DRINKING WATER DETOXIFICATION

DRINKING WATER DETOXIFICATION

Edited by M.T. Gillies

NOYES DATA CORPORATION
Park Ridge, New Jersey, U.S.A.
1978

Library of Congress Catalog Card Number: 78-61889
ISBN: 0-8155-0723-2
Printed in the United States

Published in the United States of America by
Noyes Data Corporation
Noyes Building, Park Ridge, New Jersey 07656

Foreword

This book reviews the methods used to remove toxic chemicals from drinking water including toxic substances formed by chlorination. It also concerns itself with new and proposed amendments to the *Safe Drinking Water Act of 1974* and therewith presents a comprehensive exposition of the whole drinking water topic.

At issue is EPA's controversial plan to require waterworks serving more than 75,000 persons to filter their water through beds of granular activated carbon in order to remove traces of chemicals suspected of causing cancer. Most suspect are the trihalomethanes such as chloroform, which are formed from traces of natural and waste organic matter whenever water is chlorinated. Also implicated are many other organic chemicals that enter water supplies as a result of industrial activity, and from municipal and agricultural runoffs.

It should be noted that adsorption on activated carbon is a nonspecific process which will remove not only selected key contaminants, but many other poisons, some presently undetected by current analytical capability. This may result in greater protection of human health than the arbitrary setting of maximum contaminant levels for selected substances.

While the primary aim is to remove potential carcinogens, a serendipitous note is introduced by the fact that the proposed filter systems will also greatly improve the taste and odor of any potable water.

Because the information in this book is taken from federally funded studies, and other sources, it is possible that certain portions of this book may disagree or conflict with other parts of the book. This is especially true of monetary values and opinions of future potential. We chose to include these different points of view, however, in order to make the book more valuable to the reader. Cost figures provided are those given in the report cited, the date of which is always given. When the dates of the cost figures themselves are given, we have included them.

Advanced composition and production methods developed by Noyes Data are employed to bring these durably bound books to you in a minimum of time. Special techniques are used to close the gap between "manuscript" and "completed book." Industrial technology is progressing so rapidly that time-honored, conventional typesetting, binding and shipping methods are no longer suitable. We have by-passed the delays in the conventional book publishing cycle and provide the user with an effective and convenient means of reviewing up-to-date information in depth.

The Table of Contents is organized in such a way as to serve as a subject index and provides easy access to the information contained in this book. The bibliography at the end of the volume lists the important government reports as well as additional reports pertaining to topics covered in this book. A list of hard-to-find abbreviations for scientific terms, societies and agencies should be a welcome addition.

Some of the illustrations in this book may be less clear than could be desired; however, they are reproduced from the best material available to us.

Contents and Subject Index

Proposed EPA Drinking Water Regulations and AWWA Comments

DRINKING WATER AND HEALTH—RECOMMENDATIONS OF THE NAS

> The material in this section of the chapter was taken from material from the USEPA published in the *Federal Register*, Vol. 42, No. 132, of July 11, 1977.

This publication of the recommendations of the National Academy of Sciences (NAS) is made according to the requirements of the Safe Drinking Water Act (Pub. L. 93-523). The present publication consists essentially of the summary of the NAS report which was delivered to Congress on May 26, 1977. The full report was delivered on June 20, 1977.

The Safe Drinking Water Act (Pub. L. 93-523) was enacted on December 16, 1974, giving the Administrator of the Environmental Protection Agency (EPA) the power to control the quality of the drinking water in public water systems through regulation and other means. The Act called for a three-stage mechanism for the establishment of comprehensive regulations for drinking water quality:

(1) Promulgation of National Interim Primary Drinking Water Regulations;
(2) A study to be conducted by the National Academy of Sciences, within 2 years of enactment, on the human health effects of exposure to contaminants in drinking water; and
(3) Promulgation of Revised National Primary Drinking Water Regulations based upon the NAS report.

National Interim Primary Drinking Water Regulations, pursuant to section 1412(a) of the Act, were promulgated on December 24, 1975, and July 9, 1976, to become effective on June 24, 1977. These were based on the Public Health Service Drinking Water Standards of 1962, as revised by the EPA Advisory Committee on the Revision and Application of the Drinking Water Standards, and contain maximum contaminant levels and monitoring requirements for microbiological contaminants, 10 inorganic chemicals, 6 organic chemicals, radionuclides and turbidity.

Introduction

The high quality of drinking water in the United States is recognized throughout the world, and popularly endorsed by the freedom with which water is consumed. Nevertheless, mounting concern over the spread of environmental pollution and the application of increasingly sensitive methods of analysis have led to new legislation that seeks to ensure that the quality of drinking water poses no threat to public health.

The Safe Drinking Water Act of 1974 (Pub. L. 93-523) requires the Administrator of the EPA to promulgate national standards for the purity of drinking water and regulations for enforcing them. The Act also directs the Administrator to arrange with the NAS, or other appropriate organization, to study the adverse effects on health attributable to contaminants in drinking water.

The primary purpose of the study has been to assess the significance of the adverse effects that the constituents of drinking water may have on public health. The economic or technological feasibility of controlling the concentration of these constituents is outside the scope of the study. The health effects associated with some methods of disinfection have received attention, but the relative effectiveness and potential hazards associated with the various methods of water disinfection have not been evaluated.

Application of analytical methods of great sensitivity has, in recent years, expanded our knowledge of the occurrence and diversity of impurities in drinking water. However, information about the results of chronic ingestion, at low dose rates, of most of these substances is acquired slowly because the bioassays that are usually required may take two or more years to complete. Although new approaches to the problem of estimating chronic adverse health effects may, in the future, ease this difficulty, the current knowledge on which this study is based is insufficient to assess all the contaminants of drinking water. The results reported here must therefore be considered as the first contribution of an effort that should be continued.

General Considerations

Besides the known constituents of water, we have considered also some that it would be plausible to expect to be present, even though they have not yet been identified. (Certain pesticides used in large quantities fall into this category.) In our review of water constituents, we have attempted to take into account not only their identities, concentrations, and toxicities, but also have considered other questions, such as:

(1) What reason is there for concern about the material? What risks are associated with its presence in water?
(2) How does the material get into water?
(3) What sources are there other than water?
(4) What contaminants need to be controlled?
(5) Are there special places or persons at higher than average risk?
(6) Are there essential health requirements for this material? (See particularly the discussion of inorganic solutes.)
(7) In view of the data at hand, can one say that this is a material that causes temporary ill effects? permanent ill effects? reversible effects?

(8) In view of these effects, and their reversibility (or lack of it), is it possible to set "no-observed-adverse-health-effects" levels?

(9) For materials with special health benefits, what concentrations will maximize these benefits, while keeping the health risk, associated with them at an acceptably low level?

(10) What additional information is required to resolve the outstanding problems?

Many of the constituents of drinking water occur naturally, and enter water from the rocks and the soil and the air. Some are the natural waste products of men or animals. Others are artificial or synthetic materials, made and used for special purposes, that inadvertently find their way into water. Yet others occur naturally, but have become more widely distributed in populated areas as a result of industrial and agricultural activity.

Water Consumption: In this study, the average amount of water consumed per person is assumed to be two liters per day. This is also the amount used by the EPA to calculate the current interim standards. Daily consumption of water is a function of temperature, humidity, physical activity, and other factors that vary widely. Although average water consumption per capita may be estimated from the literature on human physiology and nutrition to be about 1.6 l/day, the larger volume of 2 l/day was adopted as representing the intake of the majority of water consumers.

We estimate that most of those who consume volumes larger than 2 l/day still are afforded adequate protection because the margin of safety estimated for the contaminants is sufficient to offset the increased water consumption. Nevertheless, consideration should be given to establishing some standards on a regional or occupational basis to take extremes of water consumption into account.

Special Cases: Groups of people who may be more susceptible than average for the whole population are considered in connection with the particular contaminants to which they are sensitive. This report is concerned only with water that is used for drinking. Although all contaminants may cause problems when present in water used in health care facilities, the health hazards associated with such diverse uses of water as in humidifiers, kidney dialysis units, laundries, heating and cooling equipment, or many special uses that require further treatment of tap water, have not been considered.

Safety and Extrapolation

The hazards of ingesting chemical pollutants in drinking water can be assessed in two general ways: with epidemiological studies and with laboratory studies of toxicity. The aim of both types is to provide information on the risk to man.

Most of the current knowledge of toxicity is based on observations of the effects on man and animals of doses and dose rates that are much larger than those that correspond to the usual concentrations of harmful materials in drinking water. There is, consequently, great uncertainty in estimating the magnitude of the risk to health that ingestion of contaminants in water may produce. An additional problem is presented by the combined effects of two or more contaminants.

The theoretical and experimental bases for extrapolating estimations of risk to low levels of dose have been reviewed. Some principles are proposed to guide the conduct of this and similar studies.

Large populations are repeatedly exposed to potentially toxic contaminants in the drinking water, in minute amounts over many months or years, or over whole lifetimes. Delayed, essentially irreversible effects can occur. Methods and criteria of classical, conventional toxicology do not offer reliable means for assessing long-term toxic effects such as carcinogenesis in man by extrapolation from animal data; hence, novel considerations have to be applied in assessing risks.

The insidious effects of chronic exposure to low doses of toxic agents is difficult to recognize, because there are few, if any, early warning signs and, when signs are ultimately observed, they often imply irreversible effects. For example, cancer induction in experimental animals, even with the most potent carcinogenic chemicals, requires at least several months and in many instances a whole lifetime.

There are as yet no easy, straightforward methods for extrapolating even chronic-exposure experimental data to calculate risks to large human populations. Teratogenic effects are easier to establish by animal experimentation, but there are similar uncertainties in extrapolating to human populations. Mutagenic effects are difficult to assess experimentally in mammals, and such effects are particularly insidious, in that they appear only in later generations.

Various measures used in assessing acute toxicity, such as LD_{10}, LD_{50}, and maximal tolerated dose, are generally found to be quantitatively similar among most animals. On the basis of dose-per-unit of body surface, toxic effects in man are in the same range as those in experimental animals, such as mouse, rat, hamster, dog, and monkey.

On a body-weight basis, man is generally more vulnerable than the experimental animal, probably by a factor of 6 to 12. Comparative studies have shown generally that absorption, metabolism, and excretion of various drugs are slower, dose-for-dose, in man; that there is greater retention of such drugs; and that higher concentrations occur in body fluids and tissues in man than in small mammals. With an awareness of these quantitative differences, appropriate safety factors can be applied to calculate relatively safe therapeutic dosages for man.

These methods and principles of classical toxicology are useful for assessing toxic effects that are reversible and nonprogressive. They are much less useful in dealing with the problems of chronic irreversible toxicity or the effects of long-term exposure. This subject has not been considered widely in the past.

From the review of available information, two major questions emerge: "What types of experimental-assay procedures are required for a valid assessment of chronic toxicity of chemicals in experimental animals?" "How can such data be extrapolated to estimate risks in humans?" In dealing with these questions, our recommendations are restricted to a specific risk, namely, cancer, with the understanding that the same considerations will apply at least partially to the problems of mutagenesis and teratogenesis. Furthermore, we consider only carcinogens whose mechanisms involve somatic mutations. Some principles that underlie efforts to assess the irreversible effects of long-continued exposure to carcinogenic substances at low dose rates are outlined below.

Principle 1: Effects in Animals, Properly Qualified, Are Applicable to Man – This premise underlies all of experimental biology and medicine, but because it is continually questioned with regard to human cancer, it is desirable to point out that cancer in men and animals is strikingly similar. Virtually every form of human cancer has an experimental counterpart, and every form of multicellular organism is subject to cancer, including insects, fish, and plants. Although there are differences in susceptibility between different animal species, between different strains of the same species, and between individuals of the same strain, carcinogenic chemicals will affect most test species; and there are large bodies of experimental data that indicate that exposures that are carcinogenic to animals are likely to be carcinogenic to man, and vice versa.

Evidence that circumstances leading to cancer induction in humans are also applicable to experimental animals stems from the very first observation of chemical carcinogenesis, the appearance of cancer of the scrotum in chimney sweeps, observed by the British surgeon, Percival Pott, in 1775. It was not until modern times that a substance implicated in human cancer was found to be carcinogenic in animals, when the Japanese scientists, K. Yamagiwa and K. Ichikawa, found in 1915 that extracts from coal tar cause cancer when applied to the skin of experimental animals.

Many pure carcinogenic chemicals have since been isolated from a wide variety of "tars" derived from incomplete combustion of organic matter, such as coal, wood, and tobacco. There is little doubt that these and other chemicals, alone or in combination, are responsible for the greatly increased incidence of lung cancer among smokers. With the possible exception of arsenic and benzene, all known carcinogens in man are also carcinogenic in some species, although not in all that have been tested.

Principle 2: Methods Do Not Now Exist to Establish a Threshold for Long-Term Effects of Toxic Agents – With respect to carcinogenesis, it seems plausible at first thought, and it has often been argued, that a threshold must exist below which even the most toxic substance would be harmless. Unfortunately, a threshold cannot be established experimentally that is applicable to a total population. A time-honored practice of classical toxicology is the establishment of maximal tolerated (no-effect) doses in humans based on finding a no-observed-adverse-effect dose in chronic experiments in animals, and to divide this dose by a "safety factor" of, say, 100 to designate a "safe" dose in humans. There is no scientific basis for such estimations of safe doses in connection with carcinogenesis.

For example, even if no tumors are obtained in an assay of 100 animals, this means only that at a 95% confidence level, the true incidence of cancer in this group of animals is less than 3%. Even if we were to carry out the formidable task of using 1,000 animals for assay and no tumors appeared, we could only be 95% sure that the true incidence were less than 0.3%. Obviously, 0.3% is a very high risk for a large human population.

In fact, there are no valid reasons to assume that false-negative results of carcinogenicity tests are much less frequent than false-positive ones. To dismiss all compounds that did not induce tumors in one or two mouse and rat experiments as noncarcinogenic is wrong. Labeling as "carcinogens" all substances that gave rise to increased incidence of tumors is justified only if there is conclusive evidence of a causal relationship. The "relative risk" of compounds that are not found to induce tumors in animal experiments must also be considered. But this requires evaluation of data other than those collected in chronic toxicity studies on rodents.

Experimental procedures of bioassay in which even relatively large numbers of animals are used are likely to detect only strong carcinogens. Even when negative results are obtained in such bioassays, it is not certain that the agent tested is unequivocally safe for man. Therefore, we must accept and use possibly fallible measures of estimating hazard to man. This reasoning leads us to the statement of Principles 3 and 4.

Principle 3: The Exposure of Experimental Animals to Toxic Agents in High Doses Is a Necessary and Valid Method of Discovering Possible Carcinogenic Hazards in Man – The most commonly expressed objection to regulatory decisions based on carcinogenesis observed in animal experiments is that the high dosages to which animals are exposed have no relevance in assessment of human risks. It is therefore important to clarify this crucial issue. Practical considerations in the design of experimental model systems require that the number of animals used in experiments on long-term exposure to toxic materials will always be small, compared with the size of the human populations similarly at risk.

To obtain statistically valid results from such small groups of animals requires the use of relatively large doses so that effects will occur frequently enough to be detected. For example, an incidence as low as 0.01% would represent 20,000 people in a total population of 200 million and would be considered unacceptably high, even if benefits were sizable. To detect such a low incidence in experimental animals directly would require hundreds of thousands of animals. For this reason, we have no choice but to give large doses to relatively small experimental groups and then to use biologically reasonable models in extrapolating the results to estimate risk at low doses.

Several methods of making such calculations have been considered and used, but we think that the best method available to us today is to assume that there is no threshold, and that the incidence of tumors is directly proportional to dose. However, it is important to recognize that such calculations may give either too small or too large an estimate risk. The actual risk to humans might be even greater over a human lifetime, because it is about 35 times that of a mouse; and there is evidence that the risk of cancer increases rapidly with the length of exposure.

Moreover, experimental assays are conducted under controlled dietary and environmental conditions with genetically homogeneous animals, whereas humans live under diverse conditions, are genetically heterogeneous, and are likely to include subpopulations of unusual susceptibility.

It should be emphasized that these general considerations give only a minimal estimate of human risk; moreover, they do not take into consideration differences in susceptibility between species. For example, β-naphthylamine is well established as a human carcinogen on the basis of epidemiologic studies of occupationally exposed workers, whereas experiments have not shown it to be carcinogenic in the hamster, which is relatively resistant.

Not all substances that cause a given incidence of cancer in the rat are equally carcinogenic for man. This means that chronic-toxicity studies, which are imperfect assay systems for carcinogenicity testing, should not be used as the sole criterion in the assessment of risk.

Principle 4: Material Should Be Assessed in Terms of Human Risk, Rather than as "Safe" or "Unsafe" – The limitations of the current experimental techniques do not allow us to establish safe doses, but with the help of statistical methods we may be able to estimate an upper limit of the risk to human populations. To calculate such a risk, we need data to estimate population exposure; a valid, accurate, precise, and reproducible assay procedure in animals; and appropriate statistical methods.

Several general guidelines may be presented. First, no rigid, generally applicable procedure can be recommended for testing all toxic agents. Substances differ too much in their overall effects, and the design of appropriate assays will ultimately have to be left to the well-informed judgement of expert investigators. If substances that affect large populations are found to be carcinogenic, experiments of much wider scope may have to be conducted to obtain more detailed information on their possible effects in humans.

As a pragmatic guideline, it would be desirable to test a compound for carcinogenicity in at least two species, such as the mouse and the rat, and the strains of animals used should have a rather low incidence of spontaneous tumors under the conditions of the test. It is important to include "positive" controls, with known carcinogens, under the same conditions used for the test animals. This has been a point of considerable controversy.

Experiments should be conducted over as much of the lifetime of the experimental animal as possible. The highest dose should be the maximum that is tolerated without shortening the life span through causes other than cancer. Every animal, whether it dies during the exposure period or is sacrificed at the end of the experiment, should be examined grossly and microscopically, and all toxic effects (not only cancer) should be noted.

Risk constitutes but half the essential comparison that should be made in the assessments of human hazard; the other half is benefit to the exposed population of the agent for which hazard has been identified. Decisions cannot involve merely the risk. Statements of benefits should include the nature, extent, and recipient of the benefit. Technology has always been associated with some risk. But the acceptability of risk should depend on the specific benefits derived, the nature of the population exposed, and the availability of practical alternatives.

It is not possible to guarantee a risk-free society; nor is a risk-free society necessarily the best society. It is often necessary to accept the risks of chemicals, such as drugs and pesticides, when the benefits warrant their use. Risks imposed on persons who gain no benefits are generally not acceptable. Personal choice and personal values enter into the risk-benefit comparison. For major benefits, for example, in the treatment of otherwise incurable or incapacitating diseases, much higher risks are allowable than otherwise. An important principle in risk-benefit assessment is that each person must be allowed the widest possible choice, supported by full information on risks, as well as benefits, so that intelligent choices may be made.

The benefit portion of the equation should be well defined by knowledgeable experts, and based on data at least as good as the risk data. It is important, therefore, that the benefit-risk comparisons be established with the active cooperation of people who are qualified to assess the usefulness of a substance and the consequences to those in need of it, as well as to the population at large.

Finally, mankind is already exposed to many carcinogens whose presence in the environment cannot be easily controlled. In view of the nature of cancer, the long latent period of its development, and the irreversibility of chemical carcinogenesis, it would be highly improper to expose the general population to an increased risk if the benefits were small, questionable, or restricted to limited segments of the population.

Estimation of Risk

Chronic low-dose-rate toxicity was assessed quantitatively by different procedures, the method chosen depending on the character of the experimental evidence and whether the substance in question was judged to be carcinogenic or not.

Assessments of the toxicity of noncarcinogenic substances are described in the sections on Inorganic Solutes and Organic Solutes. These are expressed as estimates of maximal "no-observed-adverse-effect" concentrations in water, and are based on the assumption that, for these noncarcinogenic materials, there are threshold doses below which no adverse effects on health are likely to occur.

Risks of exposure to radionuclides and carcinogenic organic compounds were estimated by methods that involve an assumption that there are no thresholds in the dose-response relationships. In the case of radionuclides, the estimates were based, in large measure, on the report of the Advisory Committee on the Biological Effects of Ionizing Radiation (National Academy of Sciences-National Research Council, 1972); the method used for organic compounds is described below.

After a substance had been identified as a carcinogen, the risk to man was expressed as the probability that cancer would be produced by continued daily ingestion over a 70-year lifetime of 1 liter of water containing a standard quantity (1 μg/liter) of the substance in question. Estimates expressed in this form may then be used to calculate risk due to the concentrations actually found in drinking water and the daily consumption.

To make such estimates from the results of animal feeding studies, two steps are necessary. The first involves conversion of the standard human dose to the physiologically equivalent dose in the animal. This was performed on the basis of relative surface area. The second step requires use of a risk model relating dose to effect. The model used for this purpose is:

$$P(d) = 1 - e^{-(\lambda_0 + \lambda_1 d + \lambda_2 d^2 + \ldots \lambda_k d^k)}$$

where P(d) is the lifetime probability that dose d (total daily intake) will produce cancer, k = the number of events in the carcinogenic process, and λ_0, λ_1,

λ_2, etc. are nonnegative parameters. At low doses, the higher-order terms in d^2, d^3, etc., may be neglected and

$$P(d) \cong 1 - e^{-(\lambda_0 + \lambda_1 d)} \cong \lambda_0 + \lambda_1 d$$

λ_0 representing the background rate. When two or more sets of results of lifetime animal feeding studies were available, experimental values of P(d), the fraction of test animals developing cancer, and d, the total daily dose, were fitted to the equation to determine how many of the terms λ_0, $\lambda_1 d$, $\lambda_2 d^2$, etc., were necessary to give the best fit. Corresponding values of λ_0, λ_1, or λ_0, λ_1 and λ_2, etc., were used to calculate P(d) for the low-dose of interest, namely the animal dose that was physiologically equivalent to the standard dose for man. If the animal experiments involved only one dose level, the $\lambda_1 d$ term, alone, was used in the calculation. Upper confidence limits in the estimated low-dose risk were also calculated by use of maximum likelihood theory and these values were tabulated.

Since the animal data were obtained from lifetime feeding studies, the risk estimates calculated from them for the low-doses that were estimated to be physiologically equivalent to the human dose were taken to represent the lifetime risks for man. The background rate, obtained from the cancer incidence in the control groups of experimental animals and represented by the parameter λ_0, was excluded from the tabulated values of P(d), which therefore represent the incremental risks due to ingestion of the compounds in water.

Solid Particles in Suspension

Materials suspended in drinking water include inorganic and organic solids as finely divided particles of sizes ranging from colloidal dimensions to more than 100 micrometers. Such particles may also have other substances and microorganisms attached to them. Small particles of some materials, such as the asbestos minerals, may have the potential to affect human health directly when they are ingested, and there is widespread concern over the biological effects of such substances.

Many kinds of particles, though apparently harmless in themselves, may indirectly affect the quality of water by acting as vehicles for concentration, transport, and release of other pollutants. Water treatment can often be effective in removing most of the suspended particles but conventional methods of detecting the presence of particulate material by measurement of turbidity have serious deficiencies.

Direct Effects on Health: Particles of asbestos and other fibrous minerals occur in raw water, usually in a range of sizes from fractions of a micrometer to a few micrometers. Generally, there are fewer than 10 million fibers per liter, but waters are found with from less than 10,000 to more than 100 million fibers per liter. Some of the highest counts have been found near some cities. Fibers in drinking water are typically less than 1 micrometer long and fibers longer than 2 micrometers are uncommon.

Identification and counting of fibers is difficult and time-consuming, usually requiring the transmission electron microscope. The reported counts are highly variable, often differing from one count to the next by a factor of 10 or more.

Epidemiological studies of workers exposed to asbestos by inhalation have shown an increase in death rates from gastrointestinal cancer. With respiratory exposure it is likely that more fibers are swallowed than remain in the lungs. The workers studied were exposed to asbestos with a large range of fiber lengths. It is not clear whether fiber length is pertinent to the development of cancer in the digestive tract in humans.

Epidemiological studies of cancer death rates in Duluth, Minnesota, where the water supply has been contaminated with mineral fiber, have so far not revealed any increase of gastrointestinal cancer with time, in comparison with death rates in other areas. Contamination of the Duluth drinking water began less than 20 years ago, however, and since many cancers have long latency periods, these negative epidemiological findings do not exclude the possibility that an increase may appear within the next five to fifteen years.

Animal deposition model studies have shown that fiber length and diameter affect the carcinogenic response seen, the long thin fibers appearing to be the active ones. However, the relevance of these experimental models to the human experience is not clear. While some animal studies have shown penetration of the gastrointestinal epithelium, others have not.

It is not known whether other inorganic particulates that occur in water produce any direct effects on human health.

Indirect Effects on Health: The concentration of inorganic, organic and biological pollutants is usually much higher in the suspended solids and sediments of streams and lakes than in water. Clay and organic are chiefly responsible for such concentrations. Clay and organic particulates have large surface areas and strongly adsorb ions, polar, and nonpolar molecules, and biological agents. Occurrence of these materials in water is a consequence of natural events, as well as human activity, and is common in many waters that people drink. Although many of the clay or natural organic particulates, in themselves, may not have deleterious effects when ingested by humans, they may exert important health effects through adsorption, transport, and release of inorganic and organic toxicants, bacteria, and viruses.

The clay or organic complex with a pollutant may be mobilized by erosion from the land, or complexes may form when eroded particulate matter enters a stream containing pollutants. The atmosphere is also an important pathway. In the adsorbed state, organic and inorganic toxicants may be less active; however, the possibility exists that the toxicants may be released from the particulate matter in the alimentary tract and then exert toxic effects. It is not clear how complexes of particulate matter with viruses and bacteria behave in the gut. It is known, however, that some enzymes retain their activity when adsorbed on clay, and that viral-clay particulates are infectious in tissue culture and in animal hosts.

Turbidity as an Indicator: A high turbidity measurement is an indication that a water may produce an adverse health effect; however, a low turbidity measurement does not guarantee that a water is potable. Turbidity measurements do not indicate the type, number, or mass of particles in a water supply. Where particulates in water are suspected of being harmful, the particulate content should be identified and counted by more specific techniques. Such techniques may include biological, organic, inorganic, and fibrous-particulate surveys.

Turbidity measurements are valuable for process control in water-treatment plants. However, the results obtained with present instruments, procedures, and units of measurement are not well correlated with particle concentrations and size distributions. The test itself must be standardized and refined to facilitate its use for this and other purposes.

Conclusions and Recommendations: Certain mineral fibers found in water are suspected of being harmful upon ingestion. The available data with respect to asbestos orally ingested through drinking water do not suggest an immediate hazard to public health. They do suggest that additional research, both experimental (using animals) and epidemiological, is required to determine the degree of hazard. Until new results become available, contamination of drinking water by mineral fibers should be kept to a minimum through the use of effective coagulation and filtration processes and other appropriate measures.

Because particulates are vehicles for concentration, transport, and release of pollutants, they may have indirect effects on health. Coagulation and filtration are effective methods of reducing particulate concentrations. Measurement of particulate content by turbidimetry is imprecise and cannot be relied upon as a sole indicator of the safety of an uncharacterized drinking-water source.

Inorganic Solutes

The Interim Primary Drinking Water Regulations list maximum allowable concentrations for six metallic elements, barium, cadmium, chromium, lead, mercury, and silver. Ten additional metals were reviewed in this study, beryllium, cobalt, copper, magnesium, manganese, molybdenum, nickel, tin, vanadium, and zinc. Sodium, which is also a metal, was considered separately, because the problems it poses are quite distinct from those associated with the other metallic substances.

Eight of these metals are known to be essential to human nutrition: chromium, cobalt, copper, magnesium, manganese, molybdenum, tin, and zinc. Nickel and vanadium probably are essential also, and it is possible that barium can be beneficial under certain conditions. The metals, beryllium, cadmium, lead, mercury, and silver are believed to be not essential to humans.

Elements that are beneficial in small quantities often exhibit toxic properties when ingested in excessive amounts or concentrations. In assessment of the adverse health effects of such materials it is important not to overlook deficiency problems that might be encountered if the substances were to be completely eliminated from water supplies.

Trace metals, usually in the form of ions, occur in water both as a result of natural processes and as a consequence of man's activities. Ground waters, because of long contact with rocks and mineralized soils, usually contain greater concentrations of trace metals than surface waters. There is considerable temporal and spatial variation in concentrations of trace metals in surface waters. Generally, the trace metal concentrations of rivers tend to increase from source to mouth, and to vary inversely with discharge.

Of the 16 metals studied, the relative contribution of man's activities to the concentrations found in water supplies can be rated roughly as follows: Very great—

cadmium, chromium, copper, mercury, lead, and zinc; high–silver, barium, molybdenum, and tin; moderate–beryllium, cobalt, manganese, nickel, and vanadium; low–magnesium.

Other important sources of trace metals in drinking water are chemicals used in water-treatment processes and pickup of metallic ions during storage and distribution. Although a large fraction of the United States population continues to receive water from ground sources or from impounded upland sources without treatment other than disinfection, most large surface supplies are subjected to treatment that includes coagulation, sedimentation, filtration, and disinfection.

Should trace metals occur in the raw water supply, these normal water-treatment processes have either no effect or an uncertain one on the usual low-level concentrations of these metals. Moreover, probable trace metal impurities in the technical-grade chemicals used to treat water may introduce additional amounts.

A wide variety of materials including several metals, alloys, cements, plastics, and organic compounds are used in the pumps, pipe fittings, and reservoirs of distribution and plumbing systems. Reactions, particularly of soft, low-pH waters, with materials of the distribution system often have produced much greater concentrations of iron, copper, zinc, lead, and cadmium at the tap than those at the treatment plant.

Adverse health effects associated with trace metals depend upon the total intake from all sources, food, air and water. As a general rule concentrations of trace metals in foodstuffs greatly exceed those found in drinking waters. Because the diet of most of the United States population is increasingly varied and comes from diverse geographical sources as a result of modern food-distribution practices that counterbalance local excesses or deficiencies, the dietary intake of trace metals exhibits relatively small variation throughout the United States. This factor is helpful in evaluation of maximum no-observed-adverse-health-effect concentrations for drinking water.

Airborne exposure to trace metals other than lead is largely occupational, occurring through the inhalation of industrial dusts or fumes. At present there is more general exposure to lead from motor-exhaust fumes although evidence for acute and chronic health effects is derived from occupational exposures. Because the data relate primarily to healthy adults, caution must be observed in extrapolating these data to the general public.

All the trace metals studied are known to exhibit toxic effects at some level of intake. Many of these effects are observed, however, only at concentrations greater than the maxima that have been found in drinking water. To include such materials in primary drinking-water standards, with the accompanying requirement for mandatory surveillance, does not confer any health benefit. Augmentation of the natural concentrations of these trace elements to values of concern can be avoided most readily by preventing discharge of the contaminants into water sources.

In addition to the trace metals mentioned above, the effects on health of several other inorganic constituents of drinking water were also studied. These include sodium, arsenic, selenium, fluoride, nitrate and sulfate. The relationship between water hardness and health was also considered. The findings on these topics are summarized individually below.

Barium: It is rare to find barium in drinking water at a concentration in excess of 1 mg/l because of the low solubility of barium sulfate. Natural and treated waters usually contain sufficient sulfate so that more than 1 to 1.5 mg/l of barium cannot be maintained in solution.

Acid-soluble barium salts are very toxic, whereas insoluble compounds are benign. There has been no determination of the chronic effects of low levels of barium ingested over a long period of time.

The Interim Primary Standard of 1mg/l for barium has been based on extrapolation from effects of industrial exposure to dusts of soluble barium salts. Insufficient data are available to estimate maximum no-observed-adverse-health-effect concentrations on the basis of water intake. The limit of 4 mg/l of the USSR is based on organoleptic factors. International and European Standards of the World Health Organization (WHO) do not list barium.

It is recommended that animal studies involving long-term low-level ingestion of barium salts in water be carried out to determine possible health effects.

Beryllium: Because the oxide and hydroxide are relatively insoluble at the usual range of pH, beryllium is unlikely to occur in drinking water at more than trace concentrations. The sulfate and chloride are very soluble, but they hydrolyze quickly to the insoluble hydroxide.

Beryllium produces acute or chronic toxicity in animals when ingested continuously as beryllium sulfate in food in amounts greater than 10 to 20 mg/kg/day, or in water at concentrations greater than 5 mg/l. Soluble beryllium has been shown to be transported in the bloodstream to bone, and to be able to induce bone cancer in animals, but the data are insufficient to allow estimation of risk.

Prolonged inhalation of dusts containing beryllium is known to produce pulmonary sarcoidosis. However, increased incidence of lung cancer has not been found among workers exposed to dusts containing beryllium.

No maximal allowable concentration for beryllium has been listed in the Interim Primary Drinking Water Regulations, nor has the WHO recommended a maximum limit. The USSR, however, has set a limit of 0.2 μg/l. Until now the maximum concentration of beryllium found in U.S. surface waters has been 1.2 μg/l and in finished U.S. drinking waters has been 0.17 μg/l. Only 1.7% of drinking water supplies examined have been found to contain beryllium.

Additional studies of the frequency of occurrence and concentrations of beryllium in natural waters are needed to determine the extent to which it presents a hazard to health.

Cadmium: Cadmium is not known to be an essential or beneficial element. It has been found in 2 to 3% of U.S. surface waters, generally in concentrations not exceeding a few micrograms per liter because solubilities of cadmium carbonate and hydroxide are low at pH greater than 6. Only 0.1% of the supplies in the Community Water Supply Survey showed cadmium in excess of 10 μg/l. In addition to its geological sources, cadmium enters water from the discharge of plating wastes and by corrosion of plumbing.

Food is the primary source of cadmium intake. Total daily intake from air, water, food and tobacco ranges from 40 μg/day for the rural nonsmoker on a low cadmium diet to 190 μg/day for the urban smoker on a high cadmium diet. Drinking water contributes only a small fraction (<5%) to this total intake.

Chronic ingestion of cadmium at levels greater than 100 μg/day, in combination with several other necessary predisposing factors, was found to be responsible for the onset of "Itai-Itai" disease in Japan. Dietary intake of amounts in excess of a milligram per day is needed for appearance of acute toxicity. Major toxic effects are on the kidney; data indicate that the toxicity of cadmium is related to the zinc:cadmium ratio within the organs. Both zinc and calcium may protect against cadmium toxicity.

Persons deficient in these elements, and especially lactose-intolerant persons, who are also likely to be calcium deficient, may constitute a high-risk group relative to cadmium. Some animal studies have shown carcinogenic and teratogenic effects, but dose-response relationships are unknown. Cadmium has also been implicated as a factor in hypertension.

Insufficient data are available for establishment of a maximum no-observed-adverse-health-effect value. It may be noted, however, that consumption of two liters per day of water containing 10 μg/l of cadmium would contribute only about 20% of the normal total daily adult intake. Both the WHO and the USSR have set the maximum allowable limit for cadmium at 10 μg/l.

Chromium: Microgram amounts of chromium, derived primarily from food, are essential for maintenance of normal glucose metabolism. But chromium(VI) is known to be toxic, principally on the basis of information from respiratory occupational exposures. Increased risk of lung cancer among those occupationally exposed to chromium(VI) has been established.

Although inhaled hexavalent chromium may cause cancer of the respiratory tract, a working group of the International Agency for Research on Cancer concluded "there is no evidence that nonoccupational exposure to chromium constitutes a cancer hazard."

Concentrations of chromium found in natural waters are limited by the low solubility of chromium(III) oxides. A study of more than 1,500 surface waters showed a maximum chromium content of 0.11 mg/l, with a mean of 0.01 mg/l.

Little information is available about the average total daily intake of chromium in the United States, although it appears to be in the range of 60 to 280 μg/day. It has been suggested that diets containing mostly processed foods may be chromium deficient. Tissue chromium in U.S. adults has been shown to decline with age.

In addition to the beneficial effect of chromium on glucose metabolism, some animal studies indicate that chromium deficiency may induce atherosclerosis.

Toxicity of chromium depends on the valence. No toxic effects were observed in rats when drinking water contained 25 mg/l of trivalent chromium for a year or 5 μg/l for life. Acutely toxic doses of trivalent chromium fall in the range of grams per kilogram of body weight. Hexavalent chromium was also tolerated at the 25 mg/l level for a year by rats. Dogs showed no effects with 11 mg/l over a 4-year period. Higher doses are toxic, however, producing erosion of the gastrointestinal tract and kidney lesions.

The maximum limit of the Interim Primary Drinking Water Regulations, 0.05 mg/l, is only one-hundredth of the maximum no-observed-adverse-health-effect concentration. The European Standards of the WHO and Japanese Standards give the same limit as acceptable, but set it in terms of hexavalent chromium only. The USSR has limits of 0.1 mg/l chromium (VI) and 0.5 mg/l total chromium, based on organoleptic factors.

More informationi is needed on the carcinogenic potential of ingested chromium(VI) and chromium(III). If it becomes clear that highly toxic or carcinogenic effects occur only with chromium(VI), and a suitably sensitive analytical technique is available, then the standard might be set for chromium(VI) alone. In view of the trend in the United States toward dietary chromium deficiency, and the suggestion that chromium protects against atherosclerosis, it seems advisable to determine whether concentrations greater than that prescribed by the current drinking-water regulations are without adverse health effects, as some animal experiments suggest.

Cobalt: Cobalt is an essential element as a component of vitamin B_{12}. Excessive intake of cobalt may be toxic, however, as shown by the association of congestive heart failure with the consumption of beer containing about 1.5 mg/l of cobalt.

Cobalt has been observed in natural waters only in trace amounts. Most waters contain no detectable cobalt, and values greater than 10 μg/l are rare. The maximum concentration recorded in several extensive studies was 99 μg/l.

The major source of cobalt is food; concentration in green leafy vegetables may be as great as 0.5 mg/kg dry weight. Normally, less than 1% of total intake of cobalt is derived from aqueous sources.

Acute toxic effects in animals have been observed only at daily doses greater than several mg/kg of body weight. Chronic cobalt toxicity has been observed in children taking cobalt preparations to correct anemia at daily doses of 1 to 6 mg/kg body weight.

The Interim Primary Drinking Water Regulations do not list cobalt, nor has the WHO recommended a limit in its International or European standards. The USSR has set a limit of 1 mg/l.

Because the maximum no-observed-adverse-health-effect concentration is more than an order of magnitude greater than that found in any natural water or drinking water supply, there appears to be no reason at present to regulate the concentration of cobalt in drinking water.

Copper: Copper is an essential element for both plants and animals; it is a component of several enzymes that perform important biological functions.

Copper is a minor constituent of natural waters. In a survey of 1,600 surface waters of the United States, the concentrations were 1 to 280 μg/l. Corrosion of copper piping may increase concentrations in drinking waters to several milligrams per liter. Copper may also be released into water in industrial discharges, and has been used for algal control in reservoirs at a few tenths milligram per liter.

Average total intake of copper is about 2.5 mg/day, so that when water contains more than 1 mg/l of copper, the intake from water may equal or exceed that from food.

The general health hazard from copper intake at a few milligrams per liter appears to be small, but a few people are adversely affected by ingestion of even trace amounts of copper. This disorder of copper metabolism, called Wilson's disease, can be arrested by the use of chelating agents. Individuals with deficiency of glucose-6-phosphate dehydrogenase may be sensitive to copper.

The USPHS Drinking Water Standards (1962) recommended a limit for copper of 1 mg/l based on organoleptic rather than health effects. Because no general adverse health effects have been observed at the organoleptic limit and because the few individuals with metabolic deficiency are at the mercy of total copper intake rather than copper in water, there is no hygienic reason to impose a limit lower than the presently accepted secondary standard.

Lead: No beneficial effects of lead on human or animal development have yet been found. Although acute lead poisoning is rare, chronic lead toxicity is severe and occurs even with low daily intake ($<$1 mg) because of its accumulation in bone and tissue.

The natural lead content of surface waters is generally small. In a survey of nearly 1,600 raw surface waters, 20% were found to contain detectable concentrations of lead and these had a mean value of 0.023 mg/l. The lead concentration in municipal supplies at the tap may be much greater, however, for soft, low-pH (aggressive) waters dissolve lead from service connections, lead-lined household piping or soldered joints. Lead concentrations in excess of the interim level of 0.05 mg/l were found in 1.4% of the water systems examined in the Community Water Supply Survey. The maximum value was 0.64 mg/l.

The mean concentration of lead in U.S. drinking waters has been estimated to be 0.013 mg/l. Consumption of 2 liters per day per capita gives a mean daily intake of 26 μg.

Lead intake from food varies greatly; mean daily values are estimated to be 100 to 300 μg per capita for adults. Average intake in water is considerably less than that from food, but when the concentration in water is close to or exceeds the interim level of 0.05 mg/l, intake in water approaches that from food.

Absorption of lead from dietary sources, either food or water, is estimated to be about 10% for adults. Daily lead absorption from food is, then, 10 to 30 μg, while absorption from water ranges from an average of 3 to 10 μg or more, when water containing 0.05 mg/l or greater is ingested at two liters per day.

The daily intake from air also ranges widely, and is greatest among city dwellers. For a daily inspiration volume of 20 m^3 for adults and a lead concentration of 3 $\mu g/m^3$ in urban air, the per capita daily intake is 60 μg. Absorption from air is about 40%, however, so that the daily quantity absorbed is 24 μg, a value comparable with the dietary absorption. The sum of the estimated absorptions from the various routes, 50 to 60 μg/day, is already at the maximum no-observed-adverse-health-effect value of 50 to 60 μg/day.

Children, and especially inner-city urban children, are a special risk group with regard to lead toxicity. A primary reason is that absorption of lead from food and water is 40 to 50% for 2- to 3-year old children, rather than the 5 to 10% characteristic of adults. Also, water intake per kilogram of body weight is considerably greater for young children than for adults. Moreover, lead concentrations in urban air increase with proximity to the ground, so that urban children tend to have increased intake from this source. Young children also have the added risk of ingestion of flaking lead-based paints especially in depressed, older, urban areas.

Dietary lead intake for a 2-year-old child (12 kg) has been estimated to be 100 μg/day (8.3 μg/kg/day); with water at the present 0.05 mg/l limit and a consumption of 1.4 l/day, and with air intake about 18 μg/day, the estimated total intake for a 2-year-old would be close to 190 μg/day, not including other possible sources.

Major chronic adverse effects of lead are produced in the hematopoietic system, central and peripheral nervous systems, and kidneys. Disturbance in heme synthesis is considered to be the most sensitive effect. There is a detectable increase in red-cell protoporphyrin in women and children with blood lead concentrations greater than about 25 to 30 μg/dl (micrograms per deciliter). For men occupationally exposed, the maximum no-observed-adverse-health-effect level appears to be somewhat greater at 50 to 60 μg/dl.

Results of studies in the Boston area indicate that increased blood levels of lead occur in children when the water supply contains 0.05 to 0.1 mg/l of lead. Thus, the interim limit of 0.05 mg/l may not provide a margin to safeguard the high-risk population in urban areas. The WHO recommendation of 5 μg of lead per kg per day as a safe total daily intake cannot be met for a 12 kg child when the water supply contains as much as 0.05 mg/l. It is concluded that the no-observed-adverse-health-effect level cannot be set with assurance at any value greater than 0.025 mg/l.

Manganese: Manganese resembles iron in its chemical behavior and occurrence in natural waters, but is found less frequently and usually at lower concentrations than iron. Manganese, like iron, is an essential trace nutrient for plants and animals. It is not known whether human manganese deficiency occurs in the United States.

The solubility of the several oxidation states of manganese (II, III, and IV) depends upon pH, dissolved oxygen, and the presence of complexing agents. Occasionally, deep lakes or impounding reservoirs that contain organic sediments under anaerobic reducing conditions can distribute several mg/l of Mn^{+4} throughout the water body during "turnover" mixing. Normally, however, the concentration of manganese in natural surface waters is less than 20 μg/l.

Manganese can be absorbed by inhalation, ingestion, and through the skin; the consequences of this have been recently reviewed in depth by the National Academy of Sciences. It has been known that the occupational inhalation of manganese dusts results in a disease of the central nervous system resembling Parkinsonism, and a form of pneumonia.

Ingestion of manganese in moderate excess of the normal dietary level of 3 to 7 mg/day is not considered harmful. A reported outbreak of manganism in Japan was attributed to drinking well water containing about 14 mg/l of manganese.

The maximum concentration of manganese found in the 1975 Survey of Interstate Water Supply Systems was 0.4 mg/l except for samples from two Alaskan airports which showed 1 and 1.1 mg/l. A total of 669 supplies were examined. Similarly, the maximum concentration found in the 1969 Community Water Supply Survey was 1.3 mg/l from 969 supplies. Both these maximum concentrations are an order of magnitude less than minimum concentrations at which adverse health effects are observed. Moreover, even with manganese at 0.4 mg/l the intake of manganese from water would be only about 15% of the normal total dietary intake of manganese.

Because concentrations of manganese found in water supplies are much less than those at which adverse health effects have been observed and because the regulation of manganese for esthetic and economic reasons is also far more stringent than would be required for reasons of health, there seems little need to establish a maximum no-observed-adverse-health-effect value.

Magnesium: Magnesium is an essential element in human, animal, and plant nutrition. It is geologically ubiquitous and its salts are widely used industrially. The average U.S. adult ingests between 240 to 480 mg/day of magnesium. Magnesium intake from 3.6 to 4.2 mg/kg of body weight is believed to be adequate to maintain magnesium balance, which is closely regulated by normal kidneys. The median concentration of magnesium in the water of the 100 largest U.S. cities was reported at 6.26 mg/l with a maximum of 120 mg/l. It can be greater, especially in arid western states.

An excess of magnesium in the diet is seldom harmful, for it is generally excreted promptly in feces. High concentrations of magnesium sulfate in drinking water have a cathartic effect on new users, but a tolerance is soon acquired. Excessive magnesium in body tissues and extracellular fluids occurs only as a result of severe kidney malfunction.

Magnesium deficiency in humans may occur in alcoholics, persons performing hard labor in hot climates (because magnesium is excreted in perspiration), those with certain endrocrine disturbances, and patients using potent diuretics. Such deficiencies can best be overcome by oral administration of magnesium compounds.

The National Interim Primary Drinking Water Regulations contain no limit for magnesium, nor did the 1962 USPHS Drinking Water Standards. The USSR has set no limit, but the WHO has recommended a maximum of 150 mg/l. In view of the fact that concentrations of magnesium in drinking water less than those that impart astringent taste pose no health problem and are more likely to be beneficial, no limitation for reasons of health appears needed.

Mercury: Mercury is a comparatively rare element. Its inorganic compounds are relatively insoluble and can exist in solution only in extremely small concentrations under natural conditions. Recent measurements show that only 4% of water supplies contain mercury at concentrations greater than about 1 μg/l and only one of these exceeded the current standard of 2 μg/l. Concentrations in these supplies range from 0.1 to 10 μg/l. Industrial use of mercury has resulted in increased environmental contamination. The health effects of populations occupationally exposed to mercury and mercury compounds have long been recognized, but contamination of the general environment is of recent origin.

Inorganic mercury in bottom sediments can be transformed biochemically to injurious methylmercury or other organic mercurial compounds. The organic form readily enters the food chain with concentration factors as great as 3,000 in fish.

Several investigators have estimated the blood levels of mercury at which identifiable symptoms of mercury intoxication occur. These levels may be obtained with a steady mercury intake of from 4 to 14 μg/kg/day. This would be 240 to 840 μg/day for adults and 80 to 280 μg/day for children.

It is estimated that the normal diet contributes about 10 μg/day of mercury. With daily intake of 10 μg from food and 4 μg from water it appears that there is considerable margin of safety. However, those individuals regularly consuming fish from contaminated areas may exceed the normal intake by a factor of three or more and thus constitute a high-risk population.

There is no indication that concentrations of mercury in drinking water or air have contributed in any significant way to methylmercury intoxication of the general population. The interim level limits the daily intake to 3 to 4 μg/day. Nearly all public water supplies in the United States contain less than 1 μg/l of mercury. The WHO has set no limit and the USSR has a maximum permissible concentration of 5 μg/l.

Molybdenum: Soluble molybdate ions are present in trace concentrations in many surface waters, primarily as a result of discharge of industrial wastes but also as a product of natural weathering of molybdenum-bearing soils. Both suspended insoluble molybdenum disulfide and soluble molybdates are present in streams draining areas where molybdenum ore is mined and processed, especially in Colorado and New Mexico.

Typical diets contain on the order of 100 to 1,000 μg/kg, whereas typical surface waters (except those draining mining areas) contain less than 100 μg/l, with median values about 10 μg/l. Hence, in most locations, water is a minor factor in the total molybdenum intake. However, some finished waters are reported to contain as much as 1.0 mg/l, and so may provide as much as 2,000 μg/day of molybdenum. More information is needed about adverse health effects of molybdenum at these levels to deal properly with such supplies.

Molybdenum poisoning has rarely been observed in humans. Although it has been implicated for gout in Armenia and for a bone-crippling disease in India, more information is needed to establish cause-and-effect relationships.

Molybdenosis in livestock is a significant toxicological problem in many areas of the world. Consumption of molybdenum-rich forage by cattle and sheep causes severe diarrhea (scouring), which sometimes results in death. It can be prevented or alleviated by the administration of copper.

The USSR has established a limit for molybdenum of 0.5 mg/l in open waters, but the WHO has not promulgated a limit.

Nickel: Nickel may occur in water from trace concentrations of a few micrograms per liter to a maximum of 100 μg/l. At these levels the daily intake of nickel from water ranges from less than 10 μg/day to a maximum of 200 μg/day, as compared to a normal food intake of 300 to 600 μg/day. Available information indicates that nickel does not pose a toxicity problem because absorption from food or water is low. The principal reason for considering nickel stems from epidemiological evidence that occupational exposure to nickel compounds through the respiratory tract increases the risk of lung cancer and nasal-cavity cancer. There is difficulty in separating the effect of nickel from the effects of simultaneous inhalation of other carcinogens including arsenic, chromium, and cobalt.

Because of the generally low concentration of nickel in drinking water and its reported low oral toxicity, there is no present need to set primary health effect limits for nickel in water. WHO and the USSR have set no standards for nickel in drinking water.

Silver: Trace amounts of silver are found in some natural waters and in a few community water supplies. It has not been detected at levels exceeding the interim standard of 50 μg/l. Colloidal silver consumed in large doses, several hundred milligrams per kilogram of body weight can cause anemia and possibly death. The main chronic effect in man is "argyria." Argyria is a cosmetic defect once caused through medical or occupational exposure to silver preparations. Dosages of from 1 to 5 g of silver are sufficient to produce this syndrome.

On the assumption of 50% absorption of silver, consumption of 2 l/day of water containing 0.005 mg/l of silver would result in an accumulation of 1 g of silver over 55 years.

Silver ion has not been detected in water supplies in concentrations greater than half the no-observed-adverse-health-effect level.

Tin: There is some indication that tin may be a beneficial micronutrient, although it has not been conclusively demonstrated that tin is an essential trace element in human nutrition. Inorganic tin is relatively nontoxic, but organotin compounds can be toxic at high concentrations. Indeed, they are used as fungicides, insecticides, and anthelminthics.

Tin has seldom been determined in natural or municipally treated water. The few available data generally show concentrations of the order of 1 to 2 μg/l. In contrast, tin is present in most natural foods, and especially in canned products, to the extent that the normal human ingestion varies from 1.0 to 30 mg/day which is three or more orders of magnitude higher than the probable amount in a liter of tap water.

EPA has not set a limit for tin in its National Interim Primary Drinking Water Regulations. In view of the foregoing considerations, no regulation seems necessary.

Vanadium: Vanadium is a trace metal which has been introduced into the environment in large quantities. Fresh surface waters show concentrations in the 2 to 300 μg/l range, but with low frequency of detection. The data are limited on concentrations in finished drinking waters, but vanadium concentrations of up to 19 μg/l have been reported.

Occupational exposure to pentoxides and trioxides of vanadium leads to ear, nose and throat irritation and generally impaired health. The consequences of exposure to vanadium in air, water and food have been reviewed recently. There is no evidence of chronic oral toxicity. Vanadium is considered a beneficial nutrient at μg/l levels, and has been suggested as protective against atherosclerosis.

Zinc: Concentrations of zinc in surface water are correlated with man's activities and with urban and industrial runoff. The solubility of zinc depends upon the pH of the water. Concentrations ranging from 2 to 1,200 μg/l were detected in 77% of 1,577 surface water samples and 3 to 2,000 μg/l in 380 drinking waters.

Zinc is relatively nontoxic and is an essential trace element. Recommended minimum intake levels are 15 mg/day for adults and 10 mg/day for children over one year of age. A wide margin of safety exists between normal intake from the diet and doses likely to cause oral toxicity. Concentrations of 30 mg/l or more impart a strong astringent taste and a milky appearance to water. Some acute adverse effects have been reported from consumption of water containing zinc at 40 to 50 mg/l. There are no known chronic adverse effects of low-level zinc intake in diet, but human zinc deficiency has been identified. The proposed EPA secondary maximum contaminant level is 5 mg/l.

Sodium: Sodium ion is an ubiquitous constituent of natural waters. It is derived geologically from the leaching of surface and underground deposits of salts such as sodium chloride, from the decomposition of sodium aluminum silicates and similar minerals, from the incorporation of evaporated ocean spray particles into rainfall and from the intrusion of seawater into fresh water aquifers. Sodium chloride used as a deicing agent on roads enters water supplies in runoff from roads and storage depots. This added sodium chloride amounting to 9 million tons in 1970, is distributed throughout the snow belt of the northern U.S. and is most heavily concentrated in metropolitan areas.

A survey of 2,100 supplies, covering approximately 50% of the population of the U.S., was carried out in 1963 to 1966. The concentrations of sodium ion found ranged from 0.44 to 1,900 mg/l. About 42% of the supplies had sodium ion concentrations in excess of 20 mg/l and nearly 5% had concentrations greater than 250 mg/l.

Few studies of habitual sodium ion intake by healthy adults in the U.S. have been reported. Such data as have been reported are based on measurement of sodium excretion in 12- or 24-hour urine collections. Wide variations occur among individuals and in the same individual from day to day. One study reported a mean 24-hour excretion of 4,100 mg with a range from 1,600 to 9,600 mg in 71 working adult males in New York. Another study reported a mean sodium excretion near 2,800 mg/24 hr in 171 black women ranging in age from 35 to 44 years. Infants have been estimated to excrete 69 to 92 milligrams per kilogram per day.

Sodium chloride is added to many foods during processing. Additional sodium chloride is often added during cooking, and again at the table. None of this is essential, for habitual intake of sodium bears no relationship to physiological need. Healthy individuals have been shown to maintain sodium balance on an intake of less than 2,000 mg/24 hr while sweating 9 l/day. A variety of preindustrial societies, in widely divergent habitats (for example, tropical jungle, desert, arctic) subsist for generations on sodium intake less than 1,000 mg/day and show no evidence of sodium deprivation. Requirements for sodium in growing infants and children are estimated at less than 200 mg/day.

It thus appears that habitual intake of sodium in adults in the United States often exceeds body need by tenfold or more. Evidence that this excessive intake may have harmful consequences is summarized in the detailed report.

Specification of a "no-observed-adverse-health-effect" level in water for a substance such as sodium, for which the effect is associated with total dietary intake and for which usual food intake is already greater than a desirable level, is impossible.

Since adult fluid intake averages 1.5 to 3 l/day, sodium intake from drinking water represents less than 10% of the habitual total intake of 3,000 to 4,000 mg so long as the sodium content of the water does not exceed 200 mg/l. Adverse health effects may be anticipated with sodium concentrations in water greater than 20 mg/l only for that special risk group restricted to total sodium intake of 500 mg/day, because it is not feasible to reduce intake from food below 440 mg/day. For this group, whose diets must be medically supervised, knowledge of the sodium ion concentration of the drinking water permits prescription of bottled water low in sodium when necessary.

Reduction in hypertension for a small segment of the U.S. population who are on severely restricted diets requires a total intake of sodium less than 500 mg/day. These persons need water containing less than 20 mg/l sodium ion.

A larger proportion of the population, about 3%, is on sodium-restricted diets that require sodium intake of less than 2,000 mg/day. The fraction that can be allocated to water varies, depending on medical judgment in individual instances. Knowledge of the sodium ion content of the water supply and maintenance of it at the lowest practicable concentration is clearly helpful in arranging diets with suitable sodium intake. In many diets allowance is made for water to contain 100 mg/l of sodium.

It appears that at least 40% of the population would benefit if total sodium ion intake were maintained at less than 2,000 mg/day, provided that sodium ion concentration in the water supply were less than 100 mg/l, the contribution of water to the desired total intake of sodium would be 10% or less at a daily consumption of two liters.

Arsenic: Arsenic is not known to be essential to humans, nor are there known beneficial effects from its ingestion in any form, even though a number of arsenic compounds, principally organic, have been used medicinally for treatment of a number of diseases. Minimization of intake is, therefore, desirable.

Trace concentrations of arsenic are rather widely distributed in natural waters of the United States. Surface water surveys have indicated that 20 to 25% contain arsenic in excess of the detection limit of 10 μg/l, and that concentrations as great as 1,000 μg/l occur. Concentrations as great as 1,400 μg/l have been reported for ground waters. Enhanced values for arsenic content have been encountered in the vicinity of smelters, and in connection with dumping or spills of arsenical pesticides.

Other sources of human intake of arsenic include residues of arsenical insecticides on fruits and vegetables, naturally occurring arsenic in food products such as shellfish, residual dietary organic arsenicals in pork and poultry, and inhalation of dusts containing arsenic from occupational or environmental contamination. The median total intake of arsenic from all sources in the United States has been estimated to be 137 to 330 μg/day.

The toxicity of arsenic depends greatly on chemical form, route of exposure, and the rate and duration of exposure. Arsines and trivalent inorganic arsenic (arsenite) are the most toxic forms. The lethal oral dose of sodium arsenite lies in the range of 1 to 25 mg/kg; arsenic trioxide is one-third to one-tenth as toxic, and pentavalent arsenic and organic arsenicals are less than one-tenth as toxic.

Chronic or subacute toxicity of arsenic has been observed with ingestion of a few milligrams per day for two weeks or longer. Initial symptoms are skin erythema, edema and pigmentation, gastrointestinal and neurological disturbances. Similar symptoms have been observed in several populations that use water containing 100 to 1,000 μg/l of arsenic. Other conditions attributed to excessive human intake of arsenic include neuropathy, increased heart attacks, and vascular injury leading to gangrene and "Blackfoot." Industrial exposures, by inhalation or skin contact, sufficient to cause serious effects on health, have been reported in the United States and several other countries.

Human exposure to inorganic arsenic compounds has been linked to development of cancer of the skin, respiratory system, and gastrointestinal tract. However, animal studies have not shown arsenic compounds to be carcinogenic even when administered at the maximally tolerated dosages for long periods of time. This absence of positive results from controlled animal studies makes it impossible to estimate quantitatively a risk of cancer from intake of arsenic in any form of concentration.

Arsenic compounds are fetotoxic in animals at high doses, and teratogenic at lower doses. They have also been found to be mutagenic and are associated with chromosomal aberration in man.

There is speculation that interactions between arsenic and heavy metals or between arsenic and irritating substances, such as sulfur dioxide, may be important in determining overall effects on humans exposed to mixtures of these environmental contaminants. Arsenic has been found to protect against selenium poisoning in some circumstances, but under other conditions selenium and arsenic appear to be additive in toxicity.

The maximum no-observed-adverse-effect-level for arsenic in water is less than 100 μg/l. The current mandatory U.S. drinking water limit of 50 μg/l provides only a meager margin of safety. Intake from 2 l/day of water containing 20 μg/l is slightly greater than 10% of the median total intake of arsenic. The present WHO limit is 50 μg/l, as it is in the USSR.

A research program should include:

(1) Improvement of analytical techniques and methodology for better adaptability to water and foods. (Definition of chemical form is required.)
(2) Epidemiological and analytical studies of the distribution of the various forms of arsenic in water at low concentrations, and relationship to disease patterns.
(3) Development of a suitable animal model for long-term studies of arsenic toxicity at low levels.
(4) Intensive studies on the metabolism of arsenic in mammalian systems.
(5) Studies on the interaction of arsenic with other trace elements in the environment, such as Se, Cu, Zn.

Selenium: Either a deficiency or an excess of selenium can result in adverse responses. Selenium is an essential nutrient, part of the enzyme glutathione peroxidase, and may have a role in other biologically active compounds. It is a detoxifying agent for heavy metals, especially cadmium, and in some circumstances acts antagonistically to arsenic. On the other hand, chronic exposure to excess selenium results in dermatitis, central nervous system and gastrointestinal disturbances. Large doses cause acute toxicity or death.

Most natural waters contain only minute concentrations of selenium, less than 10 μg/l, but in regions with seleniferous soils concentrations in water may reach several hundred micrograms per liter, particularly for well water. One surface water receiving irrigation drainage from seleniferous soils has been found to contain 2,000 μg/l.

Most selenium intake normally is from food. Concentrations in foodstuffs vary widely, depending on the type and the selenium content of the soil in which the crop was grown. Cereals, meats, and seafoods are likely to be major contributors, with average concentrations of a few tenths of a mg/kg. Minimum nutritional requirements for selenium have been estimated to be 1 mg/month.

Industrial exposure to selenium may occur in copper refining, in the mining and milling of lead, zinc, phosphate, or uranium, in the manufacture of glass, ceramics, electronic devices and pigments, and as a result of coal or oil combustion. Atmospheric pollution and general respiratory intake may also occur in the neighborhood of these industries.

Both inorganic and organic forms of selenium are readily absorbed from the gastrointestinal tract of animals. Selenite and selenate are distributed largely to the liver, kidneys, muscle mass, gastrointestinal tract, and blood. The principal route of excretion of selenium is in the urine, mainly as trimethyl selenonium ion.

Most indications of the health effects of selenium are derived from animal studies; the number of reports of industrial or accidental exposures to toxic levels is limited. The severity of response depends on the chemical form of selenium, hydrogen selenide being most toxic. Symptoms of selenium toxicity in animals include gastroenteritis, myocardial damage, hydrothorax, pulmonary edema, and renal and liver damage.

Sodium selenite is toxic to rats at concentrations of 6 to 9 mg/l in drinking water, concentrations less than 1 mg/l are without observed toxic effect.

Cited evidence for carcinogenic effects of selenium is tenuous because of poor experimental design or protocol, and has not been confirmed in properly conducted studies. Epidemiological and demographic studies tend to suggest a protective effect of selenium against certain types of cancers, as do statistical data comparing sheep on selenium-supplemented diets with those on normal diets. There are no reports of mutagenicity of selenium.

Although the WHO limit on selenium, like, the EPA-proposed maximum contaminant level, is 10 μg/l and the USSR limit is 1 μg/l as SeO_3, most evidence indicates that there is greater overall potential for selenium deficiency than for selenium toxicity at current levels of selenium intake. The maximum no-observed-adverse-health-effect level for selenium in water is at least 100 μg/l and appears to be as great as 500 μg/l. A concentration of 20 μg/l just barely provides a minimum nutritional amount of selenium with a consumption of 2 l/day.

A research program should include:

(1) Development of more rapid, accurate and reproducible analytical methods to provide qualitative and quantitative assays of chemical forms, oxidation state, and solubility in water.
(2) Improved systems for monitoring selenium in the environment (water, air, food).
(3) Molecular transformations of selenium compounds in mammalian systems.
(4) Interactions between selenium, mercury, cadmium, arsenic, and other trace elements and heavy metals in the biosphere and in animal organisms.
(5) Determination of natural and industrial emissions and cycling of selenium in the environment.
(6) Effects on animal systems of long-term, low levels of selenium, alone and in combination with other trace elements.
(7) Baseline data on selenium levels in humans in health and disease.
(8) Effects of deficiency or excess of selenium on the development of animal tumors.
(9) Studies of the variation in human nutritional requirements for selenium.

Fluoride: Fluoride is found widely in water supplies, but the concentration is usually not great enough to be undesirable. The maximum concentration found for the 969 supplies studied in the 1969 Community Water Supply Survey was 4.4 mg/l. Most supplies that were not intentionally fluoridated had fluoride concentrations less than 0.3 mg/l.

A more extensive survey by the Dental Health Division of the U.S. Public Health Service showed that more than 2,600 communities with a population of 8 million people have water supplies with more than 0.7 mg/l of naturally occurring fluoride. Most of these communities are in Arizona, Colorado, Illinois, Iowa, New Mexico, Ohio, Oklahoma, South Dakota, and Texas. Of these, 524 communities representing 1 million people had supplies with fluoride concentrations greater than 2 mg/l.

Small amounts of fluoride, on the order of 1 mg/l, depending on the environmental temperature, in ingested water and beverages, are generally conceded to have a beneficial effect on the rate of occurrence of dental caries, particularly among children.

Two forms of chronic toxic effects are recognized generally as being caused by excess in intake of fluoride over long periods of time. These are mottling of tooth enamel or dental fluorosis, and skeletal fluorosis. In both cases, it is necessary to consider the severity since the very mild forms are considered beneficial by some. The most sensitive of these effects is the mottling of tooth enamel, which, depending on the temperature, may occur to an objectionable degree with fluoride concentrations in drinking water of only 1.5 to 2 mg/l. These observations were made a number of years ago and there have been no recent studies to determine if these levels still cause mottling.

Apparently there has been little systematic investigation of the degree to which consumers of drinking water with several milligrams per liter of fluoride regard the resultant mottling as an adverse health effect.

Skeletal fluorosis has been observed with use of water containing more than 3 mg/l. It now appears that some probability of objectionable dental mottling and increased bone density exists with long-term consumption of water containing fluoride in excess of 1 mg/l in patients with long-standing renal disease and polydipsia. Increased bone density, however, has often been regarded as a beneficial rather than an adverse effect. This, therefore, makes the implications of such changes unclear. Intake of fluoride for long periods in amounts greater than 20 to 40 mg/day may result in crippling skeletal fluorosis.

Other reported adverse health effects of intake of milligram-per-liter levels of fluoride in drinking water, including mongolism, cancer mortality, mutagenic or birth effects and sensitivity have either been unconfirmed or found lacking in substance. There is also no evidence that there is any difference between the effects of naturally occurring or intentionally added fluoride.

Epidemiological studies where the water is naturally high in fluoride have shown no adverse effects other than dental mottling except in rare cases. Controlled studies with fluoridation at the 1 mg/l level have reported no instances of adverse effects. Available evidence does not suggest that fluoridation has increased or decreased cancer mortality rates.

Additional studies of mottling and skeletal fluorosis need to be done in communities with several milligrams per liter fluoride in their water supplies to ascertain whether the no-adverse-health-effect level for fluoride is greater or less than 1 mg/l. In addition, sociological studies are needed to ascertain the extent to which dental mottling is regarded as an adverse effect.

Nitrate: All sources of combined nitrogen must be regarded as potential sources of nitrate, for there is a tendency for all nitrogenous materials in natural waters to be converted into nitrate. Major point sources of combined nitrogen in water are municipal and industrial wastewaters, refuse dumps, animal feedlots and septic tanks. Diffuse sources include runoff or leachate from manured or fertilized agricultural lands, urban drainage and biochemical nitrogen fixation. Small amounts of combined nitrogen occur in rainfall from solution of atmospheric ammonia and oxides of nitrogen.

In the Community Water Supply Survey of the Bureau of Water Hygiene in 1969, the range of nitrate concentrations found was 0 to 127 mg/l. Nineteen systems, about 3% of those examined for nitrate, had concentrations in excess of the recommended limit of 45 mg/l.

Ordinarily, the major human intake of nitrate is from food rather than from water. The mean food intake in the United States has been estimated to be nearly 100 mg/day, most of it coming from vegetables such as spinach, lettuce, and root vegetables, which may contain several thousand mg/kg of nitrate.

Nitrate is secreted in the saliva, the mean value being about 40 mg/day, of which about 10 mg/day is reduced to nitrite and found in that form. These quantities, although internally derived, also represent inputs to the gastric system.

Two health hazards are related to the consumption of water containing large concentrations of nitrate (or nitrite): induction of methemoglobinemia, particularly in infants, and possible formation of nitrosamines, some of which may be carcinogenic.

Acute toxicity of nitrate occurs as a result of reduction to nitrite, a process that can occur under specific conditions in the stomach, as well as in the saliva. Nitrite acts in the blood to oxidize hemoglobin to methemoglobin, which does not perform as an oxygen carrier. Consequently, anoxia and death may ensue.

Healthy adults are reported to be able to consume large quantities of nitrate in drinking water with relatively few effects, if any. Acute nitrate toxicity is almost always seen in infants rather than adults. This increased susceptibility of infants has been attributed to high intake per unit weight, to the presence of nitrate-reducing bacteria in the upper gastrointestinal tract, to the condition of the mucosa, and to greater ease of oxidation of fetal hemoglobin.

Assessment of maximum nitrate levels in water exhibiting no adverse health effects has been based principally on a study of known cases of methemoglobinemia. No cases of methemoglobinemia were found in the original studies in which the water contained less than 10 mg/l nitrate as nitrogen. Later, a small fraction of total cases was found in which the nitrate concentration of the drinking water was somewhat less than 10 mg/l as nitrogen. Only one case in the United States has been associated with a public water supply regardless of nitrate content.

Studies supplementary to the previous ones, in which levels of methemoglobin in the blood of infants were related to concentrations of nitrate in the water being fed, showed elevation of methemoglobin levels in infants supplied with water containing nitrate as nitrogen only slightly in excess of 10 mg/l.

It can be concluded that from the viewpoint of induction of methemoglobinemia, the maximum concentration of nitrate in water exhibiting no significant adverse health effects is close to the interim standard of 10 mg/l as nitrogen. However, there appears to be little margin of safety for some infants with the standard at this concentration.

The other health hazard proposed for nitrate in water, that it may act as a procarcinogen, is more speculative. A series of reactions is involved by which it is proposed that nitrate in water may be converted to N-nitroso compounds that may be carcinogenic. The steps in the reaction sequence are: (1) reduction of nitrate to nitrite; (2) reaction of nitrite with secondary amines or amides in food or water to form N-nitroso compounds; and (3) carcinogenic reaction of N-nitroso compounds.

Reaction of nitrites and secondary amines or amides to form N-nitroso compounds occurs readily in acidic solution, and particularly at the normal pH of 1 to 5, that is characteristic of gastric contents after a meal.

However, the relation of nitrate concentrations in water supplies to the presence of nitrite in the digestive tract is much more problematic. The major source of nitrite to the stomach, at least for healthy individuals, is the saliva, normally containing 6 to 15 mg/l of nitrite. Little reduction of nitrate to nitrite occurs in the human stomach unless the gastric pH is greater than 4.6. Thus, the pH for formation of nitrite is quite different from that required for ready formation of N-nitroso compounds, pH 3.5 or less.

Epidemiologically, correlations have been shown between incidence of gastric cancer and concentration of nitrate in the drinking water. An unusually high incidence of stomach cancer in certain mountainous areas of Columbia is associated with high concentration of nitrate in the drinking water. The findings, however, are preliminary and only suggestive. They provide no firm evidence of a causal link between incidence of cancer and high intake of nitrate. They do indicate a need for caution in assessing lack of adverse health effects even at 10 mg/l nitrate as nitrogen and a need for continued intensive study on the metabolism and effects of nitrate in man.

In conclusion, available evidence on the occurrence of methemoglobinemia in infants tends to confirm a value near 10 mg/l nitrate as nitrogen as a maximum no-observed-adverse-health-effect level, but there is little margin of safety in this value. There is little scientific basis to support conclusion on the hazard of any concentration of nitrate in water with regard to carcinogenic potential.

Sulfate: No adverse health effects have been noted for concentrations of sulfate in drinking water less than about 500 mg/l. Diarrhea is the only physiological effect observed at concentrations greater than 1,000 mg/l.

The taste threshold for sulfate in water lies between 300 and 400 mg/l for most persons, but some are able to detect as little as 200 mg/l.

Water Hardness and Health: A large body of scientific information indicates that certain inorganic or mineral constituents of drinking water are correlated with increased morbidity and mortality rates. These constituents are not usually considered to be "contaminants" since they are often associated with the level of "hardness" of drinking water, and occur naturally or are picked up from water treatment or distribution systems. Hardness is due primarily to the presence of ions of calcium and magnesium and is expressed as the equivalent quantity of calcium carbonate ($CaCO_3$). Water containing less than 75 mg/l $CaCO_3$ equivalent is generally considered to be soft, and above 75 mg/l as hard.

The literature suggests that in the United States and other developed nations, the incidence of many chronic diseases, but particularly cardiovascular diseases (heart disease, hypertension, and stroke), is associated with various water characteristics related to hardness. Most of these reports indicate an inverse correlation between the incidence of cardiovascular disease and the amount of hardness. A few reports also indicate a similar inverse correlation between the hardness of water and the risk of several noncardiovascular causes of death as well.

Several hypotheses have been proposed to account for the correlations; these mostly involve either a protective action attributed to some elements found in hard water or harmful effects attributed to certain metals often found in soft water.

The hypothetically protective agents include calcium, magnesium, vanadium, lithium, chromium, and manganese. The suspected harmful agents include cadmium, lead, copper, and zinc, all of which tend to be found in higher concentrations in soft water as a result of its relative corrosiveness. However, there is disagreement over the magnitude, or even the existence, of a "water factor" in the risk of cardiovascular disease; the identity of the specific causal factors; the mode of action; and the specific pathological effects.

The wide spectrum of alleged associated effects, the lack of consistency in theorized or reported etiologic factors, the very small quantities of the suspected elements in water in comparison with other sources, and the discrepancies between studies, raise serious questions as to whether drinking water serves as a vehicle of causal agents, is an indicator of something broader within the environment, or represents some unexplained spurious associations.

Despite these uncertainties, the evidence is sufficiently compelling to treat the "hard water hypothesis" as plausible, particularly when the number of potentially preventable deaths from cardiovascular diseases is considered. In the United States, cardiovascular diseases account for more than one-half of about two million deaths that occur each year. On the assumption that water factors are causally implicated, it is estimated that optimal conditioning of drinking water could reduce this annual cardiovascular disease mortality rate in the United States by as much as 15%.

In view of this potential health significance, it is essential to ascertain whether water factors are causally linked to the induction of cardiovascular or other diseases and, if so, to identify the specific factors that are involved. Much more definitive information is needed in order to identify what kinds of remedial water treatment, if any, can be considered.

Organic Solutes

The organic compounds that have been identified in drinking water make up a small fraction of the total organic matter present. About 90% of the volatile organic compounds have been identified and quantified, but these represent no more than 10% by weight of the total organic material. Only 5 to 10% of the nonvolatile organic compounds, that comprise the remaining 90% of the total organic material, have been identified. (In this context, volatile signifies that the compound is detectable by gas chromatography.)

The compounds selected for review in this study included 74 nonpesticides of the approximately 309 volatile organic compounds so far identified in drinking water, and 55 pesticides. Some of the pesticides studied have not yet been detected in drinking water, but were included because they are or have been used in large quantities. A compound was selected for consideration if any of the following criteria applied:

(1) experimental evidence of toxicity in man or animals, including carcinogenicity, mutagenicity, and teratogenicity;
(2) identified in drinking water at relatively high concentration;
(3) molecular structure closely related to that of another compound of known toxicity;
(4) pesticide in heavy use; potential contaminant of drinking water supplies; and
(5) listed in the Safe Drinking Water Act or National Interim Primary Drinking Water Regulations.

Toxicological information about the compounds of interest was variable in quality and quantity and, in some instances, inadequate for a proper assessment of toxicity. In evaluating the potential effects on health of these organic compounds the principal concern was to assess their carcinogenicity. At the concentrations found in drinking water, none of the compounds would be expected to produce acute toxicity, but the effects of long continued ingestion of the carcinogens might well become a serious public health problem.

The risks associated with ingestion of compounds that were identified as carcinogenic (to man or animals, confirmed or suspected) were calculated, to the extent that data were available, by the method described in the section on Safety and Extrapolation. The results of these assessments are given in Table 1.1.

TABLE 1.1: CATEGORIES OF KNOWN OR SUSPECTED ORGANIC CHEMICAL CARCINOGENS FOUND IN DRINKING WATER

Compound	Highest Observed Concentrations in Finished Water, μg/liter	Upper 95% Confidence Estimate of Lifetime Cancer Risk per μg/liter
Human carcinogen		
Vinyl chloride	10	4.7×10^{-7}
Suspected human carcinogens		
Benzene	10	I.D.
Benzo[a]pyrene	D.	I.D.
Animal carcinogens		
Dieldrin	8	2.6×10^{-4}
Kepone	N.D.	4.4×10^{-5}
Heptachlor	D.	4.2×10^{-5}
Chlordane	0.1	1.8×10^{-5}

(continued)

TABLE 1.1: (continued)

Compound	Highest Observed Concentrations in Finished Water, μg/liter	Upper 95% Confidence Estimate of Lifetime Cancer Risk per μg/liter
DDT	D.	1.2×10^{-5}
Lindane (γ-BHC)	0.01	9.3×10^{-6}
β-BHC	D.	4.2×10^{-6}
PCB (Aroclor 1260)	3	3.1×10^{-6}
ETU	N.D.	2.2×10^{-6}
Chloroform	366	1.7×10^{-6}
α-BHC	D.	1.5×10^{-6}
PCNB	N.D.	1.4×10^{-7}
Carbon tetrachloride	5	1.1×10^{-7}
Trichloroethylene	0.5	1.1×10^{-7}
Diphenylhydrazine	1	I.D.
Aldrin	D.	I.D.
Suspected animal carcinogens		
Bis(2-chloroethyl) ether	0.42	1.2×10^{-6}
Endrin	0.08	I.D.
Heptachlor epoxide	D.	I.D.

I.D. = Insufficient data to permit a statistical extrapolation of risk.
D. = Detected but not quantified.
N.D. = Not detected.

Chronic toxicity of the compounds that were judged not to be carcinogenic was assessed by calculating, from such experimental results as were available, Acceptable Daily Intakes (ADI). These values are given in Table 1.2, together with estimates of maximal no-observed-adverse-health-effect concentrations in water that were derived from them. Compounds that could not be assessed for lack of experimental evidence are listed in Tables 1.3 and 1.4.

TABLE 1.2: ORGANIC PESTICIDES AND OTHER ORGANIC CONTAMINANTS IN DRINKING WATER—CONCENTRATION, TOXICITY, ADI, AND SUGGESTED NO-ADVERSE-EFFECT LEVELS

Compound	Maximum Observed Concentrations in H_2O, μg/l	Maximum Dose Producing No-Observed-Adverse-Effect, mg/kg/day	Uncertainty Factor*	ADI,** mg/kg/day	Suggested No-Adverse-Effect Level from H_2O,*** μg/l 1	2
2,4-D	0.04	12.5	1000	0.0125	87.5	4.4
2,4,5-T	—	10.0	100	0.1	700	35.0
TCDD	—	10^{-5}	100	10^{-7}	7×10^{-4}	3.5×10^{-5}
2,4,5-TP	detected†	0.75	1000	0.00075	5.25	0.26
MCPA	—	1.25	1000	0.00125	8.75	0.44
Amiben	—	250	1000	0.25	1750.0	87.5
Dicamba	—	1.25	1000	0.00125	8.75	0.44
Alachlor	2.9	100	1000	0.1	700.0	35.0
Butachlor	0.06	10	1000	0.01	70.0	3.5
Propachlor	—	100	1000	0.1	700.0	35.0

(continued)

TABLE 1.2: (continued)

Compound	Maximum Observed Concentrations in H_2O, µg/l	Maximum Dose Producing No-Observed-Adverse-Effect, mg/kg/day	Uncertainty Factor*	ADI,** mg/kg/day	Suggested No-Adverse-Effect Level from H_2O,*** . . . . µg/l 1	2
Propanil	–	20	1000	0.02	140.0	7.0
Aldicarb	–	0.1	100	0.001	7	0.35
Bromacil	–	12.5	1000	0.0125	87.5	4.4
Paraquat	–	8.5	1000	0.0085	59.5	2.98
Trifluralin (also for Nitralin and Benefin)	detected	10	100	0.1	700.0	35.0
Methoxychlor	–	10	100	0.1	700.0	35.0
Toxaphene	–	1.25	1000	0.00125	8.75	0.44
Azinphosmethyl	–	0.125	10	0.0125	87.5	4.4
Diazinon	–	0.02	10	0.002	14.0	0.7
Phorate (also for Disulfoton)	–	0.01	100	0.0001	0.7	0.035
Carbaryl	–	8.2	100	0.082	574	28.7
Ziram (and Ferbam)	–	12.5	1000	0.0125	87.5	4.4
Captan	–	50	1000	0.05	350	17.5
Folpet	–	160	1000	0.16	1120	56.0
HCB	6.0	1	1000	0.001	7	0.35
PDB	1.0	13.4	1000	0.0134	93.8	4.7
Parathion (and Methyl parathion)	–	0.043	10	0.0043	30	1.5
Malathion	–	0.2	10	0.02	140	7.0
Maneb (and Zineb)	–	5.0	1000	0.005	35	1.75
Thiram	–	5.0	1000	0.005	35	1.75
Atrazine	5.1	21.5	1000	0.0215	150	7.5
Propazine	detected	46.4	1000	0.0464	325	16.0
Simazine	detected	215.0	1000	0.215	1505	75.25
Di-n-butyl phthalate	5.0	110	1000	0.11	770	38.5
Di(2-ethylhexyl) phthalate	30.0	60	100	0.6	4200	210
Hexachlorophene	0.01	1	1000	0.001	7	0.35
Methyl methacrylate	1.0	100	1000	0.1	800	35.0
Pentachlorophenol	1.4	3	1000	0.003	21	1.05
Styrene	1.0	133	1000	0.133	931	46.5

*Uncertainty factor—The factor of 10 was used where good chronic human exposure data was available and supported by chronic oral toxicity data in other species, the factor of 100 was used where good chronic oral toxicity data were available in some animal species, and the factor 1000 was used with limited chronic toxicity data.

**Acceptable daily intake (ADI)—Maximum dose producing no observed adverse effect divided by the uncertainty factor.

***Assumptions—Average weight of human adult = 70 kg. Average daily intake of water for man = 2 liters.

1. 20% of total ADI assignment to water; 80% from other sources.
2. 1% of total ADI assigned to water; 99% from other sources.

†Detected but not quantified.

TABLE 1.3: ORGANIC PESTICIDES AND OTHER ORGANIC CONTAMINANTS FOUND IN DRINKING WATER, WITH INSUFFICIENT DATA ON CHRONIC TOXICITY TO CALCULATE AN ADI

Compound	Highest Concentration in Finished Water, μg/l
Acetaldehyde	0.1
Acrolein	ND
Bromobenzene	D
Bromoform	D
Carbon disulfide	D
Chloral	5.0
Chlorobenzene	5.6
Cyanogen chloride	0.1
1,2-Dichloroethane	21.0
2,4-Dichlorophenol	36.0
2,4-Dimethylphenol	D
ε-Caprolactam	D
Hexachloroethane	4.4
o-Methoxyphenol	D
Methyl chloride	D
Methylene chloride	7.0
Phenylacetic acid	4.0
Phthalic anhydride	D
Propylbenzene	<5.0
t-Butyl alcohol	0.01
Tetrachloroethane	4.0
Tetrachloroethylene	<5.0
Toluene	11.0
Trichlorobenzene	1.0
1,1,2-Trichloroethane	D
Nicotine	3.0
Methomyl	ND
Cyanazine	D
Xylene	<5.0

ND = Not detected in finished water.
D = Detected but not quantified.

TABLE 1.4: ORGANIC CONTAMINANTS FOUND IN DRINKING WATER WITH NO AVAILABLE INFORMATION ON CHRONIC TOXICITY

Compound	Highest Concentration in Finished Water, μg/liter	Highest Concentration in Raw Water, μg/liter
1,2-Bis(chloroethoxy) ethane	0.03	–
Bis(2-chloroisopropyl) ether	1.58	–
Bromochlorobenzenes	detected	–
Bromodichloromethane	116	11
Butyl bromide	detected	–
Chloroethyl methyl ether	detected	–
Chlorodibromomethane	100	1.4
Chlorohydroxybenzophenone	detected	–
Chloromethyl ethyl ether	detected	–

(continued)

TABLE 1.4: (continued)

Compound	Highest Concentration in Finished Water, μg/liter	Highest Concentration in Raw Water, μg/liter
Chloropropene	detected	–
Crotonaldehyde	5.0	–
Dibromobenzene	detected	–
Dibromodichloroethane	0.63	–
1,3-Dichlorobenzene	<3.0	–
Dichlorodifluoroethane	detected	–
Dichloroiodomethane	0.5	–
1,1-Dichloro-2-hexanol	1.0	–
1,2-Dichloropropane	<1.0	–
1,3-Dichloropropene	<1.0	–
1,2-Dimethoxybenzene	detected	–
4,6-Dinitro-2-aminophenol	detected	–
Dioctyladipate	20.0	–
Hexachloro-1,3-butadiene	0.07	–
Isodecane	5.0	–
Metachloronitrobenzene	detected	–
Methylstearate	detected	–
Nonane	4.0	–
Octyl chloride	detected	–
Pentachlorophenyl methyl ether	0.1	–
1,1,3,3-Tetrachloroacetone	1.0	1.0
2,4,6-Trichlorophenol	detected	–
Trimethylbenzene	6.1	–

The ADI represents an empirically derived value that reflects a particular combination of both knowledge and uncertainty concerning the relative safety of a chemical. When there is more confidence about data derived from animal experiments or observations on humans the uncertainty factor is smaller than when little is known about the potential toxicity of a chemical.

These numbers are not meant to represent a guaranteed safety level, but rather to indicate a level at which exposure to the single chemical in question is not anticipated to produce an observable toxic response in man. The ADI values do not consider interactions (e.g., synergism, antagonism) among the many possible contaminants. Furthermore, the ADI values do not represent safe levels in drinking water, because they do not take into account what fraction of the potential contaminant intake may come from water.

Suggested no-observed-adverse-health-effects concentrations in water have been calculated under two different assumptions: (1) That 20% of total intake of a material is from water and 80% from other sources, and (2) that 1% of total intake is from water and 99% from other sources (see Table 1.2). Similar calculations can be made for other materials discussed in this report using such data as may be available with regard to concentration of the contaminant in food or other sources.

Because the experimental data on the effects of many substances are inconsistent, "no-observed-adverse-health-effect" levels cannot be firmly specified for all organic contaminants. Most of the materials considered have not been studied sufficiently to firmly establish their carcinogenic potential with certainty. The risk assessments do not take into account interactions such as additive toxicity, synergism, and antagonism. What ultimately may be most important is the interaction of these compounds with each other and with other material in contributing to the total body burden resulting from multiple sources of contaminant exposure. For these reasons the ADI is intended to be used only as a guide for assessment of toxicity from chronic exposure. Furthermore, an ADI is not meant to provide a basis for the continuing discharge of a compound into the environment.

In the present limited state of our knowledge concerning structure-activity relationships for carcinogenic and other toxic effects, one cannot consistently and accurately extrapolate these properties from one compound to another. Nevertheless, in certain instances (for example, the substitution of bromine for chlorine in a halogenated methane) it is presumed that the relationship is sufficiently strong to justify the suspicion that the related compounds may be similarly toxic.

The potential for existing concentrations of organic pesticides and other organic contaminants in drinking water to adversely affect health, cannot be answered with certainty at this time. The key issue is whether or not certain organic chemicals found in very low concentrations can cause or increase the rate of cancer development in man. Even though several of these chemicals have demonstrated carcinogenicity in laboratory animals, the extrapolation of such results to man remains difficult for a number of reasons.

Because bioassays that have been used to establish carcinogenicity of certain organic chemicals are conducted at doses which are hundreds to thousands of times greater than the levels at which these chemicals occur in water, the risks at these low levels must be obtained by extrapolation from higher doses. There is no hard evidence that low level oral exposure to any of these chemicals produces cancer.

An argument has been made that the dose levels used to establish carcinogenicity are so high that they overwhelm normal detoxification or repair mechanisms or both, and produce cancer by some mechanism that does not operate under low dose conditions. Experimental animals subjected to such high doses could be considered a population different from those exposed to lower doses that do not produce pathological alterations and changes in pharmacokinetic parameters, or biochemistry.

Extrapolating from laboratory animals to man would be more meaningful if comparative metabolic information between the different species were available. Some species do not metabolize a parent compound to its activated form so that use of these species in toxicological bioassays is inappropriate if the compound undergoes activation in man. The converse situation also is true. Differences may also occur with respect to other parameters such as rates of biotransformation, absorption, excretion, and biological half life.

Risk assessments based on extrapolations which fail to consider species differences with respect to sensitivity, tissue susceptibility, kinetics, pathology, or biotransformation pathways may be inappropriate. This kind of information is not presently available.

In light of such uncertainties, a cautious approach must be adopted when dealing with potentially harmful chemicals. Even more uncertainty exists when one considers the possibility that some of these chemicals may also be mutagenic or teratogenic. The methodologies used to establish these effects are even less applicable to man than cancer bioassays.

For many of the organic compounds identified in drinking water, virtually no toxicity data are available. Ideally, all of these agents (as well as any new ones) should be subjected to an extensive battery of toxicity tests including chronic bioassay. In practice, there is a need to determine those agents for which the generation of data is most pressing. Several criteria are important for the development of an order of priority for testing. The main factors identified in the assignment of priorities are:

(1) the relative concentrations of the compounds and the number of people likely to be exposed as well as the identity of defined subpopulations exposed;
(2) the number of water systems in which they occur;
(3) positive responses to in vitro mutagenic screening systems;
(4) positive responses to in vitro carcinogen prescreens (mammalian cell transformations)
(5) similarity of chemical structure of the test compound to those of other compounds having defined toxic properties (i.e., structure-activity relationships); and
(6) relationship of dose from water to total body burden.

A number of assays using bacteria and yeast have shown promise in yielding high correlations between mutagenic activity and known carcinogenic activity for certain classes of materials. These may prove to be useful in establishing a first level screen for potential carcinogens.

Conclusions: Carcinogenicity – Table 1.1 lists the organic contaminants for which positive data on carcinogenesis exist. For these compounds, where adequate (lifetime) feeding studies were available, a statistical extrapolation of risk was performed. The method is described in the section on Safety and Extrapolation. The numbers in Table 1.1 are upper 95% confidence estimates of cancer risk to man from a lifetime of exposure to a particular compound. These estimates have been corrected for interspecific differences (that is, between the experimental animal and man) on the basis of relative surface area.

Bacterial Mutagenicity – In addition to examining data from animal feeding studies for the identification of suspect carcinogens, data for mutagenesis in bacteria, or other test systems were also examined. Available data are summarized as follows: (1) benzo[a]pyrene, chlorodibromomethane, Captan, and Folpet have been found to be mutagenic; (2) bromoform and vinyl chloride, weakly mutagenic; and (3) carbon tetrachloride, bromobenzene, nicotine, DDE, dieldrin, carbaryl and trifluralin, nonmutagenic.

Teratogenicity – Data on teratogenic potential exist for 24 of the compounds under study. Hexachlorophene, nicotine, the phthalate esters, 2,4-D, 2,4,5-T, and Folpet have been shown to be teratogens, while benzene, benzo[a]pyrene, carbon tetrachloride, PCBs, Captan, carbaryl, chlordane, DDT, Kepone, malathion, methylparathion, mirex, paraquat, and parathion have been reported to be nonteratogenic. Nowhere is the paucity of toxicologic data more evident than in the data on teratogenesis.

Noncarcinogenic Toxicity – For 45 compounds there were sufficient data to calculate ADI's. These are summarized in Table 1.2. Occasionally the ADI was calculated from partial lifetime exposure studies when no other data were available. Toxicity was measured by various responses.

The health effects of many compounds of interest could not be assessed because toxicological information about them was inadequate or unavailable. These compounds are listed in Tables 1.3 and 1.4 together with their reported occurrence in drinking water in the United States.

Definitions

The Safe Drinking Water Committee adopted the following working definitions prior to its review of the scientific literature of organic contaminants. The term carcinogen is used in its broad sense, because in most of the current human epidemiologic approaches and certain animal bioassays it is not possible to differentiate clearly between initiating agents, promoting agents, and certain modifying factors. Any factor or combination of factors which increases the risk of cancer in humans is of concern regardless of its mechanism of action. The criteria listed here apply only to chemical agents.

A malignant neoplasm is composed of a population of cells displaying progressive growth and varying degrees of autonomy and cellular atypia. It displays, or it has the capacity for, invasion of normal tissues, metastases, and causing death to the host. Benign neoplasms are a less autonomous population of cells and exhibit little or no cellular atypia or invasion of normal tissues and do not metastasize. In particular cases, however, benign neoplasms may endanger the life of the host by a variety of mechanisms, including hemorrhage, encroachment on a vital organ, or unregulated hormone production.

The cytologic and histologic criteria utilized in determining whether a lesion is benign or malignant differ depending upon the tissue in which the neoplasm arises. Evaluation of whether a specific lesion is benign or malignant should therefore, follow standard criteria used by experimental oncologists and pathologists with the emphasis on correlation of the histopathologic pattern with the biologic behavior of the lesion or type of lesion. In equivocal cases, the diagnosis of a specific lesion may require a panel of experts; recognizing that they may not always agree.

Depending upon the particular case, benign neoplasms may represent a stage in the evolution of a malignant neoplasm and in other cases they may be "end points" which do not readily undergo transition to malignant neoplasms.

Criteria in Human Studies: An agent, which may comprise a combination of chemicals, is carcinogenic in man if it increases the incidence of malignant neoplasms (or a combination of benign and malignant neoplasms) in humans to levels that are significantly higher than those in a comparable group not exposed (or exposed at a lower dose) to the same agent. If all of the induced neoplasms are benign, rather than malignant, then the agent must be considered a possible carcinogen and it should, therefore, be very carefully evaluated as a health hazard.

Types of evidence suggesting that an agent is carcinogenic in humans include: neoplastic response directly related to exposure (both duration and dose); incidence and mortality differences related to occupational exposure; incidence and mortality differences between geographic regions related to different exposures rather than genetic differences and/or altered incidence in migrant populations; time trends in incidence or mortality related to either the introduction or removal of a specific agent from the environment; case control studies; and the results of retrospective-prospective and prospective studies of the consequences of human exposure.

Clinical case reports may also provide early warning of a potential carcinogen. Negative epidemiologic data may not establish the safety of suspected materials. Negative data on a given agent obtained from extensive epidemiologic studies of sufficient duration are useful for indicating upper limits for the rate at which a specific type of exposure to that agent could affect the incidence and/or mortality of specific human cancers.

Criteria in Experimental Animal Studies: The carcinogenicity of a substance is established when the administration to groups of animals in adequately designed and conducted experiments results in increases in the incidence of one or more types of malignant neoplasms (or a combination of benign and malignant neoplasms) in the treated groups as compared to control groups maintained under identical conditions but not given the test compound. The increased incidence of neoplasms in one or more of the experimental groups should be evaluated statistically for significance, and the only major experimental variable between the control and the experimental group should be the absence or presence of the single test agent.

Such increases may be regarded with greater confidence if positive results are observed in more than one group of animals or in different laboratories. The demonstration that the occurrence of neoplasms follows a dose-dependent relationship provides additional evidence of a positive result.

The occurrence of benign neoplasms raises the strong possibility that the agent in question is also carcinogenic since compounds that induce benign neoplasms frequently induce malignant neoplasms. In addition, benign neoplasms may be an early stage in a multistep carcinogenic process and they may progress to malignant neoplasms.

For these reasons, if a substance is found to induce benign neoplasms in experimental animals it should be considered a potential human health hazard which requires further evaluation. In experiments where the increased incidence of malignant neoplasms in the treated group is of questionable significance, a parallel increase in incidence of benign tumors in the same tissue adds weight to the evidence for carcinogenicity of the test substance.

Mutagen – A chemical that is capable of producing a heritable change in genetic material. These changes may be either point mutations or chromosomal mutations and can occur in either somatic or germ cells.

Teratogen – An agent which acts during pregnancy to produce a physical or functional defect in the developing offspring.

Organoleptic Test – The use of odor and taste thresholds to establish permissible levels of exposure to chemicals.

Adverse Response – With increasing dosage in the continuum of the dose-response relationship, the region is generally entered where the effects are clearly adverse. Thus, adverse effects may be defined as changes that:

(1) occur with intermittent or continued exposure and that result in impairment of functional capacity (as determined by anatomical, physiological, and biochemical, or behavioral parameters) or in a decrement of the ability to compensate for additional stress;
(2) are irreversible during exposure or following cessation of exposure if such changes cause detectable decrements in the ability of the organism to maintain homeostasis; and
(3) enhance the susceptibility of the organisms to the deleterious effects of other environmental influences. (Quoted from the NAS publication, *Principles for Evaluating Chemicals in the Environment,* 1975.)

Toxicity – The intrinsic quality of a chemical to produce an adverse effect. The term includes capacity to induce teratogenic, mutagenic, and carcinogenic effects.

Safety – Safety is the practical certainty that injury will not result from the substance when used in the quantity and in the manner proposed for its use. (Quoted from *Evaluating the Safety of Food Chemicals,* NAS, 1970.)

Evaluation of Safety – An estimation of the potential of the substance to cause injury and review and evaluation of sufficient data to warrant a conclusion that the conditions of proposed use will provide an intake so low in relation to the toxic dose that there is a practical certainty no harm can result. (Quoted from FDA Papers, November 1971.)

For the purpose of this study the proposed use was limited only to exposure from drinking water.

Safety Factor or Uncertainty Factor – A number that reflects the degree or amount of uncertainty which must be considered when experimental data in animals are extrapolated to man. When the quality and quantity of data are high the uncertainty factor is low and when data are inadequate or equivocal, the uncertainty factor must be larger. The following general guidelines have been adopted in establishing the uncertainty factors:

(1) Valid experimental results from studies on prolonged ingestion by man, with no indication of carcinogenicity. Uncertainty factor = 10.
(2) Experimental results of studies of human ingestion not available or scanty (e.g., acute exposure only). Valid results of long-term feeding studies on experimental animals or in the absence of human studies, valid animal studies on one or more species. No indication of carcinogenicity. Uncertainty factor = 100.
(3) No long-term or acute human data. Scanty results on experimental animals. No indication of carcinogenicity. Uncertainty factor = 1,000.

These uncertainty factors are used in every case as a divisor of the highest reported long-term dose that is observed not to produce any adverse effect.

Carcinogens: Categories in Table 1.1 –

(1) Human Carcinogen: Based on strong epidemiological evidence and toxicological studies in animals.
(2) Suspected Human Carcinogens: Based on limited epidemiological evidence in man and equivocal toxicological studies in animals.
(3) Animal Carcinogens: Based on toxicological studies in at least one species of animal.
(4) Suspected Animal Carcinogens: Based on equivocal toxicological studies in animals or on a structural similarity to a known carcinogen.

Radioactivity in Drinking Water

Everyone is exposed to some natural radiation that comes from both cosmic rays and terrestrial sources. Although there are large geographic variations in the amount of natural background radiation, the average background dose in the United States is about 100 mrem/year. A small proportion of this unavoidable background radiation comes from drinking water that contains radionuclides.

By far the largest contribution to the radioactivity in drinking water comes from potassium-40, which is present as a constant percentage of total potassium. Only a small percentage of the total potassium-40 body burden, however, comes from drinking water. The total body dose from other possible radioactive contaminants of water constitutes a small percentage of the background radiation to which the population is exposed.

Although the amounts of individual radioactive contaminants fluctuate from place to place, calculations made for a hypothetical water supply that might be typical for the United States have shown that a total soft-tissue dose of only 0.24 mrem/year would be contributed by all the radionuclides found in the water. Even with rather wide fluctuations in the concentrations, the total contribution of the radionuclides will remain very small.

However, bone-seeking radionuclides, such as strontium-90, radium-226, and radium-228, account for a somewhat larger proportion of the total bone dose. This is particularly true for the two isotopes of radium because they emit high-linear-energy-transfer (LET) radiation, and because certain restricted localities have been found to have rather high concentrations of radium in drinking water.

Nevertheless, in the hypothetical typical water supply, less than 10% of the annual background dose comes from such radiation. It has also been estimated that the total population exposed to levels of radium greater than 3 pCi/l is about a million people. About 120,000 people are exposed to radium at levels greater than 9 pCi/l.

Risk estimates were made of three kinds of adverse health effects that radiation could produce: developmental and teratogenic effects, genetic effects and somatic (chiefly carcinogenic) effects.

Developmental and Teratogenic Effects: The developing fetus is exposed to radiation from radionuclides in drinking water for nine months. Thus, the total dose accumulated by the fetus will be very small. Furthermore, although the fetus is sensitive to the effects of radiation in some stages of development, these periods are sharply limited and extremely short. For this reason, too, the total dose administered that could have possible developmental and teratogenic effects would be extremely small.

Current concentrations of radionuclides in drinking water lead to doses of about one five-thousandth of the lowest dose at which a developmental effect has been found in animals. Therefore, the developmental and teratogenic effects of radionuclides would not be measurable.

Genetic Effects: It has been estimated that there are about 94,400 genetic diseases per million live births in the United States. The maximum permissible dose of man-made radiation for the general population (170 mrem/year) has been estimated to increase this number in the first generation by 170 to 215, with an unlikely upper limit of 4,250. On the basis of a 30-year generation and 3.6 million live births per year in the United States, we would expect the 0.24 mrem soft-tissue dose, or gonad dose, to lead to 0.0098 additional cases of genetic disease per million live births per year or 0.035 additional cases of genetic disease in the United States per year.

Even at the unlikely extreme upper limit of possible genetic effects of radiation of around 4,000 extra cases in the first generation, there would still be less than one additional case in the 94,400 x 3.6 = 340,000 live births with genetic defects. The wide fluctuation in bone dose caused by fluctuations in the radium concentration of drinking water would not have any sensible effect on the genetically significant dose, because radium is predominantly a bone-seeker and will deliver very little radiation to the gonads.

Somatic and Carcinogenic Effects: The natural background of radiation can be estimated to cause 4.5 to 45 cases of cancer per million people, depending on the risk model used. The amount of whole-body radiation from radionuclides in typical drinking water contributes less than 1% of this amount, and thus, for cancers other than those in bone, may cause a negligible increase in the total. Radium, however, can contribute somewhat less than 7% of the total bone dose received from background radiation in areas of "normal" radium concentration.

The average carcinogenic risk associated with skeletal irradiation by radium in a population with a typical distribution of ages, is estimated to approximate 0.2 fatal cases of bone cancer per million persons per year per rem. Therefore, over a period from 10 to 40 years after the beginning of skeletal irradiation, the average risk attributable to natural background radiation is estimated to range from 0.6 per million persons per year, under typical conditions, to as much as 4.2 per million per year, in regions where 25 pCi/l of radium-226 are found in the drinking water.

It has been noted that in the United States 120,000 people are estimated to drink water containing between 9 and 25 pCi/l of radium-226 and only a small number lie near the upper end of this range. The number of excess cancers in this group would therefore lie between 0.16 and 0.43 per year. Since not all the 120,000 people drink water containing 25 pCi/l of radium-226, the latter number is inordinately high.

Conclusions: The radiation associated with most water supplies is such a small proportion of the normal background to which all human beings are exposed, that it is difficult, if not impossible, to measure any adverse health effects with certainty. In a few water supplies, however, radium can reach concentrations that pose a higher risk of bone cancer for the people exposed.

Future Needs: The precision of estimation of the health risks associated with radioactivity in drinking water could be enhanced if several water systems were analyzed to determine the complete distributions of beta and alpha radiation that constitute the gross counting measurements.

Because the precise ratio of radium-228 to radium-226 in water has not been measured extensively, an attempt should be made to determine the ratio in several ground and surface waters whose content of radium-226 is known. The waters to be analyzed should range from about 0.1 to 50 pCi/l. The percentage of the daughter radionuclides present should be determined.

Because radon is a noble gas that is quickly released from water, it is possible that, in some areas of high radon content, water vapor containing radon might constitute an inhalation hazard when such water is used, for example, in humidifiers or for showers. A study should be made to determine whether or not radon emanations from water do indeed constitute an inhalation hazard.

The models used in this report do not take into account the possibility that the finely divided solid particles that occur in water may alter the uptake of radionuclides. The effects of the solids in drinking water on the metabolism and uptake of radionuclides merit investigation.

INTERIM PRIMARY DRINKING WATER REGULATIONS—CONTROL OF ORGANIC CHEMICAL CONTAMINANTS

> The material in this section was provided by the Environmental Protection Agency, and includes a proposed amendment to the NIPDWR. It was published in the *Federal Register*, Vol. 43, No. 28, on February 9, 1978.

This proposed amendment to the National Interim Primary Drinking Water Regulations is intended to protect the public health from organic chemical contaminants in drinking water. It consists of two parts:

(1) a Maximum Contaminant Level (MCL) of 0.10 mg/l (100 parts per billion) for total trihalomethanes (TTHM's) including chloroform, which occur in drinking water as the result of the interaction of chlorine applied for disinfection and other purposes, with the organic substances which naturally occur in raw water, and

(2) a treatment technique requiring the use of granular activated carbon in the treatment process for the control of synthetic organic chemicals associated with industrial pollution and urban and agricultural runoff contaminating drinking water supplies.

The Administrator has determined that the presence in drinking water of chloroform and other trihalomethanes, and synthetic organic chemicals may have an adverse effect on the health of persons and that, therefore, human exposure to these chemicals should be reduced.

The MCL for total trihalomethanes (TTHM's) is initially applicable only to community water systems serving a population of greater than 75,000 people and which add a disinfectant to the water in any part of the treatment process. Prior to the effective date, these water systems are required to conduct monitoring under section 1445 of the Safe Drinking Water Act.

Community water systems serving populations between 10,000 and 75,000 are only required to monitor for the level of TTHM's in their systems for one year. Community water systems serving fewer than 10,000 persons are not required to comply with the MCL or conduct monitoring under this regulation.

Additional microbiological monitoring is required for systems that change their treatment practices as a result of this regulation to ensure that no degradation in bacteriological quality of the drinking water occurs as a result of any such modification.

In addition, this proposed regulation limits the application of two of the disinfectants, chlorine dioxide and chloramines, which might be utilized in place of chlorine to reduce THM generation.

The treatment technique regulation initially requires community water systems with populations greater than 75,000 people to use granular activated carbon in their drinking water treatment systems. Community water systems which demonstrate that this treatment is not necessary to protect the health of persons due to the nature of their raw water source may be granted a variance from the treatment technique requirements.

A system may also be granted a variance to use an alternative treatment technique if it can demonstrate to EPA's satisfaction that its proposed alternative is at least as effective as granular activated carbon in reducing synthetic organic chemicals in drinking water.

Unless a community water system subject to the treatment technique is granted a variance, it is required to design, construct and operate a treatment system which uses granular activated carbon (GAC) to reduce the level of synthetic organic chemicals to the maximum extent feasible and is designed to meet specified criteria.

In some cases, these criteria may be achieved through addition of GAC in the existing system with only the carbon reactivation frequency being affected. In many cases, however, economics or practical feasibility will require the design and construction of postfiltration contactors and on-site reactivation facilities. In the latter cases, the regulation provides 3½ years after the effective date for design and construction of filters and furnaces.

Optimal design specifications, including contact time, type of carbon and regeneration frequency to meet the effluent criteria are to be determined on a case-by-case basis for each system, with approval by EPA or the State Director where the State has primary enforcement responsibility. The State is further authorized to impose such monitoring, operation and maintenance, and reporting requirements as it deems necessary to assure that the system, once completed, is being operated to minimize the presence of synthetic organic chemicals in the drinking water.

In addition, as an interim measure, some systems are required to replace the media in the existing filters of their treatment plants with GAC.

As additional operating and technical experience is gained with respect to the use of alternative disinfectants and the use of GAC in the treatment of drinking water, the scope of these regulations will be expanded to include smaller community water systems. In accordance with the requirements of the SDWA, Revised Primary Drinking Water Regulations will be forthcoming which will impose additional MCL's on organic chemicals as well as require further reductions in the levels of THM as additional toxicological, technological and analytical information becomes available.

Background

Under Section 1401(a)(1) of the Safe Drinking Water Act (SDWA), 42 U.S.C. §300(f) et seq, EPA is required to prescribe National Primary Drinking Water Regulations for those contaminants which the Administrator determines may have an adverse effect on human health. In the House Report, No. 93-1185, which accompanied the passage of the Act, Congress strongly emphasized the preventive philosophy underlying the Act.

Conclusive proof of an adverse effect is not a prerequisite to regulation. This is particularly important with respect to the Agency's authority to control organic chemicals in drinking water since these chemicals are most often found in only very small amounts. Even at these low levels, however, they are believed to pose a significant health risk to consumers of drinking water as a result of long-term exposure. Particularly, there is evidence to suggest that such exposure may increase the risk of human cancer and other chronic effects.

Thus, although it is not possible to quantify the harm caused by these chemicals, the Administrator has determined, based upon the toxicological and epidemiological data presently available, that the presence of organic chemicals in drinking water may have an adverse effect on human health and, therefore, should be controlled.

Under Section 1412(a)(2) of the SDWA, Interim Primary Drinking Water Regulations are required to protect health to the maximum extent feasible using treatment methods which are generally available (taking costs into consideration) when the Act was enacted. On December 24, 1975, EPA promulgated Interim Primary Drinking Water Regulations, 40 CFR Part 141, 40 FR 59556 et seq, which became effective on June 24, 1977. These regulations only controlled organic chemicals to the extent of establishing Maximum Contaminant Levels for six organic pesticides based upon data then available to the Agency.

An MCL for carbon-chloroform extract (CCE) was proposed by EPA during the rulemaking process but was subsequently deleted from the final regulations because it was determined that CCE had many failings as an indicator of health protection from organic chemical contamination (see 40 FR at 59568-59569). Nevertheless, at the time it promulgated the Interim Primary Regulations in December 1975, the Agency noted that when sufficient information concerning organic chemicals was derived from ongoing monitoring and other research programs, the Interim Primary Regulations would be amended as authorized under Section 1412(a)(1) of the Act.

Concurrent with the promulgation of the Interim Regulations, EPA published Special Monitoring Regulations under Section 1445 of the Act, aimed at gathering a comprehensive data based on the occurrence of the organic chemicals in drinking water, and on means for determining their presence. This resulted in the National Organic Monitoring Survey (NOMS).

On July 14, 1976, EPA published an Advance Notice of Proposed Rulemaking (ANPRM), titled "Control Options for Organic Chemicals in Drinking Water" 41 FR 28991 et seq. The ANPRM described the many facets of the issue of organic chemicals in drinking water including the legislative background, health effects data, the state of available control technology and costs. Advantages and disadvantages of various regulatory and nonregulatory options were examined, and the ANPRM solicited comments and information regarding the problem and options presented.

The ANPRM addressed the problem of organic contaminants in general, placing emphasis on the monitoring and control of specific chemical groups, particularly chloroform and the other trihalomethanes, as well as the treatment approach for controlling organic chemicals in general, including the use of GAC.

Over 100 comments were received on the ANPRM, and the consensus on those comments was that when sufficient information becomes available, organic contaminants should be controlled by means of establishing Maximum Contaminant Levels (MCLs) rather than by establishing required treatment techniques, and that surrogate parameters, if available, would be preferred over limits for individual compounds.

Many of the comments contained expressions of concern over the lack of data in support of regulation of the broad spectrum of organic contaminants. The comments generally supported the control of contaminants or groups of contaminants for which evidence of hazard to health could be presented.

Since the appearance of the ANPRM, a considerable amount of additional data on the occurrence and treatment of synthetic organic chemicals, and on possible human health effects from consumption of drinking water has become available. The NAS report *Drinking Water and Health* [see the previous section of this chapter, and chapter 3] was released in June 1977 and provides the most comprehensive compilation of data presently available on the health effects of organic chemicals in drinking water.

Additional toxicological information, particularly on drinking water concentrations, as well as epidemiological studies, suggests the possibility of a drinking water variable related to a human health risk. Although granular activated carbon was a treatment technology available at the time the Act was passed, its use by the American waterworks industry was very limited. Pilot and full-scale demonstration projects utilizing GAC and other adsorbants have given further assurance of the practical feasibility of the application of GAC in drinking water treatment plants.

The Administrator has concluded that the regulation of both THMs and synthetic organic chemicals in drinking water should be initiated at this time by amending the National Interim Primary Drinking Water Regulations. Consideration was given to the issuance of these requirements as part of the Revised Primary Drinking Water Regulations since there is no question that Congress expected the Revised Regulations to include such controls. However, it is equally clear that the establishment of Interim Regulations for these chemicals was left to the Administrator's discretion based upon the availability of data which would provide sufficient support for such Agency action.

Both in the House report and the floor debate accompanying the passage of the Act, Congress expressed concern with the presence of organic chemicals in drinking water. Due to their increasing influx into the environment and their extreme variation, Congress anticipated that organic chemicals would be regulated as a group. A comprehensive approach to standard-setting was encouraged. As a first step toward comprehensive control of organic chemical contamination of drinking water, it is consistent with the purpose of the Act to propose these regulations as amendments to the Interim Regulations.

The Agency's failure to include more comprehensive requirements governing the control of organic chemical contaminants in drinking water in the Interim Regulations promulgated in December 1975 has been the subject of litigation and continued Congressional concern. A decision in the court suit brought by the Environmental Defense Fund in the United States Court of Appeals for the District of Columbia Circuit seeking more comprehensive control of organic chemicals under the Interim Regulations is still pending.

During the Congressional debates over the amendments to the Safe Drinking Water Act in July 1977, the Agency was also criticized for its failure to control organic chemicals. It was noted that the 1962 Public Health Service standards had indicated such a need and indeed had included a standard for carbon-chloroform extract (CCE). The Congressional floor managers urged EPA to update its regulations to include more comprehensive controls for organic chemicals. The authorization for EPA to conduct additional studies on organic chemicals in drinking water was accompanied by the explicit caveat that such studies should not delay the standard-setting process.

Based upon further information presently available, it is necessary to proceed with efforts to reduce the level of organic chemicals in drinking water in the earliest feasible time frame. Such controls take time to implement and it would not be prudent to delay the imposition of controls while the means are available. The Revised Regulations will be promulgated concurrent with the establishment of recommended maximum contaminant levels and will seek to establish long-range requirements based upon the most current toxicological, occurrence and technological information available at that time.

By amending the Interim Primary Regulations as authorized by the Act to include more comprehensive control of organic chemicals, the Agency is clearly acting consistent with the Congressional intent to protect the public health to the maximum extent feasible using control methods which have been available since the Act's passage.

Organic Chemicals in Drinking Water

More than 700 specific organic chemicals have been identified in various drinking water supplies in the United States. These compounds result from such diverse sources as industrial and municipal discharges, urban and rural runoff, and natural decomposition of vegetative and animal matter, as well as from water and sewage chlorination practices. Compositions and concentrations vary from virtually nil in protected ground water to substantial levels in many surface waters and contaminated ground waters.

Organic chemical contaminants in drinking water can be divided into two major classes: those of natural origin and those of synthetic origin. The natural substances represent by far the greatest portion and consist primarily of undefined humus and fulvic materials and others produced by normal organic decomposition or biotic transformation and are not known to be harmful in themselves.

The synthetic chemicals in water can be subdivided into two groups. The first group consists of those chemicals that result from water treatment practices (e.g., trihalomethanes). Recent EPA studies indicate that, except for certain cases, the trihalomethanes constitute the largest portion of the identifiable synthetic chemicals in drinking water. Unlike other synthetic chemicals, chloroform and other trihalomethanes are formed during the treatment process. They are thus found in virtually every drinking water supply that is disinfected with chlorine, and not uncommonly at concentrations of several hundred parts per billion (ppb or micrograms per liter).

Studies by Sontheimer and Kuhn indicate that the THMs may represent only a portion of the total halogenated products of chlorination of water. Methods are being developed to quantify the total halogenated organic compounds produced during chlorination; however, for other than the chlorinated phenols and a few other substances, identification is very difficult.

Halogenated organics such as carbon tetrachloride, chloroform and hexachloroethane have also been detected at parts-per-million levels as contaminants in chlorine. Chlorine manufactured by the graphite-anode process is more likely to have higher levels of organics than chlorine from metal anodes, particularly if the chlorine has not been "scrubbed" properly after manufacture. Chlorine used for potable water disinfection should be of the highest purity to avoid introduction of these by-products into drinking water.

The second group of synthetic chemicals consists of those chemicals introduced as a result of point and nonpoint sources of pollution. Nationally, both surface waters and to a lesser degree ground waters are contaminated with a variety of these pollution-related synthetic organic chemicals ranging generally from the lower molecular weight halogenated hydrocarbons and monocyclic aromatic compounds to higher molecular weight pesticides, polycyclic aromatic compounds, and pesticide-like compounds.

These classes of compounds have been found in drinking water using gas chromatography or gas chromatography/mass spectroscopy. However, the large bulk of organic matter (primarily natural products but also higher molecular weight synthetics) in water is not amenable to detection by these commonly used methods. Those organic contaminants which have been identified in drinking water constitute only a small percentage of the total amount of organic matter present.

Because the chemicals thus far identified in drinking water account for only a small fraction of the total organic content, the possibility, and indeed the probability, exists that additional substances of equal or greater toxicological significance may be present but remain undetected by present monitoring capabilities.

In 1974 monitoring of 10 cities for a broad range of organic compounds was conducted as part of the National Organics Reconnaissance Survey (NORS). Contamination of finished water supplies revealed the presence of 129 organic compounds, attributable to industrial, agricultural and municipal sources. A ground water source was also found to be contaminated.

The most recent comprehensive data on the presence of organic chemicals in drinking water can be found in the National Organics Monitoring Survey (NOMS) in 1976 and 1977. The NOMS was intended to provide a more comprehensive survey of synthetic organic contaminants in finished drinking water by monitoring for approximately 20 specific organic compounds as well as such general organic indicators as Total Organic Carbon (TOC) and Carbon Chloroform Extract (CCE).

The NOMS data clearly demonstrated that the THMs have the greatest occurrence in finished water. A large number of other synthetic chemicals were also detected at lower levels. Some apparent correlation could be found between TOC and THM, but no correlations could be made for the remaining organics and any of the surrogate indicators used.

As the sensitivity of analytical methods has improved, greater numbers of organic compounds have been found. Ongoing studies being conducted by EPA's National Screening Program will further help to identify those water systems vulnerable to contamination by organic chemicals.

Part I: Trihalomethanes

The trihalomethanes found in drinking water are members of the family of organohalogen compounds which are named as derivatives of methane, where three of the four hydrogen atoms have been replaced by three atoms of chlorine, bromine or iodine. Ten distinct compounds are possible by various combinations of three halogenated atoms, one hydrogen and carbon atom. Current analytical technology applied to drinking water has thus far detected chloroform (trichloromethane), bromodichloromethane, dibromochloromethane, bromoform (tribromomethane) and dichloroiodomethane.

The principal source of chloroform and other trihalomethanes in drinking water is the chemical interaction of the chlorine added for disinfection and other purposes with the commonly present natural humic and fulvic substances and other precursors produced either by normal organic decomposition or by the metabolism of aquatic biota.

Recent EPA studies of the chlorination of certain algae in culture indicate the formation of THM. The actual human exposure potential, however, will vary depending upon the season, contact time, water temperature, pH, type and chemical composition of raw water and treatment methodology.

Since the natural organic precursors are more commonly found in surface water, water taken from a surface source is more likely than ground water (with notable exceptions) to produce high THM levels. Generally, the THM-producing reaction is as follows: chlorine + (bromide ion or iodide ion) + precursors = chloroform + (other trihalomethanes).

Chloroform is the most common trihalomethane found in drinking water and it is also usually present in the highest concentration. In a number of cases the concentrations of the brominated trihalomethanes were found to far exceed the chloroform concentrations. The mixed trihalomethanes appear to form by way of an initial oxidation of bromide ion in solution by added chlorine, followed by rapid and apparently preferential bromination of the organic precursors. Bromine may also be introduced as a contaminant of chlorine.

Iodination probably occurs by a similar process, however, the details are less well understood. Dichloroiodomethane has been detected in drinking water along with the chlorinated and brominated THMs but it is not included in the present regulation because of analytical and quantification difficulties. Chloroform is also a common contaminant in chlorine.

Chloroform and other trihalomethanes were first reported in drinking water in late 1974. EPA initiated the National Organics Reconnaissance Survey (NORS) of 80 water utilities, which confirmed that THMs were being formed during chlorination in the drinking water treatment process. Water samples were collected at the treatment plant and iced for shipment but not dechlorinated. Concentrations in finished water appeared to be roughly related to the amounts of natural chemicals present in the water.

In late 1975, EPA initiated the National Organics Monitoring Survey (NOMS) in 113 cities throughout the United States, and this study is now virtually complete. The NOMS expanded on the NORS by including analyses for THMs throughout the year in three phases at each utility, as well as quantifying a large number of other synthetic chemicals found in the water. THM samples were collected at water treatment plants and also in the distribution systems.

Phase I analyses in the NOMS were conducted similarly to the NORS. Phase II analyses were performed after the THM-producing reactions were allowed to run to completion (terminal). Phase III analyses were conducted on both dechlorinated samples and on samples that were allowed to run to completion (terminal).

The NOMS demonstrated that considerable amounts of THMs could form in the water after it has entered the distribution systems on the way to the consumer's tap. It also showed that THMs far exceeded the concentrations of other synthetic organic contaminants in finished drinking water, and that brominated THMs could also exceed the chloroform concentrations. Part of the THM concentration variation between Phases I, II, and III might also be attributable to seasonal effects. [Additional information on these studies is contained in chapter 2 of this book.]

Human Health Considerations: Past use of some of the trihalomethanes, particularly chloroform and bromoform in anesthesia and medicinal preparations, respectively, has provided an extensive catalog of health effects data from high levels of human exposure.

Chloroform has been shown to be rapidly absorbed upon oral and peritoneal administration and subsequently metabolized to carbon dioxide and unidentified metabolites in urine. Chloroform is metabolized in vitro to 2-oxothiazolidine-4-carboxylic acid and methylene chloride. The metabolic profile of chloroform in animal species such as mice, rats, and monkeys is qualitatively similar to that in man. Trihalomethanes other than chloroform may be expected to be absorbed rapidly by ingestion and metabolized in a similar manner to chloroform in both humans and experimental animals because of their structural and chemical similarities with chloroform.

Human exposure to chloroform may result from several sources including ambient air, residues in food (from its use as a fumigant or from natural sources), toothpaste and cough medicines (until recently), and occupational exposures, as well as from drinking water. The Food and Drug Administration prohibits the use of chloroform in human drugs and cosmetics, and has proposed its prohibition in food packaging, because of the potential risk associated with its use. Depending upon the ranges of chloroform (and trihalomethanes) concentrations that have been detected in air, food and water (which are junctions of location, urbanization and industrialization), drinking water may contribute from none to 90% or more of the total daily intake.

Mammalian bioeffects following exposure to chloroform include its effects on the central nervous system, hepatotoxicity, nephrotoxicity, teratogenicity, and carcinogenicity. These responses are discernible in mammals from exposure to high levels of chloroform ranging from 30 to 350 mg/kg; the intensity of response was dependent upon the dose. Short term exposure to the low levels of chloroform commonly found in drinking water supplies are not known to manifest acute toxic effects. The potential for human effects from chronic lifetime exposure at low concentrations is the basis for this regulation.

Evidence of the carcinogenicity of chloroform in mammals has been confirmed by several studies. These data are extensively reviewed in the report of the National Academy of Sciences titled *Drinking Water and Health* (June 1977) and in the *Statement of Basis and Purpose* incorporated by reference herein (see chapter 2).

Chloroform produced malignant and metastatic neoplasms (cancers) in at least one feeding study in mice, and has produced tumors in both rats and mice in other studies. Malignancies have been produced in a dose-related fashion in both rats and mice in the studies by the National Cancer Institute and the latency period for the carcinogenic effect decreased as the dose increased in the male animals. Those studies established that chloroform was carcinogenic to the animals under the test conditions and therefore, might also present a carcinogenic risk to humans.

Data on the oncogenic (tumor causing) effect of other trihalomethanes are very limited, but several of them will be tested in the NCI Bioassay Screening Program. Bromoform generated excess lung tumors in preliminary pulmonary tumor induction tests with strain A mice, but chloroform did not. Brominated compounds have exhibited more carcinogenic analogs in several cases in similar studies. Tests of the mutagenicity of a number of halogenated compounds including trihalomethanes, in the *Salmonella typhimurium* bacterial system in vitro have demonstrated that, in general, brominated compounds are more active than chlorinated compounds. Although chloroform was not mutagenic in that test, other trihalomethanes containing bromine and iodine were mutagenic in the bacterial test system.

Thus, although less toxicological information is available for the brominated THMs than for chloroform, toxicity, mutagenicity, and carcinogenicity have been detected in some test systems. Because of the liability of organically-bound bromine, physiological chemical activity would be expected to be greater for the brominated THMs than for chloroform. For these reasons, as well as the chemical structural similarity and common route of formation, the brominated THMs are also included with chloroform in this regulation.

Epidemiological evidence relating THM concentrations or other drinking water quality factors and cancer morbidity/mortality is not conclusive, but suggestive of a health risk. Positive statistical correlations have been found in several studies, but causal relationships cannot be established on the basis of those epidemiological studies. One preliminary epidemiological study investigated the association between chloroform and TTHM and cancer mortality and obtained some positive correlations particularly with bladder cancer.

The correlation was stronger for the brominated THMs than with chloroform. Several other preliminary studies have been conducted on the relationship between chlorination, sources of drinking water supplies and cancer morbidity/mortality. The evidence from these studies thus far is incomplete due to limitations in the scope of the studies, the small sample sizes, the lack of water quality data and the trends and patterns of association not having been fully developed.

When viewed collectively, however, the epidemiological studies provide sufficient evidence for maintaining the hypothesis that a potential health risk may exist and that the positive statistical correlations may be due to some association with drinking water quality. More definitive studies are being conducted by EPA and the National Cancer Institute. In addition, EPA has requested the National Academy of Sciences to provide an independent assessment of those epidemiology studies already completed.

Although it is generally agreed that it is not possible to project with accuracy from risk estimates to absolute numbers of cancers in a human population exposed to a given agent, the approximate impact of the standard has been estimated using statistical extrapolation models derived from animal data. Statistical assessments of risk have been attempted using several mathematical models, and they indicate that the excess risk for lifetime exposure at the regulation level (0.10 mg/l) would be on the order of 10^{-4} to 10^{-5}; that is, an excess lifetime (70 years) cancer risk in the range of 1 in 10,000 to 1 in 100,000. This will be reduced considerably as actual average levels below 0.10 mg/l are achieved and as the standard itself is reduced in the future.

Those risk estimates are computed on the basis of chloroform alone, and do not consider the presence of other substances which might be introduced concurrently with chloroform or pollution-related contaminants and which would contribute to a higher level of risk.

Control of Trihalomethanes: There are three basic ways of controlling trihalomethanes in drinking water: (1) use of a disinfectant that does not generate trihalomethanes in water; (2) treatment to reduce the precursor concentration prior to chlorination; and (3) treatment to reduce the trihalomethane concentration after formation.

Each of these general approaches can be further divided into other control options, depending upon individual circumstances. However, these should not be considered as mutually exclusive options. Indeed, in many cases some combination of all three may be necessary to simultaneously minimize organics and optimize pathogen control.

Since any of the disinfectants or their corresponding by-products may have some undesirable properties, a fundamental principle should be to apply whatever treatment is needed to produce water of high quality and low chemical content prior to the application of the disinfectant. Thus, the chemical disinfectant demand of the water will be minimized and pathogen control will be maintained while disinfectant use and by-product formation will be minimized.

Disinfectants other than free chlorine or hypochlorite, such as ozone, when used in conjunction with a disinfectant residual such as a chloramine or chlorine dioxide, will eliminate or greatly reduce the formation of THM. Ozone in combination with a chlorine residual may be usable in some waters, although preliminary data indicate that ultimate THM precursor concentrations are not reduced.

More efficient operating control of chlorination at the treatment plant may also result in decreased trihalomethane formation. Use of a chlorine residual in a less active form such as chlorine combined with ammonia (chloramine) will significantly reduce trihalomethane formation, however, chloramines are much less potent disinfectants than free chlorine. In fact, early studies, subsequently confirmed, demonstrated that chloramines required approximately a 100-fold increase in contact time to inactivate coliform bacteria and enteric pathogens as compared to free available chlorine.

For this reason, chloramines are not recommended for use as primary disinfectants in drinking water treatment. Chloramine treatment finds its widest application in maintenance of chlorine residuals in the distribution systems, after primary disinfection with free available chlorine or other disinfectant. The health effects of water treatment with chloramine have not been studied in detail.

Although these disinfectants in the absence of chlorine do not produce trihalomethanes, questions have been raised on the issue of the toxicology of chlorine dioxide and other disinfectants and their by-products. Chlorine dioxide introduces chlorite ion upon partial reduction.

Studies with cats have shown that chlorite has a deleterious effect on red blood cell survival rate at chlorine dioxide concentrations above 10 mg/l. Therefore, a limit of 1.0 mg/l is necessary to prevent potential adverse effects on sensitive individuals, particularly children. Additional studies are under way to clarify this matter.

The Administrator is authorized to restrict the use of alternative disinfectants such as chloramines and chlorine dioxide under Section 1401(1)(D) of the Act which provides that in addition to a maximum contaminant level, a Primary

Drinking Water Regulation must include "criteria and procedures to assure a supply of drinking water which dependably complies with such maximum contaminant levels; including quality control and testing procedures to insure compliance with such levels and to insure proper operation and maintenance of the system."

Congress anticipated that additional safeguards would be necessary to assure that public water systems dependably supply safe drinking water. While the Administrator is prohibited by Section 1412(b)(6) of the Act from requiring the addition of any substance for preventive health care purposes, he is not precluded from restricting the use of chemicals which he finds might jeopardize the provision of a safe supply of drinking water.

Because the establishment of an MCL for trihalomethanes attributable to chlorination practices will cause many water suppliers to turn to alternative disinfectants which are either less effective than chlorine or have undesirable side effects of their own, it is necessary to restrict the use of such alternatives in order to assure that the most healthful drinking water is being provided to consumers while efforts are being made to comply with the specific MCL requirements for TTHM and microbiological contaminants.

High Total Organic Carbon (TOC) levels in water can be indicators of potentially high disinfectant demand because of consumptive chemical processes (e.g., THM production from chlorine). Ozonation of high TOC waters can also lead to aftergrowth of flora in distribution systems. Thus, particular care must be taken in the selection of the total treatment scheme, especially in waters with TOC concentrations above approximately 2 mg/l, to avoid the potential problems that can be introduced by excessive application of any disinfectant.

The trihalomethanes precursors, which are probably complex mixtures of humic and fulvic substances and may also include simpler low molecular weight compounds containing the acetyl group ($CH_3{-}C{=}O$), can be reduced to some degree by effective coagulation and filtration. The maximum benefit from this technique for precursor removal is not achieved unless the chlorine is added only at that point in the treatment process where precursor concentration is lowest (i.e., after coagulation and settling or after filtration).

Use of fresh granular activated carbon (GAC) is the best technique among all of the control options because it is effective both for precursor and trihalomethane removal (to a lesser degree) as well as for general removal of contaminating organic chemicals. The versatility of GAC application for general organic chemical removal is important because Revised National Primary Drinking Water Regulations will contain MCLs or treatment techniques for additional organic chemicals. Thus, the use of GAC for THM control could obviate the need for additional treatment modifications to control other contaminants and to provide added protection from other undesirable organic chemicals such as from upstream pollution.

Thus, EPA recommends that public water systems drawing water from sources subject to contamination with synthetic organic chemicals should use adsorbents like GAC in their treatment processes in order to provide the best level of protection of the public health from organic chemical contamination. This is consistent with the accompanying proposed treatment regulation for synthetic organic chemicals.

Some operational problems with GAC use include its relatively short life for THM control in some waters and bacterial growth on the adsorbers. In some water treatment procedures, in German practice, biological activity on the GAC, "biological activated carbon (BAC)," is encouraged as an effective means of reducing organic chemical loadings and extending the time between reactivation of the carbon.

A typical treatment train utilizing the BAC approach might include raw water chlorination and coagulation followed by ozonation, contact with deep beds of granular carbon, then application of a disinfectant such as chlorine dioxide for maintenance of a residual. EPA encourages the use of adsorbents such as GAC where appropriate to reduce THM precursors as well as other contaminants and our research program is actively studying all aspects of its application.

Two techniques have been tested for removal of THM from water after formation. These include aeration with a high gas-to-water ratio and the use of an adsorbent specifically designed for THM removal. Neither of these techniques would alter the continued generation of THM in the distribution system from the reaction of remaining precursors and a free chlorine residual.

The NORS and NOMS studies have shown some degree of correlation between concentrations of trihalomethanes and Nonpurgeable Total Organic Carbon (NPTOC) measurements on finished water. NPTOC and TOC measurements can now be made readily and at low cost per sample. Thus, TOC or NPTOC analyses conducted along with the THM monitoring might provide some insights to the water system operators as to water quality factors influencing THM formation. NPTOC or TOC measurements might ultimately be useful for predicting THM formation potential in a water system using a particular treatment method.

Discussions of the various treatment techniques in detail, including the estimated cost of treatment, specific suggestions for modifications of several different types of water treatment plants, and recommendations for operational monitoring, as well as of NPTOC analyses will be found in chapters 4 through 9 of this book.

Regulatory Approach: The Safe Drinking Water Act mandates that substances in drinking water that may have adverse human health effects should be controlled to the extent feasible using available technology and taking costs and other factors into consideration. In the Interim Primary Drinking Water Regulations promulgated December 24, 1975, 40 CFR, at 5956B, the Administrator stated that as soon as sufficient information was developed from the monitoring programs and related research, the regulations would be amended to deal with additional organic chemicals beyond the six pesticides regulated at that time.

The Administrator has now determined that a maximum contaminant level for chloroform and related trihalomethanes under the Interim Primary Regulations is warranted and that means are available to larger water systems to control trihalomethanes without increasing the risk of microbiological contamination. This decision was reached based on the following factors:

(1) The health effects data, including bioassay, monitoring, mutagenicity, and epidemiology are further developed for chloroform and by implication for THMs than for other organic chemicals in drinking water.

(2) THMs are generally present in greater concentrations than other organic chemicals in drinking water, thus facilitating feasible monitoring and analytical techniques.
(3) THMs are more ubiquitous contaminants of drinking water than other synthetic organic chemicals.
(4) Several treatment procedures are available for THM control in drinking water.

These regulations should not be construed as compromising the absolute principle of providing drinking water of the highest microbiological quality, or of discouraging the use of chlorine (or hypochlorite), but rather as encouraging treatments under optimal conditions so as to minimize formation of undesirable byproducts, be they organic or inorganic. This philosophy encourages selection of the best available raw water source and application of the most appropriate physical and chemical processes to produce water of very high quality (and low disinfectant demand) prior to application of the disinfectant.

This amendment to the Interim Regulations will initially affect those situations where excessive quantities of trihalomethanes are being generated by current treatment practices. In addition, the techniques that will be utilized to reduce trihalomethane concentrations will also concurrently result in reductions of many other contaminants that are undesirable by-products of the treatment process or are contaminants in raw water.

The causes and incidences of water-borne disease outbreaks demonstrate the essentiality of water disinfection and the continued use of chlorine. Except for the demonstrated public health benefits of chlorination, trihalomethane concentrations might have been controlled at a lower level sooner. At least in the near term, chlorination must be continued to prevent infectious water-borne disease and thus, at this time THM concentrations are being reduced by technologically and economically feasible methods rather than being eliminated.

The proposed MCL of 0.10 mg/l should not be construed as a "safe" level. The current scientific thought on human exposure to substances which have been demonstrated to be carcinogens in animals in appropriate tests is that they be considered potential carcinogenic risks to humans. The presumption is that health risk is related to the extent of exposure and that no threshold exposure level without risk can be demonstrated for a genetically diverse population.

Thus, translated into a regulatory philosophy, exposure to those substances should be minimized so as to minimize risks. Therefore, water systems should strive to reduce THM concentrations to levels as low as is economically and technologically feasible, without compromising the microbiological quality of the drinking water.

Based upon these considerations, an MCL of 0.10 mg/l for TTHM has been proposed under interim regulations as a reasonable level providing health protection to the extent feasible. This should not lead to complacency on the part of systems with a THM of 0.09 mg/l, for example, but should be construed only as a starting point which will, over time, be lowered progressively as technologic, economic, and practical constraints permit.

Within 2 years after the effective date of this regulation, EPA will review the implementation experience. Assuming successful implementation, it is expected that the MCL will be reduced and/or coverage of impacted public water systems will be expanded to include additional smaller systems. As soon as sufficient practical experience has been gained so as to optimize the application of these or other technologies, EPA expects that levels as low as 0.010 mg/l or less will be attainable in many cases.

Selection of a Maximum Contaminant Level of 0.10 mg/l for THM was based on a balancing of public health considerations and feasibility of achievement. This included such practical considerations as availability of control technology, monitoring methods and trained personnel, economic considerations, new risks which may be introduced from changes in current treatment practice and the resource-limited technical assistance available from EPA.

Establishment of a specific MCL is therefore not a license for those systems with lower THM levels to allow concentrations to increase to the MCL. Rather, all water systems should strive to minimize THM concentrations by technologically and economically feasible methods that will not compromise protection from water-borne infectious disease. EPA expects that application of the various control possibilities will probably result in many water systems achieving levels well below the MCL.

Analogous to the use of the coliform test as a simple indicator of the presence of microbial pathogens in water, THM concentrations should be treated as indicators of the presence of other undefined halogenated compounds concurrently produced during chlorination. Thus, measures that are taken to reduce THM concentrations will concurrently provide the additional benefit of reducing human exposure to the other undefined by-products.

Because of the structural, and likely biochemical and toxicological similarities of the trihalomethanes, and their common occurrence in drinking water in much higher quantities than other synthetic chemicals, and because the exposed human population is very large, EPA has determined that an MCL for trihalomethanes should be part of the first phase in its approach to synthetic organic chemical control in drinking water. Pollution-related contaminants will be more comprehensively controlled in the concurrent proposed treatment regulations for synthetic organics and in the Revised National Primary Drinking Water Regulations.

Selection and Applicability of the MCL: This part of the amended interim regulations is the first step in a phased regulatory approach to controlling potentially harmful organic chemicals in drinking water. This part consists of a Maximum Contaminant Level for total trihalomethanes and of monitoring requirements that apply to community water systems in a manner that is dependent upon population served, as follows:

(1) Systems serving populations greater than 75,000 persons–MCL for total trihalomethanes at 0.10 mg/l would be effective 18 months from the date of promulgation of the regulations. Monitoring at a minimum of 5 analyses per quarter would be effective 3 months from the date of promulgation of the regulations. Standard Plate Count analyses are required to ensure maintenance of microbiological quality when treatment changes are implemented and this stipultion also effective 3 months after promulgation.

(2) Systems serving populations between 10,000 and 75,000 persons—the MCL is not applicable. Monitoring for one year at a minimum of 2 analyses per quarter effective 6 months from the date of promulgation of the regulation would be required. These results must be periodically reported to the responsible State agency and EPA.

(3) Systems serving populations smaller than 10,000 persons are not subject to the MCL for THM or monitoring requirements at this time.

The selection of an MCL, monitoring requirements and population coverage in these interim regulations was based upon the feasibility of achieving reductions of THMs and the concern that any attempted modification of disinfection practice to control trihalomethanes must not in any way affect the microbiological quality of drinking water which would increase the possibility of transmission of infectious disease.

The interim MCL of 0.10 mg/l total trihalomethanes was selected from a number of higher and lower options that were considered. Available control and monitoring technology, national and per capita costs, population coverage, and availability of personnel, equipment and materials were among the feasibility factors influencing the decision.

The mean levels of TTHMs from Phase II and Phase III of the National Organics Monitoring Survey (NOMS) were 0.117 mg/l and 0.100 mg/l, respectively, in water samples that were allowed to react to completion. Averages between dechlorinated and terminal samples could be considered as estimates representative of likely concentrations to be found at the tap of the average consumer. These were 0.093 mg/l in Phase II and 0.077 mg/l in Phase III. Maximum THM levels ranged as high as 0.695 mg/l and 0.784 mg/l in terminal samples. Therefore, an interim MCL of 0.10 mg/l will result in substantial reduction of THM concentrations in many water systems now exceeding the MCL and provide adequate health protection.

The MCL and its associated monitoring requirements apply to community water systems serving populations greater than 75,000 persons. Water systems serving between 10,000 and 75,000 persons are required only to monitor for trihalomethanes and report the results to EPA and the State. There are 390 community water systems in the United States serving populations greater than 75,000 persons and they represent a total population of 100 million persons, or 52% of the total population served by community water systems.

Initial limitations of the MCL to systems serving populations greater than 75,000 provides an achievable level of protection and introduces a manageable demand on the supplier industries for instrumentation, raw materials, construction, activated carbon and disinfectants, and expert personnel. Control is thus technologically and economically feasible in that size range. The 2,300 community water systems serving populations between 10,000 and 75,000 persons represent a total population of 62 million persons or 32% of the total population served by community water systems.

In these communities reasonable steps to minimize THM concentrations should be evaluated and utilized where feasible under proper supervision. EPA feels that requiring water systems serving populations between 25 and 10,000 persons to meet the MCL or monitoring requirements would not be feasible at this time in most instances, and it encourages their continued use of chlorine and other necessary disinfection treatment methods while additional information is being generated by the first phase of this regulatory program.

A considerable level of protection is provided by this regulation since it applies to those locations where the likelihood of THM contamination is greatest. The great majority of larger public water systems that are covered by those regulations utilize surface water which is more likely to produce greater quantities of trihalomethanes. The majority of small systems utilize ground water, a less likely source of trihalomethanes. In addition, many of the small systems do not now chlorinate and therefore would not be producing THMs.

The almost 40,000 public water systems in the United States ranging in size from 25 to several million persons serve water to almost 200 million people in total. Because of the intimate relation of THM control to the disinfection process and the limited THM control technology experience in the United States, EPA feels it is imperative that any changes in current treatment practice must be carefully supervised and supported by technical assistance from the States and EPA.

Both EPA and State resources for this type of technical assistance are very limited. In a number of cases the changes in treatment practice attendant to this regulation will involve application of fairly sophisticated new procedures involving close cooperation of engineers, chemists and microbiologists to assure successful treatment without sacrificing microbiological quality of drinking water. It is not prudent nor feasible, therefore, to deal with all public water systems in the short term.

Rather, EPA chose as an initial step to select an MCL for TTHM that would be protective and to limit its application to a large but reasonable number of public water systems that would have the financial base and access to expertise necessary to achieve that MCL.

Using a phased regulatory approach, EPA intends to initially affect the public water systems serving the largest populations and the worst cases of high level exposure to THM. In the future, as additional experience is gained, the MCL for THMs will become more restrictive and coverage will be extended to smaller water systems as necessary to adequately protect the public health.

At this time, while EPA believes that exposure of consumers of drinking water to THMs should be reduced to the extent feasible, no less health protection is necessarily being provided to those smaller public water systems not covered by the MCL, many of which utilize ground water. Because THM levels are not expected to be high for ground water sources of drinking water, smaller water systems are being required to monitor for the presence of THMs in order to better ascertain the scope of the problem in that size category.

EPA believes that it is consistent with the Act's objective of health protection to proceed with care in controlling THMs in light of the delicate balance between THM reduction and adequate disinfection.

A number of community water systems which are not now impacted by this regulation presumably will either voluntarily or due to State action also attempt to reduce THM concentrations. While such action is encouraged, EPA recommends that this be attempted only under appropriate State supervision, and the Agency will provide technical assistance to such efforts within available resources.

THM Monitoring Requirements: THM monitoring pursuant to Section 1445(a) in systems larger than 75,000 persons should be performed at least quarterly and a minimum of 5 samples should be collected on the same day and analyzed according to EPA approved procedures, and the results arithmetically averaged. Initial sampling must begin within 3 months of promulgation of these amendments so that data will be available in time for systems to determine corrective actions before the effective date of the regulation.

The sampling locations should be selected to be representative of the THM concentrations at the consumers' taps. No more than 20% of the samples should be collected at the entry point of the distribution system and no less than 20% should be at extremes in the system. The remaining 60% should be from locations representative of population density throughout the distribution system.

Compliance with the MCL is computed by averaging the quarterly values from the preceding 12 months. This method was selected, rather than basing compliance on the highest value obtained in any single sample, because of the large and possibly uncontrollable variations of raw and finished water quality that occur in the short term. These and other factors which must be dealt with on a case-by-case basis could lead to wide variations of THM concentrations.

At this time it would be unreasonable to demand the kind of pinpoint control that would be necessary to maintain THM levels below a particular figure at all times and at all locations in the distribution system of every water system. This approach could be changed as additional experience is gained through the implementation of this regulation.

Monitoring for THMs in smaller systems (10,000 to 75,000 persons) pursuant to Section 1445(a) of the Safe Drinking Water Act should be initiated within 6 months of promulgation of these amendments and should continue for a period of one year. These data will assist EPA in assessing the magnitude of THM concentrations in these systems and in making determinations as to the need and feasibility of revised regulations in that size category.

In addition, identification of locations with high THM levels will allow the system, State or EPA to take appropriate action on a case-by-case basis when feasible. Monitoring requirements are minimal, involving only 2 analyses per quarter. Both samples should be collected at the entry to the distribution system at the same time. One sample should be dechlorinated immediately; the other sample should be stored for 7 days in the sample container at room temperature (20° to 25°C) prior to analysis to permit the chlorination reactions to go to completion.

The results from these two samples should thus reflect the extremes of possible THM concentrations in the drinking water. These data must be reported to EPA and the State. Data from the 1976-1977 NOMS has been transmitted to each of the 113 participating water systems, so a number of them already have sufficient information in hand from which to plan compliance needs. To assure that sufficient qualified laboratories will be available to perform the required analyses, EPA is developing a Quality Assurance program and will expand the current laboratory certification program. This will assure the validity of the analytical data used by water systems to determine compliance and possible treatment modifications.

Additional Microbiological Monitoring Requirements: To assure that any steps taken to reduce THM concentrations in drinking water will not increase the possibility of microbial contamination, additional microbiological monitoring is required for a water system that is modifying existing treatment practice.

Standard Plate Count (SPC) determinations must be made at least daily both at the treatment plant and in the distribution system, for one month before and six months subsequent to the treatment change to assure that no degradation of water quality occurs. Analyses prior to the change are intended to provide a baseline to which subsequent effects can be compared. The appropriate number and sampling locations of SPC analyses should be determined by the State or EPA depending on local conditions. Significant deviations from the "normal" range must be reported to the State or EPA and corrective actions taken immediately.

The normal microbiological organisms in public water supplies originate from the source water. This initial population of bacteria may be drastically reduced through chemical treatment and filtration. Disinfection will further reduce the remaining population significantly to perhaps one ten thousandth of the bacterial population in the source water. These survivors may be joined by other organisms that penetrate the treatment barrier along with turbidity or during momentary lapses in treatment.

Microbial contaminants carried by airborne dust introduced through clear well vents, installation and repair of distribution lines, cross-connections, and negative line pressure situations all contribute to a microbial population that is partially enumerated by the Standard Plate Count.

Water treatment regimes are designed to minimize the public health risk associated with organisms of sanitary significance, but the process becomes selective of certain microbial strains. Although the genera of organisms detected by the Standard Plate Count may not be harmful or dangerous to normal persons when present in drinking water in low numbers, under certain conditions, these organisms are known to produce severe or chronic human infections.

The National Academy of Sciences Report *Drinking Water and Health* underscores the usefulness of the Standard Plate Count in evaluating these organisms and recognizes that the Standard Plate Count should be used in conjunction with the total coliform procedure(s) to measure the sanitary quality of potable water. In that report, the Academy recommends use of the Standard Plate Count to:

(1) provide a method of monitoring for changes in the microbiological quality of finished water,
(2) determine whether the normal flora of a water supply may be interfering with coliform detection, and
(3) monitor the effectiveness of a disinfectant or treatment practice within the plant and distribution system and provide an indication of

filter-effluent quality deterioration and the occurrence of the positive breakthrough of pathogenic microorganisms.

Use of the Standard Plate Count in evaluating disinfectant efficiency should be coupled with total coliform determinations at various plants in the treatment process. This information will provide an indication of relative efficiency of each treatment process so that when changes are made, any deterioration of water quality can be detected. It should be clearly understood that the Standard Plate Count is not a substitute for total coliform as a measurement of the sanitary significance of potable water, but rather a sensitive tool to signal an increased potential risk of breakthrough of undetected pathogens such as viruses or protozoa.

The presence of coliform bacteria in water is still considered the most reliable indicator of possible fecal contamination and any associated enteric pathogenic microorganisms. Certain viruses and the cysts of *Entamoeba hystolytica* and *Giardia lamblia*, however, are capable of surviving in water for longer periods of time than the indicator coliform bacteria. In addition, these pathogenic organisms are also somewhat more resistant to disinfection than the indicators.

Thus, use of the Standard Plate Count at least during periods of treatment alteration will provide a more sensitive indication of any changes in the effectiveness of microbial control resulting from process changes.

State Discretion for Monitoring Frequencies: The State is granted discretion to adjust monitoring requirements after one year upon a determination that local conditions are such that THM concentrations would not be likely to approach or exceed the MCL. After THM data have been collected according to the requirements for a full year, a community water system may request that the State allow a reduction of the monitoring frequency. The State's case-by-case decision should be based on such factors as the monitoring data, the quality and source of the raw water, and the type of treatment employed.

Monitoring cannot be reduced to less than one sampling every six months. After the first year's monitoring data have been obtained this minimal additional monitoring should be sufficient to demonstrate that conditions have not changed to the extent that the water supply might later exceed the MCL. However, a change in the source of water or modification of the treatment program would require that the initial monitoring requirement be reinstated to reestablish the baseline.

Effective Dates of Monitoring Requirements: The regulations would require that baseline quarterly monitoring for THMs for the larger systems (about 75,000 persons and above) would be initiated within 3 months of promulgation of these amendments, and for smaller systems (about 10,000 to 75,000 persons) within 6 months of the date of promulgation.

The effective date for the TTHM Maximum Contaminant Level would then be 18 months after the promulgation date. This schedule is proposed so that sufficient time will be available to the affected Public Water Systems for the acquisition of equipment and personnel, the accumulation of baseline compliance data, conducting of engineering feasibility studies and the introduction of control technology. In addition, 18 months is deemed sufficient time for States with primary enforcement responsibility to incorporate the new Federal requirements into their statutory and/or regulatory authorities.

Economic Impact of THM Regulations: The economic impact was projected based on several control options available to satisfy the regulatory requirements, modifying chlorination or other treatment procedures, changing disinfectants, or using an adsorbent. The calculation of cost projections for the THM regulation requires an estimation of the number of systems which are likely to select each of these treatment strategies to comply with the regulation.

An incremental cost will accrue to all systems impacted, whether or not treatment is required, to cover monitoring expenses. Because no empirical method exists for predetermining the choice of treatment that would be made by each affected public water system, a more probabilistic and structured approach was used.

A logical sequence of decision points was designed to distribute the systems covered by the regulation according to the most likely path they would follow. The decision made at each point is consistent with certain criteria as follows:

(1) The treatments currently used: If a system does not add chlorine it will not be affected by a THM regulation, and therefore will require no new treatment.
(2) Water source used: If a system uses surface water (except the Great Lakes and some high quality mountain waters) as its primary source, it is more likely to exceed a given level of THM contamination. Hence, the number of water systems using water from ground or surface sources affects the number of systems which will exceed the MCL and therefore require treatment.
(3) Degree to which water quality exceeds MCL: If the presence of THM is only slightly in excess of the initial MCL, then minimal modifications to procedures may be adequate for compliance. As the level of contamination increases, a system must consider more significant (and costly) treatment techniques.
(4) Economic considerations: The presumption was that systems would adopt the least costly treatment strategy that satisfies the regulations.
(5) Treatment effectiveness: The presence of THMs above certain levels can probably be best controlled by the use of adsorbents. This is because of the likelihood that high disinfectant demand waters cannot be disinfected adequately without generating considerable amounts of by-products of unknown hazard. Consequently, those systems with very high levels of THMs are likely to use adsorbents.

Of the 390 public water systems that serve more than 75,000 people, 60 purchase the majority of their water from other systems that are presumed to provide treatment. Thus, a total of 330 systems would be initially affected, although 18 of these are excluded because they do not presently add chlorine. Of the remaining 312, some 86 were estimated to have TTHM levels above 0.10 mg/l and hence would require treatment changes.

The 21 systems estimated to exceed the level by less than 25% were assumed to be able to comply by the least costly method: modifying current chlorination or other procedures. The remaining 65 systems are split into those above and below 0.25 mg/l of TTHM. Of those estimated to be over the 0.25 mg/l level, 20% were assumed to be able to comply by changing disinfectant and 80% would probably use an adsorbent to achieve compliance, because changing disinfectants alone might not be an appropriate method to bring the systems into compliance with the regulation considering the quantity of other by-products that might be formed. Eighty percent of those between 0.125 and 0.25 mg/l were estimated to change disinfectant and the remaining 20% would also use an adsorbent. On the basis of the above assumptions, national costs, exclusive of monitoring costs, for compliance with the proposed regulations are as follows:

Summary of Estimated Total Costs for an MCL Regulation for Trihalomethane Concentration of 0.10 mg/l (In Millions of 1976 Dollars)

Capital expenditures, cumulative to 1981	154.4
External financing, cumulative to 1981	145.2
Operating and maintenance expenses, 1981	25.9/yr
Revenue requirements, 1981	36.0/yr
Annual per capita costs* (dollars)	2.07/person

*Revenue requirements divided by population served by the cost-impacted systems

Per capita costs vary depending upon the type of treatment selected, population and other factors and range from 28 to 93 cents for alternate disinfectants and $3.30 to $6.11 for an adsorbent assuming a 60-day regeneration cycle. More detailed cost analyses are to be found in Chapter 9.

Monitoring costs were computed based upon a survey of contract analytical laboratories currently performing TTHM analyses. Per sample costs ranged from $25 to $100. After these regulations have been promulgated, the increased volume of business and competitive factors would be expected to reduce the analytical costs to below $25 per sample.

Annual monitoring costs for a large community water system (greater than 75,000 population), assuming the minimum frequency of 5 samples per quarter and $25 per sample, amount to $500. Annual costs for a smaller water system (10,000 to 75,000 population) at the minimum frequency of 2 per quarter amount to $200. Annual national monitoring costs amount to $153,500 for the 307 out of 390 (less 18 that do not chlorinate) large systems that probably would not need to alter their current treatment practice and $464,000 for the approximately 2,300 smaller systems affected.

EPA expects that a number of community water systems will choose to purchase the equipment and monitor for THM on-site more frequently than the minimum, for operational control as well as for compliance purposes. An additional benefit from purchase and on-site analytical capability, is that the gas chromatograph is versatile and can be used to monitor for the presence of many other organic chemical contaminants besides THMs. The cost of equipping an existing laboratory with an appropriate gas chromatograph is dependent upon which analytical procedure is selected and the type of instrument.

The basic instrumentation for the "liquid-liquid" extraction method consists of a gas chromatograph with an "Electron Capture" detector and recorder; cost approximately $5,000. The basic instrumentation for the "purge and trap" method consists of a gas chromatograph, a halogen-specific detector "purge and trap" sample concentrator, and recorder; cost approximately $10,000. In either case, additional expenditures for accessories should be added. Costs would also be greater for more sophisticated instrumentation beyond the basic system.

EPA is concurrently conducting studies on analytical methods based on gas chromatography alone that would be feasible for direct monitoring and operational use by public water system personnel. This project is attempting to develop methodology and demonstrate applicability in several hundred public water systems.

Part II: Synthetic Organic Chemicals

A large number of synthetic organic chemicals have been identified in various drinking water supplies in the United States. Since these chemicals are normally present in only very small quantities, it has only been within the past few years that sufficiently sensitive analytical methods have been developed to detect and quantify such chemicals at low levels. As analytical techniques become more sensitive and as new techniques are developed, the number of chemicals detected continues to grow.

These chemicals have been found for the most part in surface water supplies which have become the repository for the waste discharges of numerous industrial facilities and municipalities as well as for urban and agricultural runoff. Ground water supplies, however, have not been spared from contamination due to the leaching of these chemicals from poor maintenance practices at industrial facilities and improper on-land disposal of wastes.

At the time of the passage of the Safe Drinking Water Act in 1974, there were more than 12,000 chemical compounds known to be in commercial use, and many new chemical compounds are being added to the list each year.

The causes of synthetic organic chemical contamination are both chronic and variable in nature. Industrial discharges from point sources are regulated by this Agency under the Federal Water Pollution Control Act's National Permit Discharge Elimination System. However, despite the control of industrial discharges through a comprehensive nation-wide permit system, a certain amount of toxic pollutants continues to be discharged into surface waters. In addition, there is always the possibility of accidental spills.

A large quantity of chemicals may thus enter a water supply system against which current water treatment practices are not presently equipped to provide protection. Finally, there are an undefinable number of nonpoint sources of pollution which contribute to the problem. Many pesticides used for agricultural purposes are known to be highly toxic. Storm water runoff carrying other potentially harmful toxic chemicals is another pollution-related source of contamination.

The National Organics Reconnaissance Survey (NORS) conducted in 1974-1975 detected as many as 129 organic compounds in finished water supplies throughout the country. Synthetic organic chemicals that were found included carbon tetrachloride and 1,2-dichloroethane.

The most recent comprehensive data on the occurrence of synthetic organic chemicals in drinking water is from EPA's National Organics Monitoring Survey (NOMS) conducted under the National Interim Primary Drinking Water Regulations. The synthetic organic chemicals identified encompass many of the known classes of compounds including halogenated aliphatic and aromatic hydrocarbons such as carbon tetrachloride, dichloroethane, vinyl chloride, and chlorobenzenes; pesticides such as dieldrin and lindane; aromatics such as benzene, toluene and styrene; polynuclear aromatics such as fluoranthene; nitrogenous compounds such as aniline and nitrobenzene; esters such as dibutylphthalate; and many others.

The three compounds most frequently found at average concentration levels in the low parts per billion range were pentachlorophenol, dichlorobenzene and trichloroethylene.

This and other occurrence data clearly demonstrate that synthetic organic chemical contamination in many of the nation's drinking water supplies is a reality. As part of the Agency's pronounced objective to reduce the public's exposure to toxic chemicals in the environment generally, this proposed regulatory action to reduce the levels of such chemicals in the nation's drinking water supplies to the maximum extent feasible was deemed necessary at the present time.

Human Health Considerations: Synthetic organic chemical contaminants have been found to cause both acute and chronic adverse effects in humans at high exposure levels. However, at the lower concentrations at which they occur in drinking water, EPA's primary concern is with their potential contribution to elevated cancer risks. At the present time, the specific cause(s) of cancer is not yet understood.

Nevertheless, there is growing agreement within the scientific community that prolonged exposure to carcinogenic contaminants in the environment, including food, air and water, contributes significantly to the incidence of this dread disease which accounts for approximately 350,000 deaths annually in the United States. Other long-term risks such as mutagenicity and teratogenicity are also of serious concern.

Although the concentrations of each synthetic organic compound, when detected in drinking water, have been generally at the parts per billion level or lower, the aggregate exposure to such chemicals from a lifetime of water consumption is significant in terms of potential risk to human health. Moreover, EPA data on the occurrence of such chemicals in drinking water indicate that unique exposure to one or a few of these compounds is never the case. Thus, there is the possibility of synergistic interactions among chemicals thereby enhancing the associated risks.

As with most pathways of exposure to cancer-causing agents in the environment, there is no direct evidence that consumption of drinking water has actually caused human cancers. However, EPA believes that such carcinogens when present in drinking water pose an unreasonable risk to public health. The precise magnitude of the risks cannot be quantified since there are many unmeasured and untested chemicals in drinking water and because the extrapolation models are imprecise and require more comprehensive national occurrence data than are currently available.

However, for one group of chemicals, the THMs, a statistical extrapolation of the data indicates that a lifelong exposure to these contaminants in drinking water may account for 200 excess fatalities per year nationwide. This projection assumes that the effects of chloroform are representative of THMs and that our limited occurrence data is representative of its presence generally in drinking water.

Other studies would indicate even higher risks. Moreover, that computation probably underestimates the total risk since it relates only to chloroform for which occurrence and dose/response data exist. It does not include risks from any of the other trihalomethanes or other products of chlorination, or the risk from any of the myriad of other synthetic organic chemicals which have been detected in drinking water.

Based on this information, EPA has concluded that synthetic organic contaminants in drinking water constitute a public health concern and that the use of granular activated carbon (GAC) in the treatment process to reduce human exposure to these chemicals where they occur is a reasonable preventive health measure. We believe that the mainstream of thinking within the scientific community will support EPA's conclusions. Since it is unlikely that the uncertainties in interpreting the toxicological data will be resolved in the near future, regulatory action to reduce these apparent risks using the best technology presently available, GAC, is deemed warranted at the present time.

Regulatory Approach: In developing National Interim Primary Drinking Water Regulations for the control of synthetic organic chemicals, EPA considered the two regulatory approaches available under Section 1401 of the Safe Drinking Water Act. These two regulatory options consist of establishing a maximum contaminant level for the contaminant and/or prescribing a treatment technique for the control of such contaminant.

For synthetic organic chemicals in drinking water, the Administrator has concluded that a treatment technique should be prescribed based upon his judgment that it would not be technologically or economically feasible to monitor for the presence of all the synthetic organic chemicals in drinking water which may have an adverse effect on human health.

As previously noted, there are thousands of organic compounds potentially present in drinking water in small quantities. Only with the application of sophisticated analytical techniques such as gas chromatography/mass spectrometry in the early 1970s have researchers been able to begin to identify and quantify many of the trace synthetic organic contaminants in drinking water. However, based upon what we presently know about these chemicals that have been identified and for which toxicological data are available, there is sufficient reason to believe that the presence of synthetic organic chemicals in drinking water, both those we can currently measure and those which may be identified in the future, pose a risk to human health.

Moreover, it is likely that this health risk may be heightened by the potential for synergistic effects among the many chemicals known to be, and which may be, present in drinking water.

Monitoring feasibility is the critical determinant as to whether a contaminant or group of contaminants should be controlled by means of an MCL or a treatment technique under Section 1401 of the Act. If an MCL(s) is specified, a public water system is required to conduct periodic monitoring to insure that the level of such contaminant in the drinking water does not exceed that established by the MCL.

The infeasibility of monitoring on a routine basis for all the synthetic organic chemicals about which there is cause for concern was a key determining factor in the Administrator's decision to prescribe a treatment technique. For the purposes of this treatment technique, approximately 60 individual compounds have been listed in Table 1.5, representing key indicators of synthetic organic chemical contamination and each of them can be and has been detected in drinking water by available methods.

TABLE 1.5: CHEMICAL INDICATORS OF INDUSTRIAL CONTAMINATION

Aliphatic Halogenated Hydrocarbons	
Methane derivatives:	Dichloromethane
	Trichlorofluoromethane
	Dichlorodifluromethane
	Carbon tetrachloride
Ethane derivatives:	1,1-dichloroethane
	1,2-dichloroethane
	Hexachloroethane
	1,1,-trichloroethane
	1,1,2-trichloroethane
	1,1,2,2-tetrachloroethane
Unsaturated hydrocarbons:	Trichloroethylene
	Tetrachloroethylene
	Vinyl chloride
	1,1-dichloroethene
	1,2-dichloroethene
	1,3-dichloropropene
	hexachlorobutadiene
	2-chlorovinyl ether
Other halogenated compounds:	1,2-dichloropropane
	Bis(2-chloroethyl)ether
	Bis(2-chloroisopropyl)ether
Cyclic Aliphatic Compounds	
Chlorinated hydrocarbons:	Lindane and BHC
	Kepone
	Toxaphene
Cyclodienes:	Hexachlorocyclopentadiene
	Aldrin
	Dieldrin
	Chlordane
	Heptachlor
	Heptachlor epoxide
	Endrin
Aromatic Hydrocarbons	
Benzenes:	Benzene
	Toluene
	Xylenes
	Ethylbenzene

(continued)

TABLE 1.5: (continued)

	Propylbenzene
	Styrene
Halogenated aromatics:	Chlorinated naphthalenes
	Chlorobenzene
	Dichlorobenzenes
	Trichlorobenzenes
	Polychlorinated biphenyls
	Bromobenzene
	DDT
	DDE
	DDD
	Chlorophenols
	Pentachlorophenol
	4-bromophenylphenyl ether
	4-chlorophenylphenyl ether
	Hexachlorobenzene
Other aromatic hydrocarbons:	Nitrobenzene
	Dinitrotoluene
	Phthalate esters
	Atrazine
	3,4-benzofluoranthene
	Benzo[k] fluoranthene
	1,12-benzoperylene
	Fluoranthene
	Indeno[1,2,3-cd] pyrene
	Benzo[a] pyrene

However, there are presently no surrogates or group analytical schemes which would permit economical routine sampling and analyses to be performed to determine whether specific maximum contaminant levels have been exceeded. Research is continuing on identifying and demonstrating methods for detecting groups of compounds.

Total organic halogen (TOH) or total organic chlorine (TOCl) is one method being developed which shows promise. TOCl could have application in assessing raw water quality as an indicator of synthetic chemical contamination, or as a measure of the performance of GAC over time in reducing those contaminants, thus indicating the need for reactivation. TOCl and other such group parameters are being actively considered for possible future use in applications related to this regulation.

Monitoring for the list of chemical indicators is feasible for the limited purpose of determining whether a public water system's raw water supply is subject to significant contamination by synthetic organic chemicals and thereby eligible for a variance from the treatment technique requirement. However, the combined costs of monitoring for even all known synthetic organic chemicals to determine specific levels of such contaminants as would be required for judging compliance with a lesser number of MCLs would be clearly excessive and thereby unreasonable.

Based upon these considerations, the establishment of an MCL(s) for synthetic organic chemicals was not deemed to be a feasible regulatory approach at this time. As additional knowledge is gained with respect to the use of surrogate parameters as a means for monitoring for synthetic organic chemicals, and as analytical capabilities are improved and methods and technologies further developed, MCLs for synthetic organics or their surrogates will be included in EPA's Revised Primary Drinking Water Regulations.

A treatment technique is also preferable to MCL requirements for the control of synthetic organic chemicals because of the flexibility which it affords as a regulatory tool. Whereas an MCL requires a public water system to provide treatment to reduce the concentration below the acceptable level, however much that MCL has been exceeded, the treatment technique requires the system to operate a treatment process using GAC which has been designed to achieve a specified set of conditions.

Once the system has been so designed and is operational, it must thereafter be operated in such a manner so as to reduce the level of synthetic organic chemicals to the maximum extent feasible, using the appropriate type of carbon, contact time and regeneration frequency. Because States with primary enforcement responsibility would have the ability to take into account diverse factors such as the nature of the raw water source, and other local conditions in issuing variances from a treatment technique, they would have the lead role in controlling synthetic organic contamination. These factors are not easily incorporated into an inflexible national standard such as an MCL.

Finally, variances may be granted from an MCL requirement only if the MCL cannot be achieved despite the application of best available technology, and such variances must be conditioned upon a compliance schedule which requires compliance as expeditiously as practicable. Variances from a treatment technique are issued under entirely different conditions in that they may be granted upon a showing that the system's raw water is of such quality that the application of a treatment technique is not necessary for the protection of the health of persons.

The treatment technique requirement proposed herein is intended to limit the public's exposure to all of the synthetic organic chemicals present in drinking water which are believed to pose a risk to human health, and not just those chemicals for which MCLs could conceivably be established. This approach is clearly consistent with the preventive philosophy of the Safe Drinking Water Act and its legislative history. There is no doubt that by requiring systems to install and properly operate a granular activated carbon system as part of their treatment process, the public will be the beneficiaries of cleaner, more healthful drinking water.

Selection and Applicability of the Treatment Technique Requirement: Under Section 1401 of the Safe Drinking Water Act, the Administrator is directed to specify each treatment technique known to him which would lead to a reduction in the level of the contaminant sufficient to protect public health to the extent feasible.

As discussed in the *Statement of Basis and Purpose,* and *Treatment Guide,* several available treatment technologies for organic chemical reduction have been examined by the Administrator. These included carbon adsorption, resin adsorption and aeration. The use of GAC has been found to be the best broad spectrum treatment technology currently available for controlling these contaminants and therefore is the only technology being specified at this time. However, although GAC is required in the proposed treatment regulation, a system which can demonstrate that an alternative treatment technique is at least as efficient as GAC in reducing the levels of a broad spectrum of synthetic organic chemicals may be granted a variance conditioned on the use of that alternative treatment technique.

In accordance with the Agency's phased regulatory approach to the control of organic chemicals in drinking water, the proposed treatment technique requirement is initially made applicable to community water systems serving a population of more than 75,000 individuals. As noted in the discussion of the MCL for total trihalomethanes, this requirement will affect 52% of the total population served by community water systems.

As with MCL for TTHMs, a considerable level of protection is provided by this regulation since it applies to those locations most likely to be subject to contamination from synthetic organic chemicals. Synthetic organic chemical contamination is mainly found in surface water sources of drinking water which receive upstream chemical discharges from industrial facilities and urban and agricultural runoff. Surface waters are the predominant source of drinking water for most large community water systems covered by this regulation, whereas most smaller systems receive their drinking water from ground water sources, less likely to be subject to synthetic organic contamination.

Furthermore, the use of granular activated carbon in the treatment of drinking water is a relatively new and sophisticated technology. It is not now standard practice in the United States, although about 40 cities now use GAC filtration to improve the taste and odor of their drinking water. Careful supervision is essential for the proper design and operation of GAC system to efficiently and consistently reduce the levels of synthetic organic chemicals in drinking water.

The larger systems generally have access to the requisite engineering experience and trained personnel to implement the GAC treatment technique. Moreover, due to the limited amount of technical assistance available from EPA and the States, it was deemed necessary to limit the initial number of systems that would be required to make treatment modifications.

Finally, because GAC is a costly technology, it was not deemed economically feasible for the smaller systems to incur the relatively higher per capita costs of GAC treatment technology at the present time. Larger systems will be better able to keep the added cost burden on the individual consumer to a reasonable level.

As additional operating and technical experience is gained with respect to the use of GAC for the treatment of drinking water, the scope of this treatment technique requirement will be expanded to include smaller community water systems. Based upon further knowledge to be gained through experience within the next two years, it is also possible that the design criteria for GAC treatment may be modified to reflect such additional experience gained. Smaller systems will thus benefit from the experience gained in the initial application in large systems.

In the meantime, all water systems should strive to minimize synthetic organic chemical concentrations to the extent technologically and economically feasible.

To this end, the smaller public water systems are strongly urged to seek technical assistance from the States and EPA to reduce these undesirable contaminants from their drinking water supplies. A few smaller systems with sufficient resources are already using granular activated carbon in their treatment processes and are encouraged to continue doing so. EPA will be actively expanding its data base with respect to the application to GAC technology to small systems.

The Treatment Technique Requirement: The proposed technique requirement ultimately requires each community water system which serves a population of greater than 75,000 individuals (and which is not granted a variance from the regulation) to operate a treatment system using granular activated carbon (GAC) no later than 3½ years after the effective date of this regulation. Since the effective date will be 18 months after promulgation of this regulation, a system would actually have 5 years to comply with this ultimate treatment technique requirement. The treatment system using GAC would have to be designed to meet designated criteria to ensure maximized reduction of synthetic organic chemicals in drinking water.

This lengthy time frame was deemed necessary in order to provide systems a reasonable amount of time to gather the information prerequisite to the granting of a variance where appropriate, to conduct pilot studies to determine the specific optimal design parameters of each GAC treatment system and to construct the system.

The Agency is fully aware that this treatment technique requirement involves a significant departure from existing water treatment practice. It is therefore desirable to accord systems subject to the regulation sufficient time to proceed with care in designing a treatment process which will provide the most effective reduction in the level of synthetic organic chemicals in drinking water for maximizing public health protection.

On the other hand, it was also deemed necessary to prescribe a time frame which would require action with all deliberate speed. Exemptions may be granted from this time frame according to Section 1416 of the Safe Drinking Water Act.

Important consideration was also given to the necessity of providing States with primary enforcement responsibility adequate time to incorporate these new regulations into their State enforcement programs. In order for a State to maintain primacy, its statutory and regulatory authority must satisfy the requirements of Section 1413 of the Safe Drinking Water Act which requires that State regulations be no less stringent than those federal regulations which are in effect. Eighteen months was deemed to be a reasonable amount of time to allow States to amend their regulations and/or enabling legislation in conformance with this requirement.

Once the new regulations have been incorporated, States with primary enforcement responsibility will have the authority to grant or deny variances from the treatment technique and to ensure that systems ultimately subject to the treatment technique requirements comply with the regulations, including the intermediate milestones contained herein.

The treatment technique requirement as set forth in Subpart F of 40 CFR Part 141 requires that the treatment system using GAC be designed to meet three specific criteria to insure that the system has been designed to optimize the reduction of synthetic organic chemicals in drinking water. These three criteria are related to the breakthrough of "volatile" halogenated organic compounds and total organic carbon (TOC) and have been determined to be the current best surrogate indicators for determining whether a GAC system is providing maximum effective removal of a broad spectrum of synthetic organic chemicals generally.

In light of the need to reduce public exposure to synthetic organic chemicals in drinking water by the best available means as soon as possible, it was deemed necessary to require the use of an interim control measure in those cases where treatment systems using GAC would not be operational until from 2 to 3½ years after the effective date. Contamination of some public water supplies by synthetic organic chemicals is sufficiently significant to warrant short-term improvement of water quality while further work is under way.

It was therefore not believed to be prudent to wait as long as 5 years after the promulgation for optimal systems to be installed and operative for accomplishing any reduction in the level of synthetic organic chemicals in drinking water.

Thus, any community water system subject to the treatment technique requirement is also required to replace its existing filter media with GAC as an interim control measure not later than one year after the effective date of the regulations, unless it demonstrates to the satisfaction of the State that the failure to implement such a control measure will not result in an unreasonable risk to health, and that either of two conditions are satisfied.

First, the interim control measure may be waived by the State if the system shows that its GAC treatment system will in fact be operational in compliance with the treatment technique requirement not later than 2 years after the effective date. The interim control measure may also be waived by the State if the system demonstrates that due to compelling physical circumstances, the cost of replacing its existing filter media with GAC would be unreasonably high when considered in light of the accompanying reduction in the level of synthetic organic chemicals that would be achieved using such an interim remedial measure.

For example, a vulnerable water system subject to the treatment technique requirement may be excused from the interim control measure if it shows that the depth of its filter beds is so small that the removal capacity of that small an amount of carbon would be inconsequential and the cost unreasonably high. The case might arise that in order to achieve a reasonable removal efficiency with GAC, it would be necessary to increase the size of the existing filter beds at a cost which would be unreasonable or which would be physically impossible due to structural constraints at the treatment facility.

The decision as to whether this interim control is waived for any particular system would be made on a case-by-case basis by the State, or EPA where the State did not have the primary enforcement responsibility.

Once the existing filter media have been replaced with GAC by those systems required to implement the interim control measure, the carbon is required to be replaced or regenerated not less frequently than once every 6 months. This was deemed a reasonable time based upon a consideration of the removal efficiency of GAC being replaced in an average filter bed and cost of replacement at such a frequency. Under these operating conditions, although weakly adsorbed compounds like dichloroethane would break through the GAC, many strongly adsorbed chemicals such as polynuclear aromatic hydrocarbons would still be removed efficiently.

Also, the GAC would constitute a protective barrier from unusual events such as chemical spills into the raw water sources. Comment is solicited from the public as to whether this is a reasonable requirement.

In some cases, a system might demonstrate through monitoring data that even after 6 months, the use of GAC in its existing filter beds is allowing its treatment system to meet the three design criteria specified for the treatment technique requirement. In those cases, such a system may regenerate the carbon less frequently. Regeneration must then occur at a frequency which will not allow the criteria to be consistently exceeded.

Approximately 40 public water systems in the United States currently use GAC in their filter beds, principally for taste and odor control. The required interim control measure is therefore feasible and can be installed with minimum delay in other systems to provide some degree of interim public health protection.

Whether or not a system replaces its existing filter media with GAC as an interim control measure, by not later than 18 months after the effective date, the system must submit to the State for approval its final design plans and specifications and a construction schedule for the GAC treatment systems. Finally, by not later than 3½ years after the effective date, each system subject to the treatment technique requirement must notify the State that construction of the GAC treatment system has been completed and that it is being operated to maximize the removal of synthetic organic chemicals in drinking water.

Design Criteria for the Required Treatment Technique: The performance of a granular activated carbon (GAC) treatment system has been shown to vary considerably with the design of the system and the characteristics of the raw water. For a discussion of the scientific data on which these and the following judgments are based, see Chapters 5 and 6.

If inadequate contact time is provided in the system, then removal of organic contaminants will be less than ideal with fresh carbon. Further, as the GAC remains in the treatment system, in time its adsorptive capacity will gradually be used up, resulting in the breakthrough of some organic chemicals. Finally, different types of carbon vary in their adsorptive capacity (for which no generally accepted measurement technique exists). Thus, the design of the system must specify at least three factors: contact time, regeneration frequency, and type of carbon.

In the absence of pilot study data on the specific water to be treated, the performance of the GAC system will not in general be predictable since it can and does vary widely with the characteristics of the raw water, particularly the types and amounts of organic chemicals present. Because of this, it is not possible to specify maximum or minimum values for these parameters applicable to all systems in the proposed regulations. On the other hand, some criteria must be provided so that the requirements imposed on public water systems can be adequately specified.

To deal with this problem, the regulations specify three criteria which the GAC system should be designed to achieve. The design of each system should be based in part on pilot studies using the system's raw water to determine the configuration which would best satisfy the criteria in a particular situation. These criteria are:

(1) The concentration in the effluent of any of the volatile halogenated organic compounds (except for trihalomethanes) determinable by the purge-and-trap/gas chromatography method shall not exceed 0.5 $\mu g/l$.

(2) The removal of influent total organic carbon with fresh activated carbon shall be at least 50%.

(3) The effluent total organic carbon may not exceed the value with fresh activated carbon by more than 0.5 mg/l. Of course, the design must also assure that the requirements specified in the interim primary regulations are also met, including THM concentrations, disinfectant demand and concentrations of other regulated chemicals.

The first criterion has been chosen because the volatile halogenated organic compounds measurable by the purge-and-trap technique tend to break through GAC earlier than other heavier compounds. The volatile halogenated organics are common industrial chemicals found as contaminants in drinking water. The presence of these chemicals should also be indicative of the presence of other potentially hazardous substances which would be more difficult to detect. In addition, some of them, such as carbon tetrachloride and trichloroethylene, have been identified as carcinogens by the National Academy of Sciences.

Thus, using these compounds as the basis for one of the design criteria provides a margin of safety, in that it is unlikely that other, possibly carcinogenic, compounds will break through the carbon before these do. The value of 0.5 $\mu g/l$ has been chosen as an operational value, which is above that achievable by a system using fresh GAC with reasonable contact times and one which is really measured by the stated analytical method without excessive analytical error, thus making the breakthrough relatively unambiguous.

The system should be designed so that GAC which has been used long enough to allow these compounds to break through consistently (3-week running average) should be regenerated. In most cases this is expected to permit at least several months of operation between regenerations.

The second and third criteria are designed to deal with the situation in which a system's raw water source may not contain significant amounts of the volatile compounds measurable by the first criterion. Total organic carbon (TOC) is a measurement of all of the organic compounds present in the water. Because it tends to give excessive weight to naturally occurring high molecular weight compounds which are not believed to be hazardous, it is not suitable as the sole criterion for the design of a GAC system.

However, when a consistent significant breakthrough of TOC is observed, it could indicate that the GAC is becoming saturated in its ability to remove some of the potentially harmful synthetic organic compounds and is therefore ready to be regenerated. The target value of 0.5 mg/l allowable increase in TOC is unambiguous analytically. The second criterion of at least 50% initial removal of TOC should be easily achievable with GAC but is necessary to make the third criterion meaningful and to assure sufficient adsorptive activity of the carbon.

It should be emphasized that these criteria should not be construed as maximum contaminant levels in disguise. Rather, they are intended to be the objectives which the GAC treatment system should be designed to achieve in those public water systems subject to the treatment technique requirement. Thus, for example, a system whose raw water does not satisfy the first criterion would not automatically be denied a variance.

Moreover, the second and third criteria cannot by their nature be interpreted as MCLs. The Agency solicits comments on the appropriateness of its approach to satisfying the required treatment technique and on the particular criteria proposed. MCLs will ultimately be established in Revised Drinking Water Regulations in accordance with the SDWA for compounds for which monitoring is technologically and economically feasible.

Variances from the Required Treatment Technique: Although the GAC treatment technique is initially applicable to all community water systems serving a population of greater than 75,000 people, a system may request a variance from its requirements.

In accordance with Section 1415(a)(B) of the Safe Drinking Water Act (SDWA), to receive a variance, a system must demonstrate that the prescribed treatment technique is not necessary to protect the health of persons because its raw water source is not subject to, nor likely to be subject to, significant synthetic organic chemical contamination. These proposed regulations prescribe the demonstration which a system must make in order to satisfy the statutory requirement.

By the use of the variance procedure, only those systems whose raw water sources warrant treatment for synthetic organics will be required to install granular activated carbon. Nevertheless, each affected system has the burden of proof in demonstrating that its particular circumstance warrants the granting of a variance from the treatment technique requirement designed to protect the public health.

In order to be considered for a variance, a system must submit information of two basic types to the appropriate authority. First, the system must provide information with respect to those point and nonpoint sources of pollution that would likely be potential sources of synthetic organic chemical contamination for the system's drinking water supply. This requirement was included because upstream polluters are considered to be one of the most important contributors to chemical contamination of drinking water.

The identification of these pollution sources will serve as a significant link between drinking water programs and programs under the Federal Water Pollution Control Act. Systems are thereby encouraged as a preventive measure to look for the sources of contamination problems and to the extent possible to improve the quality of their raw water supplies. To the extent feasible, systems must provide information with respect to the location of the particular dischargers and the characteristics of their discharges, such as types and concentrations of chemicals in the normal discharges as well as the potential for accidental spills. This information is needed to give the variance-issuing authority the ability to assess the likelihood that a system's raw water supply may be contaminated by synthetic organic chemicals on a case-by-case basis based upon local conditions.

The second type of information which a system requesting a variance is required to provide is analytical data on the quality of the system's raw water. As previously discussed, although it is not feasible to require systems to monitor for the presence of all the specific synthetic organics which may pose a potential public health risk to drinking water consumers, the presence of particular compounds which have been identified and which are known to present significant health problems is deemed to be a reasonable surrogate indicator of the presence of other potentially harmful contaminants.

Variance-issuing authorities are therefore required to utilize the data provided by the variance seeker on the presence of chemical indicators in Table 1.5 as well as any other available information in making the determination of the appropriateness of issuing a variance in a given situation.

As a guide to the variance-issuing authorities, a list of chemical indicators of industrial contamination has been prepared. This list includes compounds which have been found in drinking water supplies and for which there is suspicion or confirmation of adverse health effects. This list is derived from *Drinking Water and Health* and from EPA's list of 129 priority pollutants in effluent discharges. For convenience it is divided into general classes of organic compounds as these are easier to discuss because of their chemical similarities. This list is intended to serve as a guide for the purpose of aiding in the determination for the granting of variances. (See Table 1.5.)

Testing for these compounds is feasible and their presence would be indicative of chemical pollution and therefore of the fact that the particular water source is vulnerable to synthetic chemical contamination. Obviously, many other chemicals besides those listed could be present. Therefore, the list does not comprise all possible pollutant chemicals but is intended as a surrogate for the possible presence of many other chemicals for which analysis would be technically as well as economically infeasible. This list may be modified as additional information is gathered.

The Agency expects that particular sources of drinking water are less likely to be subject to contamination by synthetic organic chemicals. These include systems whose raw water sources consist exclusively of water from deep ground aquifers, protected watersheds such as upland reservoirs, the Great Lakes where substantial dilution is known to take place, or water purchased from a supplier of water which is treating its water using granular activated carbon. Because the likelihood of potential synthetic organic chemical contamination of these sources is smaller, these systems are expected to receive variances relatively easily.

Nevertheless, variance-issuing authorities are precluded from granting a variance if, based on other information, it finds that significant contamination exists which requires the system to comply with the treatment technique to protect the public health from the adverse effects of synthetic organic chemicals. Comments are solicited as to the Agency's assumption that these raw water sources of drinking water are not likely to be subject to contamination and whether they may warrant the imposition of less stringent requirements for being granted a variance.

In addition to a variance based on water quality considerations and the need for the designated treatment to be installed, Section 1415(a)(3) of the SDWA also allows variances to be granted by EPA upon a showing that an alternative treatment not included in the treatment requirement is at least as efficient in lowering the level of contaminants being controlled.

SUPPLEMENTAL MATERIAL ON NIPDWR

The material in this section of the chapter is taken from the supplemental material published by the Environmental Protection Agency in the *Federal Register*, Vol. 43, No. 130, on July 6, 1978.

On February 9, 1978, EPA published proposed regulations for the control of organic chemical contaminants in drinking water. These proposed regulations consist of two parts: a maximum contaminant level of 0.10 mg/l for total trihalomethanes and a treatment technique of granular activated carbon for the reduction of synthetic organic chemicals present in drinking water.

The proposed regulations have generated considerable public comment including written as well as oral comments presented at the public hearings held thus far by the Agency. These comments have raised several issues with respect to the proposed regulations which have led the Agency to gather additional information in support of its initial proposal. By this supplemental notice, this additional information is presented for public comment.

The Health Basis of the Proposed Regulations

EPA has received many comments questioning whether enough information on health effects of organic contaminants exists to justify the proposed regulations. These comments have generally accepted the fact that chloroform and other compounds present in some drinking water have caused tumors in laboratory tests when fed to animals at high doses, but they do not agree that this information forms a valid basis for reducing the levels of these compounds beyond the already low levels that occur in drinking water.

A frequent comment is that the National Academy of Sciences, in its report *Drinking Water and Health*, recommended against the establishment of a maximum contaminant level (MCL) for trihalomethanes and instead urged that further research be done. EPA believes this is a misinterpretation of the report. While the Academy did identify research needs, it clearly felt that the available information on chloroform (which forms the bulk of the trihalomethanes) was sufficient to justify regulation. After the Academy summarized this data, it concluded ". . . it is suggested that strict criteria be applied when limits for chloroform in drinking water are established."

The health basis for the proposed standards has been fully supported by Federal agencies experienced in dealing with environmental carcinogens. These agencies include the National Cancer Institute, the National Institute of Environmental Health Sciences, the Food and Drug Administration, and the Occupational Safety and Health Administration.

The approach EPA is taking in the drinking water proposal is consistent with the approach which these and other health-concerned agencies have taken in regulating human exposure to carcinogens, that is, to reduce human exposure to the extent feasible, provided the costs are reasonable.

Economic Impact of the Proposed Regulations

Since the U.S. Environmental Protection Agency (EPA) proposed a regulation for the control of organic chemical contaminants in drinking water in February 1978, the Agency has received a number of comments regarding the costs and financial feasibility of compliance with the regulations. This notice presents the results of an 8-week reassessment of the economic analyses in 1977 as the regulation was being developed.

The major findings of this review process are:

(1) The unit costs for GAC installations are now being revised upward from the figures published in 1977. Capital costs are being increased generally by 50 to 80%, primarily to adjust for inflation, to allow for contingencies and higher design, legal and financing fees, and to incorporate somewhat more conservative design parameters. Operating and maintenance expense estimates remain at levels very close to the former estimates.

(2) The range of technical assumptions has been narrowed somewhat in this review with a resulting impact on the change in the national cost estimates. The lower end of the previous cost estimate has been raised because a 6-month regeneration frequency is no longer included in these conservative estimates, although it may be enough in some cases. The higher end of the range has been reduced somewhat to reflect the estimate that 11 systems affected only the THM regulation and 50 systems impacted by the GAC requirement, 61 systems in total would install GAC treatment. The national capital costs based on these assumptions and the December 1977 unit costs would be $352 to $585 million. The new unit costs presented in this report raise this range to $616 to $831 million.

(3) The estimates of local residential costs for model water systems of GAC treatment have also been increased generally by approximately 30 to 50% and could be higher for systems with significant site specific problems in implementation. The effect on residential bills for the average family of three will range from $7 to $26 per year depending upon system size, design parameters, and local conditions.

(4) Financing in the capital markets should be possible through normal financing channels for almost all of the systems under either a low or high GAC cost scenario presuming that rates are increased to cover the annualized capital costs and the O/M expenses associated with GAC treatment addition. Under the low cost scenario almost

all could finance GAC with little or no difficulty, but under the high cost scenario about half of the systems would have some difficulty and would have to phase in their financing, suffer some decline in financial strength, and/or perhaps increase revenues more than just the amount required to cover GAC capital and operating costs. A small number of systems would have major difficulty in financing and probably would have to arrange special financing or apply for relief under the Safe Drinking Water Act.

Experience in the Use of Activated Carbon

Many comments have been received stating that granular activated carbon is a new and untried technology. While GAC used for removal of organic chemicals from drinking water, with regular regeneration of the carbon, is not now common practive in the American water works industry, it is far from a new and untried technology.

It has been used for dechlorination, organic removal and turbidity control in the soft drink and beverage industry and for purification in the sugar refining industry. Its application in wastewater treatment is longstanding, particularly for the removal of organic chemicals. Even in drinking water, activated carbon has been used by a number of water utilities for taste and odor control. While the operating parameters of a GAC system for removal of organic contaminants from drinking water will differ from these applications, the basic technology has been demonstrated through experience to be a generally available treatment technique.

Alleged Adverse Effects of Granular Activated Carbon

A number of technical issues concerning the use of GAC in water treatment have been raised in some comments received to date. Questions have dealt with the possible growth of bacteria on GAC filters; the possible extraction of metals and organic compounds from carbon; and desorption or "sloughing" of chemicals off from carbon after they have once been adsorbed.

Information available to EPA ranging from the past literature to current studies has not indicated any substantial problems of these types from the use of GAC in treatment of drinking water. This information is summarized below and discussed in detail in the supplementary document "Operational Aspects of Granular Activated Carbon Adsorption Treatment."

Microorganisms and Granular Activated Carbon: A concern has been expressed about the growth of bacteria in GAC beds, suggesting that contamination with pathogenic organisms may result. The available data do not show any such danger.

Several common methods are employed in determining the extent of microbial populations in water. Coliforms have been used as indicators of the possible presence of pathogens. Standard Plate Counts (SPC) are used as an indicator of the overall control of bacterial populations and also include many nonpathogenic bacteria which may survive the treatment process or proliferate. Studies have not shown pathogens or coliforms to increase in concentration upon passage of treated water through GAC beds.

GAC removes residual disinfectant while concentrating chemical nutrients, so it is not surprising that general bacterial populations (as measured by Standard Plate Counts) may increase either on the carbon bed or in the water effluent. In fact, this phenomenon is utilized in the new "Biological Activated Carbon" treatment process to improve the efficiency of GAC and lenghten the time between reactivations.

The amount of bacterial activity in a GAC column is a function of several factors including number and type of bacteria in the applied water; nutrients in the applied water (e.g., total organic carbon); temperature; bed depth; time between backwashing; and total time in service.

Workers in the United States, England, France, Germany, and the Netherlands have studied the matter and found generally that the extent of bacterial growth on GAC filters is variable and that the bacterial populations in GAC treated water are easily controlled by a small amount of post-disinfectant.

One such study used Ohio River water which was treated by coagulation and settling, followed by either dual media filtration or GAC (10 minutes empty bed contact time). No disinfectant was added anywhere in the treatment process. Coliforms were never detected in either the dual media filter or the GAC column effluents. Even after 4 months of operations the SPC in the GAC column effluent was consistently lower than in the dual media effluent. Both systems, without disinfectant, resulted in greater than 99% reduction in bacterial counts.

Endotoxins: Endotoxins are lipopolysaccharide-protein complexes produced in Gram-negative bacteria. The possible formation of endotoxins in GAC adsorbers has been studied because of the possible adverse health effects. Studies to date have not shown increased endotoxin concentrations in effluents from GAC filters.

In the same Ohio River water study described above, EPA scientists also monitored bacterial endotoxin concentrations following GAC treatment using the Limulus lysate bioassay. Endotoxin concentrations were reduced considerably by the treatment schemes. Mean concentrations were usually lower in the GAC effluent. No increase in endotoxin activity was observed in the GAC effluent.

In a survey now underway, samples are being collected from a dozen full-scale water treatment plants using GAC filtration. Time in service ranges up to 9 years and empty bed contact time ranges from 4 to 13 minutes. Thus far, no instances have been found where endotoxin levels increased through the GAC bed.

Metals and Other Inorganic Elements: Because granular activated carbon is derived from coal and other complex materials which contain inorganic elements, it is reasonable to inquire whether those substances remain in or on processed GAC and to what degree they can be extracted from GAC into water. Tests have shown that these elements do not leach from GAC into drinking water to any substantial degree, and therefore present no health hazard.

The quality of carbon suitable for use in drinking water systems would be the same type that meets the Food Chemical Codex requirements of the National Academy of Sciences. This currently includes quality specifications for arsenic (3 ppm), lead (10 ppm), heavy metals (40 ppm), and cyanogens and aromatics as determined by extraction tests. These permissible levels are virtually the same as the USP limits for medicinal carbon. Typical carbons used in water treatment are well within those specifications.

Detailed analysis of such carbon has detected many common elements: predominantly, iron, silicon, aluminum and calcium (which are of no health concern) as well as parts per million levels of lead, mercury, arsenic and several other toxic elements. Leaching tests using drinking water under simulated treatment conditions did not detect any leaching of metals, and except for calcium, only minimal amounts of metals were extracted by boiling in distilled water. Leaching occurred when the carbon was boiled with acid, a situation not likely to occur in drinking water treatment.

Calculations show that even if all of these elements would extract in common usage, an unlikely worst case situation, resulting concentrations in finished drinking water would still only be a miniscule fraction of current drinking water standards for those substances.

STATEMENT OF THE AWWA ON THE NIPDWR AMENDMENTS

> This material is a statement of the position of the AWWA on the amendments detailed in the first part of this chapter as solicited by the USEPA in the *Federal Register*, Vol. 43, No. 28, February 9, 1978, p. 5757: "Comment is solicited on all technical, economic and policy aspects of this proposed regulation, including the selected Maximum Contaminant Level for TTHMs, proposed treatment technique requirement, population coverage and phasing concept."
>
> The American Water Works Association is a scientific and educational society with the objective of improving water service to the public. Its 28,000 members include managers, engineers, chemists, operators and other employees of public water systems, federal and state employees, consulting engineers, educators and others with an interest in water supply.

On the whole, the Association believes the Environmental Protection Agency has done a good job in implementing the Safe Drinking Water Act. It shares the agency's concern for public health and looks forward to working with EPA in the future to safeguard public health and improve water service to the people.

The Association believes, however, that EPA's proposals on organic contaminants are based far too much on speculation and not nearly enough on scientific research. It realizes that the Congress does not require EPA to gather conclusive proof of health risk before taking action. But the American people have the right to demand more than educated guesswork before agreeing to divert hundreds of millions, possibly billions, of dollars from other programs, such as education, urban renewal, health care and energy conservation.

In taking issue with these EPA proposals, AWWA offers several suggestions for the timely and cost-effective improvement of water systems throughout the nation. AWWA's recommendations include widespread health-effects research, substitution of a total trihalomethane (TTHM) goal for the proposed maximum contaminant level (MCL), elimination of the requirement of the granular activated carbon (GAC) treatment technique, extensive GAC research on a plant-scale level, and industry-wide monitoring.

Public Law 93-523 mandated the National Academy of Sciences (NAS) to determine if maximum contaminant levels should be recommended. In its report, *Drinking Water and Health,* the Academy did not recommend an MCL on TTHM or synthetic organics, but it did propose research on chemical contaminants, including:

(1) studies of the physiological and biochemical mechanisms by which the toxic substances in water produce their effects;
(2) development of rapid, inexpensive and precise tests to identify substances that may produce important toxic effects at low doses and dose rates;
(3) epidemiological studies of chronic disease; and
(4) research on statistical methods and analytical models for describing and estimating the effects of long-term exposure to low doses of toxic substances. Studies should not be limited to carcinogenesis and should consider, also, difference between species and particularly sensitive subgroups in the population.

AWWA strongly recommends that EPA implement the NAS research recommendations as rapidly as possible. The Association is aware that EPA has an extensive ongoing research program, including many studies that have just begun or are partly complete. It urges that this work be accelerated and that the results be made available to the water supply industry before any MCL is established.

Maximum Contaminant Level for Total Trihalomethanes

EPA has proposed an MCL of 100 parts per billion for total trihalomethanes in drinking water. Quite frankly, AWWA can see no basis whatever for that figure. It suspects that it represents a response to highly vocal pressure groups and individuals who have been demanding that "something be done." EPA's own advisor, the National Academy of Sciences, did not propose an MCL because it found no basis for one.

Not surprisingly, many public water system personnel fail to understand EPA's rationale for the 100-ppb MCL. They know that scientists differ greatly in their assessment of health risk associated with trihalomethanes, that may see no risk at all in the miniscule quantities currently being identified, so they find it difficult to accept the proposed regulations.

As a direction for water supply people to take while awaiting actual evidence of necessity, the Association recommends that the 100-ppb TTHM level be identified solely as a goal that is desirable, but not required. It recommends that efforts to achieve the goal be limited to improvements and modifications of existing treatment plant processes, such as relocation of the point of chlorine application. In addition, all public water systems, not just those with more than 75,000 customers, should be urged to meet this goal.

The Association does not feel that the governmental whip of regulation is necessary to obtain action. Action already is taking place. Many public water systems already are following the excellent advice that EPA has developed on testing for

TTHM and on treatment modifications to reduce TTHM concentrations. These include improved coagulation for control of THM precursors and changes in the point of chlorination. The Association will use its influence to encourage many more utilities to adopt these practices as experience and information grow.

Many treatment plants have used chloramines for decades with great success. Chloramines provide an excellent means for the control of TTHM. AWWA thinks it would be a serious mistake to limit their use now. In fact, it recommends that chloramine experimentation and use be encouraged as a means of preventing THM formation, provided that satisfactory disinfection is achieved.

Granular Activated Carbon Treatment

The Association strongly recommends the proposed requirement of a GAC treatment technique be dropped. Before such a technique or other means to control synthetic organics is again proposed, EPA should implement the NAS research recommendations to prove conclusively that a problem warranting huge expenditure of public funds does exist. Furthermore, EPA should demonstrate through plant-scale research that GAC or other techniques can indeed do the job.

As a matter of principle, the Association opposes the idea of specifying a single treatment technique as the cure for all potential water treatment problems. If a health problem is shown to exist, it prefers that an MCL be set and each affected water supply be given the flexibility to provide its own best solution. In the case of synthetic organics, the best solution may well rest in the authority granted EPA under PL 92-500 to prohibit discharges of the offending toxic chemicals into the sources of public water supply.

As a substitute for mandating the GAC treatment technique, the Association recommends that EPA finance four plant-scale facilities to demonstrate that GAC is indeed a cost-effective process that will remove the synthetic organics and not create more problems than it solves. Each facility should include two or more full-scale GAC filter units, equipment for removal, regeneration and replacement of GAC, and necessary laboratory test equipment. Also, two of the sites should substitute GAC for existing filter media and two should use GAC in postcontact filters. This type of well-managed research project makes far more sense than requiring some 60 large systems to make costly independent and poorly coordinated studies.

In brief, these are the reasons the Association recommends requirement for the GAC treatment technique be dropped in favor of further experimentation:

The Need for Such a Technique Has Not Been Demonstrated: In proposing it, EPA is asking water systems that seek a variance to demonstrate GAC isn't necessary when, in fact, EPA itself cannot demonstrate that it is. Much more research needs to be done to establish when and if GAC or other treatment should be provided.

Design Data for Synthetic Organic Removal Using GAC Are Lacking: AWWA believes plant-scale research will be invaluable if this treatment is employed extensively at some later date.

Operating Data Are Lacking: AWWA is especially concerned that GAC units may create more problems than they solve. Virtually all plants that use GAC now are controlling tastes and odors, not synthetic organics. Water suppliers have many questions about regeneration frequency and contaminant breakthrough that need to be answered.

Reliable Cost Information on Plant Construction and Operation Is Woefully Lacking: Plant-scale research is the only way to obtain this information with any reasonable degree of accuracy. Preliminary estimates made by water utilities indicate actual costs will be six to ten times higher than those announced by EPA. For example, Miami estimates its capital requirements in the range of $40 to $45 million; EPA's estimates for Miami are $3.7 million.

Data on Efficient Regeneration System Design Are Lacking: Efficient methods for removal, regeneration and replacement of GAC are essential. The water supply industry needs data based on experience, not guesswork. Among other things, the environmental impact of regeneration furnaces needs to be assessed. Energy impacts, too, need to be determined.

Data on the Performance of GAC Under "Shock" Loading Need To Be Evaluated: AWWA believes GAC may be of some help in the event of emergency spill situations, but other safeguards, such as source monitoring and provisions for bypassing, may offer greater protection to community water systems. The Association recommends all water systems develop comprehensive plans for handling emergencies such as chemical spills.

Research on Alternative Adsorption, Disinfection and Other Treatment Techniques Needs To Be Done: The Association recommends that the four plant-scale facilities be entirely financed by EPA under its research program. Moreover, it recommends the projects include pilot plant studies to establish design criteria, engineering, construction and operation data. Operation should include extensive analytical work to assess in depth the performance of both the GAC units and the regeneration systems.

Monitoring

The Association believes the monitoring program suggested by EPA in its proposed revision of Part 141.25 (Total Trihalomethanes Sampling and Analytical Requirements) should be observed with one exception. The information recorded by all water systems on TTHM should be gathered for control purposes and correlated against possible future need. However, since the systems will be working toward a goal, not an MCL, no public notification should be necessary and none should be required.

For synthetic organic chemicals, the Association recommends the establishment of a monitoring program managed, operated and financed by EPA and based on guidelines similar to those in EPA's proposed variance program, Part 141.54. Such a program should be based on cost-effectiveness, focusing on the systems identified as having the highest TTHM and synthetic organic levels. Data gathered under such a program should be published at regular intervals.

Summary

In summary, the American Water Works Association recommends a four-point program in lieu of EPA's proposal of regulations to control organic chemical contaminants in drinking water:

(1) Expanded and accelerated health-effects research on trihalomethanes and synthetic organics as recommended by the National Academy of Sciences. A scientific basis for regulation must be established before, not after, rules are proposed.
(2) Establishment of a 100-parts-per-billion level of total trihalomethanes as a goal for all public water supply systems;
(3) Elimination of EPA's proposed requirement of granular activated carbon as a treatment technique. In its place, provide at least four plant-size research projects underwritten by the government to gather financial and operating, as well as scientific, data that do not now exist.
(4) Adoption of EPA's proposed monitoring program for TTHM, except that public notification should not be required. Establishment of a monitoring program for synthetic organic chemicals financed and operated by EPA.

Results of the health-effects, GAC and other research should be communicated regularly to the water supply industry. This will build both understanding and confidence in the Safe Drinking Water Act and its various provisions. AWWA offers to arrange conferences on water quality and health and to provide the pages of its monthly Journal AWWA for articles on health-effects research and GAC performance.

Statement of Basis and Purpose for the Amendment on Trihalomethanes

The material in the body of this chapter is taken from the *Statement of Basis and Purpose for an Amendment to the National Interim Primary Drinking Water Regulations on Trihalomethanes,* prepared by the Office of Water Supply Criteria and Standards Division of the EPA and published in January 1978. It is a fairly exhaustive review of the health effects on animals and humans of ingestion of the various trihalomethanes and related compounds and an explanation of the method by which the maximum contaminant levels proposed for the drinking water regulations were derived.

INTRODUCTION

The extent and significance of organic chemical contamination of drinking water sources first came to public attention in 1972, when a report, *Industrial Pollution of the Lower Mississippi River in Louisiana* was published by the EPA. While this report did not include quantification of the pollutants found, and was directed toward locating industrial discharges responsible for the pollution, the report did include analyses of finished (treated) drinking water and provided evidence of the presence of trihalomethanes (THM) in such water.

Subsequently, a more thorough examination of finished drinking water in the New Orleans area was carried out, using the most sophisticated analytical methods available. This latter study confirmed the presence of trihalomethanes and many other organic chemicals in finished drinking water, and furthermore demonstrated that one of them, chloroform, was present in extremely high relative concentrations.

The findings in New Orleans promoted other studies, primarily for the purpose of determining how widespread and serious the organic chemical contamination of drinking water was. Impetus was added by the passage of the Safe Drinking Water Act (P.L. 93-523), which directed the EPA to conduct a comprehensive study of public water supplies and drinking water sources to determine the nature, extent, sources of, and means of control of contamination by chemicals or other substances suspected of being carcinogenic. The National Organics Reconnaissance Survey of Halogenated Organics (NORS) or "80 City Study," (see Appendix I of this chapter) was aimed primarily at determining the extent of the presence of four trihalomethanes, chloroform, bromodichloromethane, dibromochloromethane and bromoform, along with carbon tetrachloride and 1,2-dichloroethane, and at determining what effect raw water source and water treatment practices had on the formation of these compounds (refer to Table 2.1).

TABLE 2.1: ANALYTICAL RESULTS FOR WATER SUPPLIES FROM NORS AND NOMS

	Concentrations in mg/liter				
	NORS	NOMS			
		Phase I	Phase II	Phase III	
	Chloroform			Dechlorinated	Terminal
Median	0.021	0.027	0.059	0.022	0.044
Mean	-	0.043	0.083	0.035	0.069
Range	NF-0.311	NF-0.271	NF-0.47	NF-0.20	NF-0.540
	Bromoform				
Median	0.005	LD	LD	LD	LD
Mean	-	0.003	0.004	0.002	0.004
Range	NF-0.092	NF- 0.039	NF-0.280	NF-0.137	NF-0.190
	Dibromochloromethane				
Median	0.001	LD	0.004	0.002	0.003
Mean	-	0.008	0.012	0.006	0.011
Range	NF-0.100	NF- 0.19	NF-0.290	NF-0.114	NF-0.250
	Bromodichloromethane				
Median	0.006	0.010	0.014	0.006	0.011
Mean	-	0.018	0.018	0.009	0.017
Range	NF-0.116	NF-0.183	NF-0.180	NF-0.072	NF-0.125
	Total Trihalomethane (TTHM)				
Median	0.027	0.045	0.087	0.037	0.074
Mean	0.067	0.068	0.117	0.053	0.100
Range	NF-0.482	NF-0.457	NF-0.784	NF-0.295	NF-0.695

NF = not found
LD = less than detection limit

The presence of trihalomethanes in finished drinking water was confirmed, and some trend relating nonvolatile total organic carbon (NVTOC) of the raw water and the total trihalomethane concentration (TTHM) was postulated. Chloroform occurred invariably in water which had been chlorinated, while it was absent or present at lower concentrations in raw water. Water samples were collected at the treatment plant in winter and iced for shipment but not dechlorinated. Thus, these values might approximate minima for human exposure in the areas selected. Of the various trihalomethanes, chloroform was found at the highest concentrations (averaging approximately 75% of the TTHM), with progressively less bromodichloromethane, dibromochloromethane and bromoform being detected. In some cases chloroform was found at concentrations greater than 0.300 mg/l (the highest value found was 0.540 mg/l). Carbon tetrachloride and 1,2-dichloroethane were found at very low concentrations. The concentration of these two components did not increase after the chlorination process, therefore, it can be assumed that the presence of these compounds is not related to the disinfection process.

A joint federal state survey of organics and inorganics in 83 selected drinking water supplies, carried out by EPA's Region V (Chicago) provided additional evidence of the ubiquitous nature of chloroform and other trihalomethanes in chlorinated drinking water. Two conclusions reached in that study were that raw water relatively free of organic matter results in finished water that is relatively free of chloroform and related halogenated compounds, and that there is a correlation in some instances between the concentrations of chloroform, bromodichloromethane, dibromochloromethane and bromoform in finished water and the amount of organic matter found in raw water. It appeared that these compounds resulted from the chlorination of precursors in the raw water.

A more recent study, the National Organics Monitoring Study (NOMS) (see Appendix II to this chapter) was aimed not only at determining the presence of trihalomethanes in additional water supplies, but also at determining the seasonal variations in concentration of these substances.

The NOMS sample size was 113 public water systems designated by the Administrator. The study also included analyses for approximately 20 specific synthetic organic chemicals deemed to be candidates of particular concern and analyses of several surrogate group chemical parameters which are indicators of the total amount of organic contamination. Three phases of this study have been completed and the mean, minimum, and maximum values of chloroform and trihalomethanes in drinking water are reported in Table 2.1.

Phase I analyses in the NOMS were conducted similarly to the NORS. Phase II analyses were performed after the THM-producing reactions were allowed to run to completion. Phase III analyses were conducted on both dechlorinated samples and on samples that were allowed to run to completion (terminal). Again chloroform was found at the highest concentrations in most cases, however, in a few cases bromoform was found to be the highest concentration of the THMs (0.280 mg/l). The mean concentrations of chloroform were 0.043 mg/l, 0.083 mg/l, 0.035 mg/l and 0.069 mg/l for Phase I, II, III (dechlorinated) and III (terminal), respectively; the mean concentrations for total trihalomethanes were 0.068 mg/l, 0.117, mg/l, 0.053 mg/l and 0.100 mg/l for Phase I, II, III (dechlorinated) and III (terminal), respectively.

THE ROLE OF CHLORINATION

All evidence indicates that chlorination of drinking water containing organic chemicals is the major factor in the formation of halogenated organic chemicals, particularly the trihalomethanes in finished drinking water. Chlorinated organic compounds, however, can also be introduced into our drinking water from industrial outfalls, urban and rural runoff, rainfall, through polluted air or from the chlorination of sewage and industrial wastewater.

Several studies in addition to those mentioned above have demonstrated increased trihalomethane concentrations in drinking water. Works by J.J. Rook (1972) in the Netherlands, and the studies by Bellar, Lichtenberg and Kroner (1974), showed that chloroform and other halogenated methanes are formed during the water chlorination process. It should be noted that these findings came as a result of more sensitive and refined analytical techniques.

Recent work by Rook (1976, 1977) has provided some insight as to the organic precursors which might be responsible for the formation of the trihalomethanes. Studies by Sontheimer and Kuhn (1977) indicate that the trihalomethanes may represent only a portion of the total halogenated products of chlorination of water. Bunn et al (1975) have demonstrated that hypochlorite in the presence of bromide and iodide ions but not fluoride will react with natural organic matter to produce all ten possible trihalogenated methanes.

It can be concluded from the above studies and others that the trihalomethanes occur in chlorinated drinking waters, and that the concentrations of the various trihalomethanes are dependent on the type and quantity of organic precursor substances, the amount of chlorine used, and the presence of other halogen ions as well as contact time, temperature and pH.

There are a number of methods available for reducing levels of trihalomethanes in drinking water. These options include modifications of current treatment practices, such as moving the point of chlorination, the use of alternative disinfectants such as chlorine dioxide or ozone, and various methods that will reduce organic precursor concentrations such as use of adsorbents like granular activated carbon (GAC).

The two chemicals most often mentioned as alternative disinfectants, chlorine dioxide and ozone, are both well known as effective disinfectants and chemical oxidants, and some history of their practical use in water treatment has been accumulated particularly in Europe.

EPA studies of the toxicology of chlorine dioxide and chlorite ion in drinking water reveal considerable variations. These compounds have been reported to affect the hematopoietic systems by oxidative changes in hemoglobins and hemolysis of red blood cells. Other bioeffects observed include gastrointestinal disturbances. The preliminary results indicate species variability in biological manifestations. Cats and African green monkeys appear to lie at the extreme ends of the spectrum from among the species studied; cats are very sensitive to the hematopoietic effects whereas monkeys were apparently insensitive even at levels as high as 200 mg/l.

An upper limit for chlorine dioxide usage has been set primarily because of the lack of data concerning the safety of this material, and particularly its decomposition products, at higher concentrations (Musil et al, 1963; Fridlyand and Kagan, 1971). Studies with cats have shown that chlorite, which is an oxidant and can cause anemias, has a deleterious effect on red blood cell survival rate at chlorine dioxide concentrations above 10 mg/l. Therefore a limit of 1.0 mg/l is necessary to prevent potential adverse effects on sensitive individuals, particularly children.

A preliminary study by Cotruvo, Simmon and Spanggord (1976, 1977) concerning ozonation of 29 organic compounds potentially present in water supply sources indicated the formation of a number of products. These reaction mixtures were assayed for mutagenic activity employing (1) five strains of *Salmonella typhimurium* (Ames Salmonella/microsome assay) and (2) mitotic recombination in the yeast *Saccharomyces cerevisiae* D3. After very extensive ozonation in water some of the organic compounds exhibited mutagenic activity in these systems. Similar studies under extreme conditions with chlorine dioxide by-products thus far have exhibited minimal mutagenic activity.

Combining ammonia with chlorine to form chloramines has been called the chloramine process, chloramination, and combined residual chlorination. The products of this process are monochloramines, dichloramines or trichloramines (nitrogen trichloride) depending on the pH and the chlorine to ammonia ratio. The production of the latter species is referred to as breakpoint chlorination and may contribute to taste and odor problems in the finished water.

Based on the results of numerous investigations, the comparative disinfectant efficiency of chloramines ranks last when compared to ozone, chlorine dioxide, hypochlorous acid (HOCl), and hypochlorite ion (OCl^-). Early studies by Butterfield and Wattie demonstrated that chloramines required approximately a 100-fold increase in contact time to inactivate coliform bacteria and enteric pathogens as compared to free available chlorine at pH 9.5. This work was later confirmed in 1953 by Kabler and in 1962 by Clarke et al.

Results with cysts of *Entamoeba histolytica* and viruses also confirm the decreased effectiveness of chloramines as a disinfectant. Studies by Fair et al (1947) showed that additional dichloramine is about 60% and monochloramine about 22% as effective as hypochlorous acid at pH 4.5 on cysts of *E. histolytica.* Kelly and Sanderson (196) found that chloramines in the concentration of 1 mg/l at 25°C required 3 hours at pH 6, or 6 to 8 hours at pH 10 to achieve a 99.7% inactivation of polio virus. With 0.5 mg/l free chlorine at pH 7.8, by comparison, inactivation of 99.99% of polio virus can be achieved in approximately 15 minutes (Liu and McGrowan, 1973). Chloramine treatment finds its widest application in maintenance of chlorine residuals in the distributing systems. The health effects of water treatment with chloramine have not been studied in detail.

Although these disinfectants do not produce trihalomethanes, questions have also been raised on both their toxicology and the toxicology of their by-products. Studies are under way to clarify this matter and could result in the designation of maximum permissible levels for certain disinfectants when applied to drinking water. In the meantime, EPA has determined that chlorine dioxide applications should be limited to no more than one milligram per liter which is not uncommon in today's usage and that chloramines should not be used as primary disinfectants.

The use of adsorbents for trihalomethane removal has also introduced some unknown factors. Assuming that the adsorption process is effective for its intended purpose, there is always the possibility that a breakthrough of adsorbed chemicals will occur, that these substances will be adsorbed and subsequently slough off to produce contaminant concentrations intermittently, or that bacteria and/or toxins will be added to the water from growth on the adsorbent. All of these potential effects are controllable in practice, and EPA encourages the use of GAC to purify contaminated waters and to control THM precursors.

Thus, it is essential that the THM concentrations be reduced but without compromising public health from either infectious disease transmission or from the technology that is used. Outbreaks of infectious waterborne disease have been noted when there have been breakdowns in chlorination. The alternative control methods outlined previously are effective and are also being studied for their possible side effects. As soon as data becomes available EPA will make specific recommendations regarding their use.

At the present time, the best approach to reduce the organic precursors is to use adsorbents such as GAC. This approach has the benefit of reducing the concentrations of many of the organic chemicals in the water in addition to the precursors of THM and other chlorinated organics. Thus, once the organic chemical concentrations in the water have been reduced, the chemical demand for applied disinfectant will also be reduced, thus human exposure to all disinfectant chemicals, and their degradation products and by-products will be minimized.

SOURCES OF TRIHALOMETHANE EXPOSURE

McConnell et al (1975) have reported that chloroform occurs in many common foods and that while some halogenated compounds in food may result from manufacturing and pest control practices, chloroform may be introduced as the result of geochemical processes. Chlorinated compounds are the halogenated species most prevalent in food, but at least one food, Limu Kohu, a seasonal weed or alga eaten in Hawaii, contains an essential oil which is composed largely of bromoform (Burreson, et al 1976).

Chloroform has been widely used as an anesthetic, and until recently was a common ingredient in dentrifices and cough preparations. The Food and Drug Administration has taken action to halt the use of chloroform in drug products, cosmetic products, and food-contact articles (41 F.R. 145026, April 9, 1976). The EPA has issued a notice of rebuttable presumption against continued registration of chloroform-containing pesticides (41 F.R. 14588, April 6, 1976). Thus, in addition to drinking water, exposure to some or all of the trihalomethanes is complicated by other environmental sources, however, exposure from some of those sources is being reduced.

The relative contribution and uptake of chloroform can be estimated for three major sources of human exposure: atmosphere, drinking water, and the food supply. The calculations of human uptake were based on the fluid intake, respiratory volume, and food consumption data for reference man as compiled by the International Commission on Radiological Protection. The combined uptake for human adults from all three sources was estimated by multiplying estimated exposure levels times estimated intakes.

Human uptake of chloroform from air, food and drinking water is given in Table 2.2. Chloroform and trihalomethane uptake from drinking water was estimated by multiplying the chloroform and trihalomethane concentrations found in drinking water supplies from NOMS data (Table 2.1) and the average consumption of 2 liters of water per day. 100% absorption of the amount of chloroform in drinking water was assumed for these calculations. The total chloroform uptake from water was estimated at a mean value of 64 mg per year and a maximum uptake value of 343 mg per year.

TABLE 2.2: HUMAN UPTAKE OF CHLOROFORM AND TRIHALOMETHANES FROM DRINKING WATER, FOOD AND AIR

	...........Exposure Levels, mg/yr...........		
Chemical	**Drinking Water Mean (Range)**	**Food Mean (Range)**	**Air* Mean (Range)**
Chloroform	64 (0.73-343)	9 (2–15.97)	20 (0.41–204)
Trihalomethanes	85	–	–

*Calculated from data supplied by Strategies and Air Standards Division, Office of Air Quality Planning and Standards, Environmental Protection Agency, Research Triangle Park. The air samples were collected both from the rural and industrial areas during the years 1974-76. The mean value was derived from the concentrations obtained from urban industrialized areas.

In order to determine human uptake of chloroform from foods, the concentrations of chloroform in various foods was multiplied by the average consumption of each food item by human adults per year in the United States, and 100% absorption of ingested chloroform was assumed. A calculated maximum value of about 16 mg of chloroform uptake per year and a mean value of 9 mg per year from total food consumed was obtained.

The calculation for the uptake of chloroform by humans from air was based upon the assumption that an average of 63% of chloroform present in ambient air was absorbed after inhalation; the volume of air inhaled by an average adult was taken as 8.1×10^6 liters per year; 0.02 and 10 ppb (by volume) chloroform concentrations in urban air as minimum and maximum values, respectively. The minimum and maximum values for the uptake of chloroform by an adult were estimated as 0.41 and 204 mg per year respectively.

At minimum conditions from all sources of exposure the atmosphere contributes 13% of the total chloroform while the drinking water contributes 23% and food is most significant. At maximum conditions from all sources, water is the major contributor at 61%, with air at 36%. Under conditions of maximum exposure from the water and minimum exposure from the air, the major contribution by far is drinking water as a source of chloroform uptake, which is estimated to be as much as 97%. Thus, relative contributions of drinking water as a source of chloroform to the total body burden may change from a moderate to a maximum contributor as the annual exposure from water ranges from nil to 343 mg per year and from 204 to 0.41 mg per year in ambient air (see Table 2.3).

TABLE 2.3: UPTAKE OF CHLOROFORM FOR THE ADULT HUMAN FROM AIR, WATER, AND FOOD

Source	Adult (mg/yr)	Uptake (%)
Maximum Conditions		
Atmosphere	204	36
Water	343	61
Food supply	16	3
Total	563	100.00
Minimum Conditions		
Atmosphere	0.41	13
Water	0.73	23
Food supply	2.00	64
Total	3.14	100.00
Maximum Water-Minimum Air		
Atmosphere	0.41	1
Water	343.00	97
Food supply	9.00	2
Total	352.41	100.00

METABOLISM

A number of reports (Brown et al, 1974; Labigne and Marchand, 1974; Fry et al, 1972; Paul and Rubinstein, 1963; Taylor et al, 1974) have indicated that chloroform is rapidly absorbed on oral and intraperitoneal administration and subsequently metabolized to carbon dioxide and unidentified metabolites in urine. Species variation in the metabolism of chloroform has been summarized in Table 2.4 on the following page.

It is noteworthy that the mouse, a species which shows greater sensitivity to the oncogenic effect of chloroform (Brown et al, 1974; Eschenbrenner and Miller, 1945), metabolized chloroform extensively to carbon dioxide (80%) and unidentified metabolites (3%) from an oral dose of 60 mg/kg. Rats also metabolize chloroform to carbon dioxide but to a lesser extent (66%). Paul and Rubenstein (1963) recovered 4% carbon dioxide after administering 1,484 mg/kg chloroform intraduodenally to rats. The discrepancy in these two results may be dose related.

TABLE 2.4: DISPOSITION OF CHLORFORM-SPECIES VARIATION

Animal Species	Sex	Strain	Dose mg/kg	Metabolism (%) CHCl$_3$	CO_2	Urine Feces	Total Excretion
Mouse	M	CBA CF/LP C57	60 po	6	80	3	93*
Rat	M	Sprague Dawley	60 po	20	66	7	93
Rat	–	–	1484 id	70	–	–	–
Rat	M	Sprague Dawley	4710 ip	–	0.39	–	–
Monkey	M	Squirrel	60 po	78	17	2	97

*Includes radioactivity in carcas

po = orally
id = intraduodenally
ip = intraperitoneally

Dose related differences in the metabolism of compounds are known and have recently been reported for the carcinogen vinyl chloride. Squirrel monkeys, when given 60 mg/kg of chloroform orally, excreted 97% of the dose with 17% as carbon dioxide and 78% as chloroform. Fry et al (1972) recovered unmetabolized chloroform ranging from 17.8 to 66.6% of a 500 mg dose of chloroform given to human volunteers during an 8 hour time period (equivalent to about 7 mg/kg). Since the metabolism of xenobiotics is also dependent on age and sex, the widespread variation in the quantitative disposition of chloroform in human subjects may be due to the experimental protocols wherein subjects ranging from 18 to 50 years of age were used.

A related halogenated hydrocarbon, carbon tetrachloride (CCl_4), has been shown to be carcinogenic in Osborne-Mendel rats and in B6C3F1 mice at dosages ranging from 57 to 160 mg/kg and 1,250 to 2,500 mg/kg respectively. Dosages for oncogenic effects of chloroform were 90 to 200 mg/kg for rats and 138 to 477 mg/kg for mice. Metabolic similarities between those compounds include the appearance of halide ions in urine and carbon dioxide in breath.

Carbon dioxide is one of the major metabolites of chloroform in mice and rats whereas it is a minor one in carbon tetrachloride metabolism. Carbon tetrachloride also is metabolized to chloroform in trace amounts, which may in turn, be biotransformed to carbon dioxide. Carcinogenicity of carbon tetrachloride, however, has been attributed to a free radical (CCl_3) which is postulated as an intermediate in the metabolic processes.

Many carcinogens have been reported to form complexes with proteins, DNA and RNA (Miller and Miller, 1966). In some instances the first stage in chemical carcinogenesis may involve metabolism of the carcinogen to a secondary and more active compound. In the case of chloroform, Ilett et al (1973) reported covalent bonding of chloroform metabolite(s) to tissue macromolecules of mice. The covalent bonding increased or decreased when the animals were pretreated with phenobarbital or piperonyl butoxide, agents which stimulate or inhibit the metabolism of foreign compounds by drug metabolizing enzymes. This is suggestive of the involvement of chloroform metabolism in these processes.

Information regarding the metabolism of bromoform and other haloforms is not available. However, the structural similarities of these haloforms with chloroform indicate that these compounds should also be absorbed by the oral and inhalation routes of exposure and then biotransformed into carbon dioxide and halide ions. Related halogenated hydrocarbons of the dihalomethane series, i.e., dichloromethane, dibromomethane and bromochloromethanes have been reported (Kubic et al, 1974) to be metabolized to carbon monoxide; the rate of metabolism of dibromomethane was higher than that of the chloro isomer.

ACUTE AND CHRONIC HEALTH EFFECTS

Biologic responses on exposure of chloroform to mammals include its effect on the central nervous system resulting in narcosis, hepatotoxicity, nephrotoxicity, teratogenicity, and carcinogenicity. Reported LD_{50}s are as follows: for rats, 300 mg/kg administered orally (DHEW, 1973) and for mice 705 mg/kg (Plaa et al 1958).

Acute studies involving single dosage levels in animals have been reported by several researchers. Jones et al (1958) studied the effect of various doses of chloroform fed to mice and made the following observations after 72 hours of exposure: 35 mg/kg, threshold hepatotoxic effect—minimal midzonal fatty changes; 70 mg/kg, minimal central fatty infiltration; 140 mg/kg, massive fatty infiltration; 350 mg/kg, centrilobular necrosis; and 1,100 mg/kg, minimum lethal dose.

In regard to acute effects on exposure to chloroform and bromoform, species variation has been observed. Reported lethal doses for chloroform and bromoform are:

Species	Subcutaneous Lethal Dose	Values in mg/kg
Mouse	LD_{50}	704 (chloroform) 1820 (bromoform)
Rabbit	LD_{10}	800 (chloroform) 410 (bromoform)

Data on the acute toxicity of dibromochloromethane and dichloromethane are not available.

Hepatotoxicity

Plaa et al established a dose-response relationship in mice, measuring parameters indicative of hepatotoxicity. ED_{50} values of 166 mg/kg were found in mice which received chloroform subcutaneously. The inhalation exposure of chloroform by mice for 4 hours at concentrations ranging from 100 to 800 ppm resulted in fatty infiltration of the liver at all dose levels. These changes were observed at necropsy after 1 to 3 days of exposure. Like chloroform, bromoform exposure leads to fatty degeneration and centrilobular necrosis of the liver (von Oettingen, 1953). Dibromochloromethane and dichlorobromomethane may bring about similar responses.

Nephrotoxicity

Nephrotoxic effects of chloroform were studied by Plaa and Larson (1965). Median effective doses (ED_{50}) of chloroform in mice were 178 mg/kg as measured by phenolsulfophthalein excretion. Increases in urinary protein and glucose, indices of kidney damage, had an ED_{50} of 104 mg/kg for chloroform. Data concerning the nephrotoxic effect of other trihalomethanes are not available.

Teratogenicity

Teratogenic responses on oral dosing of animals to chloroform were investigated by Thompson et al (1973). Rats and rabbits were administered chloroform at 126 and 50 mg/kg respectively. No significant fetal deformities were observed. Inhalation of chloroform by Sprague Dawley rats at 30, 100 and 300 ppm for 7 hours a day on days 6 through 15 of gestation revealed significant fetal abnormalities including: acaudia, imperforate anus, subcutaneous edema, missing ribs and delayed skull ossification (Schwetz et al, 1974).

In an attempt to explain reproductive failure in laboratory animals, i.e., mice and rabbits, McKinney et al (1976) conducted a study using CD-1 mice wherein groups of mice were given tap water and purified tap water (passed through a Corning 3508 ORC and a Corning 3508 B demineralizer). The analysis of the water indicated reduced amounts of chlorinated compounds in the purified water. The study was inconclusive in relating chloroform and other chlorinated organics in tap water to reproductive failures in laboratory animals, since the concentration of chlorinated organics in water was lowest in months when reproductive failure was highest, although there did appear to be small differences in these parameters between the highly purified tap water and tap water.

In another study involving the effect of Durham tap water and purified tap water as in the above study, Chernoff (1977) did not find striking differences in the reproductive parameters of CD-1 mice. No teratogenic studies on the haloforms other than chloroform were available.

Mutagenicity

The trihalomethanes (chloroform, bromodichloromethane, dibromochloromethane and bromoform) were assayed for in vitro mutagenic activity using strains of *Salmonella typhimurium* (TA100 and TA1535). The assays were conducted in desiccators such that each compound was allowed to volatilize and only the vapor phase came in contact with bacteria on the petri plates. The activation system was tested and found not to be required for the bromohalomethanes since they were positive in the absence of activation.

The results obtained were as follows: (a) chloroform was not mutagenic in TA100 either with or without activation or in TA 1535 without activation; (b) bromodichloromethane was mutagenic in TA100 without activation, with a doubling dose of approximately 25 microliters; (c) dibromochloromethane was mutagenic in TA100 without metabolic activation, with a doubling dose of approximately 3.5 microliters; and (d) bromoform was mutagenic in TA100 without metabolic activation, with a doubling dose of approximately 25 microliters, and was also mutagenic in TA1535 with metabolic activation, with a doubling dose of approximately 100 microliters (Tardiff, 1976).

All three compounds demonstrating mutagenic activity did so in a dose-response mode. For certain classes of compounds the Ames test which utilizes *Salmonella typhimurium* bacteria correlates highly (90%) with the in vivo carcinogenicity bioassay. However, for certain chlorinated hydrocarbons the test has been shown to have limitations in detecting gene mutations (Ames et al, 1973), which can be demonstrated in other test systems.

Carcinogenicity

Prolonged administration of chloroform at relatively high dose levels to animals, specifically mice and rats, manifested oncogenic effects. An investigation conducted by Eschenbrenner and Miller (1945) revealed hepatomas in female mice (strain A) given repeated dosages ranging from 0.145 to 2.32 mg of chloroform for a period of four months. Minimum doses of 593 mg/kg chloroform per day (total of 30 doses) produced tumors in all of the surviving animals.

In a more recent study (NCI, 1976) linking chloroform with oncogenicity, rats and mice of both sexes were fed doses of chloroform ranging from 90 to 477 mg/kg. In this study, the lowest dose for observed carcinogenic effect (kidney epithelial tumors) in male rats was 100 mg/kg and for mice 138 mg/kg administered to the animals for a total period of 78 weeks. A related halogenated hydrocarbon, carbon tetrachloride, has been shown as carcinogenic in Osborne Mendel rats and in B6C3F1 mice at dosages ranging from 47 to 160 mg/kg and 1,250 to 2,500 mg/kg, respectively.

The incidence of hepatocellular tumors formed in these animals at both dose levels almost approached 100% (Table 2.5). The percent survival in mice treated with chloroform and carbon tetrachloride is depicted in Table 2.6.

TABLE 2.5: COMPARISON OF HEPATOCELLULAR CARCINOMA INCIDENCE IN CHLOROFORM- AND CARBON TETRACHLORIDE-TREATED MICE

Animal Group		Chloroform	Carbon Tetrachloride
Males	Controls	5/77	5/77
	Low Dose	18/50	49/49
	High Dose	44/45	47/48
Females	Controls	1/80	1/80
	Low Dose	36/45	40/40
	High Dose	39/41	43/45

TABLE 2.6: COMPARISON OF SURVIVAL OF CHLOROFORM- AND CARBON TETRACHLORIDE-TREATED MICE

		Chloroform			Carbon Tetrachloride		
Animal Group		Initial No.	78 Weeks	90 Weeks	Initial No.	78 Weeks	91-92 Weeks
Males	Controls	77	53	38	77	53	38
	Low Dose	50	43	37	50	11	0
	High Dose	50	41	35	50	2	0
Females	Controls	80	71	65	80	71	65
	Low Dose	50	43	36	50	10	0
	High Dose	50	36	11	50	4	1

Almost all the animals on treatment with carbon tetrachloride died between 91 to 92 weeks whereas with chloroform treatment at both dose levels, 73 and 46% of the animals survived. Miklashevskii et al (1966) fed chloroform to rats at 0.4 mg/kg apparently for 5 months and detected no histopathological abnormalities after this treatment.

A recent study on the carcinogenic effect of chloroform at dose levels of 17 mg/kg per day and 60 mg/kg per day was conducted by Roe (1976), utilizing the rat (Sprague Dawley), the beagle dog and four strains of mice (ICC Swiss, C57B1, CVA and CF/1). Comparison with the NCI study indicates that the number of animals and the duration of the experiment were essentially similar. The major differences were the dosages, which were lower than in the NCI study, and the vehicle, which was toothpaste. The only finding of neoplasia was an excess of tumors of the renal cortex in the male ICI Swiss mice at a dose level of 60 mg/kg/day. However, animals fed 17 mg/kg/day of chloroform showed no incidence of renal carcinoma.

Some renal tumors were also seen in control animals in a later study. The negative results observed in the dog experiment may be explained on the basis that either the animals were not exposed for a suitable length of time (i.e., duration of life span) or that an insufficient number of animals were tested, or that this species may not have been responsive to the oncogenic effect of chloroform. The negative results of the rat study may be explained on the basis of lack of strain sensitivity.

Much less information is available on the carcinogenicity of bromohalomethanes. Preliminary results from the strain A mouse pulmonary tumor induction technique (Theiss et al, 1977) indicated that bromoform produced a positive pulmonary adenoma response while chloroform did not. Other studies (Poirier et al, 1975) indicated that in several instances brominated compounds exhibited more carcinogenic activity than their chlorinated analogs in the pulmonary adenoma bioassay.

HUMAN HEALTH EFFECTS

NAS Principles of Toxicological Evaluation

The NAS in a recent report entitled *Drinking Water and Health,* identified several principles that are the basis of assessing the irreversible effects of long and continued exposure to carcinogenic substances on humans at low dose rates. (See chapters 1 and 3.)

On the basis of chloroform studies in animals and human toxicological data, the NAS has recommended that strict criteria should be applied for establishing exposure limits. The National Institute for Occupational Safety and Health has recommended that the occupational exposure to chloroform should not exceed 2 ppm determined as time-weighted average exposure for up to a 10 hour work day.

The human health effects as observed in accidental, habitual, and occupational exposures appear to indicate that the bioeffects on exposure to chloroform are similar to those found in experimental animals. These include the effects on the central nervous system, liver and kidney.

The symptoms observed (Storms, 1973) in a 14 year-old patient following an accidental exposure to an unknown amount of chloroform included cyanosis, difficulty in breathing and unconsciousness. Liver function tests measured by serum enzyme levels after four days of ingestion indicated very high levels of SGOT, SGPT, and LDH. The authors also noted cerebellar damage characterized by an instability of gait and a slight tremor on finger-to-nose testing. The symptoms disappeared in two weeks.

Several cases of habitual chloroform use have also been recorded by Heilbrunn et al (1965). A case study of interest was a 33 year-old male who had habitually inhaled chloroform for 12 years. The subject showed psychiatric and neurological symptoms including restlessness, hallucinations, convulsions, dysarthria, ataxia and tremor of tongue and fingers.

Lunt (1953) reported on the delayed chloroform poisoning in obstetric patients. Laboratory findings indicated renal dysfunction including: albumin, red blood cells, and pus in the urine. Chloroform exposure of humans by inhalation was studied by Lehman and Schmidt-Kehl (1936). Ten different concentrations of chloroform were used and the chloroform concentrations were determined by the alkaline hydrolysis method. Exposure at concentrations of 7 ppm for 7 minutes and at all higher levels up to 3,000 ppm caused symptoms of central nervous system depression.

Limited information is available on the controlled bioeffect studies in humans exposed to chloroform. DeSalva et al (1975) studied the effects of chloroform in humans; the subjects were given dentrifice containing 3.4% chloroform and mouthwash with 0.43% chloroform for 1 to 5 years. No hepatotoxic effects were observed at estimated daily ingestion of 0.3 to 0.96 mg/kg chloroform. Reversible hepatotoxic effects were manifested at 23 to 37 mg/kg/day chloroform ingested for 10 years in a study conducted by Wallace (1959).

Epidemiologic Studies

As of July 1977 there had been 11 different epidemiological studies with additional unpublished reports that investigated the relationship between cancer mortality and morbidity and constituents in drinking water. All of the studies were retrospective in design; nine were correlations using an indirect design, two used a case-control or direct design approach. Two studies utilized cancer morbidity or incidence rather than mortality as a measure of disease frequency. The studies vary in sample size, cancer sites considered, confounding factors selected as variables, parameters selected as indicators of water quality, and statistical analysis.

There are several problems peculiar to these studies which make it difficult to interpret their results: (1) there is a limited amount of water quality data on organics, and the data which exists covers less than a five year time period; and (2) the water quality data is often from geographic areas not conterminous with areas (usually counties) reporting cancer mortality data.

The water quality data on organics is of recent origin and it is not known the extent to which current levels may reflect past exposures. This is important, since the latent period for most types of cancer induction is measured in decades, not months or years. Comparison of the various study results is difficult because of the different approaches used.

In general, indirect, retrospective epidemiological studies are a useful methodological tool in hypothesis generation. A positive correlation cannot establish causal relationship. However, the results from these studies when viewed collectively can provide some insight into associations of potential causal relationships which need to be tested further by more highly focused direct methods, such as case-control studies.

The studies do provide evidence that there is reason for concern. When the evidence from all studies is weighed, the emphasis should be placed not only on the statistical significance of single correlation coefficients but on their consistency and patterns. When more than one independent study shows positive associations for site-specific cancers, then the association may not be due to chance alone. When the association is verified by consistent results across all four sex-race groups, the association may be due to the variable considered and the evidence should be viewed more seriously.

A large body of data provides evidence (both epidemiological and experimental) that the majority of human cancers result from multifactorial causes (Weisburger, 1977), particularly for cancers of the gastrointestinal and urinary tract. Etiologic factors, such as smoking and its association with lung cancer that result in increased relative risk greater than 5, were the first to be discovered. The etiologic factors associated with cancers of the gastrointestinal and urinary tract are more difficult to evaluate from epidemiological studies because of the lower incidence and mortality rates and because of the multifactorial interaction of environmental causes.

The increased relative risk of most potential factors associated with gastrointestinal and urinary cancers is probably less than 3. Thus, any correlation in a sound, indirect, retrospective study between drinking water and cancer mortality would most probably be weak, as is shown in the studies completed.

A number of epidemiologic studies that have been conducted did not define the water quality parameter by chemical constituents and therefore compared various sources of water supply. One investigation was performed by Page and Harris (1974). The study involved Louisiana county (parish) cancer mortality rates, 1950 to 1969, for total cancer and selected sites in white males. The parishes were categorized by the percentage of the county population drinking Mississippi River water.

The variables controlled were rural-urban characteristics, median income, population density, and proportion of employed population in the petroleum, chemical, and mining industries. An unweighted regression analysis resulted in a positive correlation between drinking water and total cancer (minus cancer of the lung, urinary tract, GI tract, and liver), gastrointestinal organs, and lung cancer mortality. These investigations suggested the possibility that there was an association between the cancer mortality rates and drinking Mississippi River water. As a result, serious questions were raised as to the safety of drinking water contaminated by suspected carcinogens, particularly various organic chemicals in the water.

Tarone and Gart (1975) reviewed the study by Page and Harris and included an additional variable, the elevation above sea level. By using a weighted regression analysis for four race-sex groups, weak, but statistically significant, positive correlations were found between the water variable and total cancer and lung cancer mortality for white males (WM), nonwhite males (NWM), and nonwhite females (NWF). The correlations were not statistically significant for white females (WF) for the same sites. Thus, there was a lack of consistency across the four sex-race groups for the aforementioned cancer sites.

Another report by Meinhardt et al (1975) commenting on the Page and Harris report, looked at the cancer mortality gradient and concluded that there was a random distribution of high and low cancer mortality rates among the river water consumers along the lengths of the Missouri and Mississippi River systems. From this study it was pointed out that the controls used might not be representative.

A second report by Page and Harris, entitled "Relation Between Cancer Mortality and Drinking Water in Louisiana," utilized independent variables and cancer sites similar to those in the first study, however, relationships for all four sex-race groups were added. Positive regression coefficients for the water variable that were statistically significant are as follows: total cancer sites, WM, NWM, NWF; all other things, WM; urinary tract, WM, NWF; and gastrointestinal, WM, NWM, WF, NWF.

DeRouen and Diem (1975) did another analysis of the relationship of cancer mortality in Louisiana and the Mississippi River as the drinking water source. An additional variable, latitude, was included, which divided Louisiana into a northern and southern section. This variable effectively resulted in an ethnic division of the population. The variables urban-rural characteristics, median income, employment characteristics, and elevation above sea level included in previous studies (Page and Harris, 1974; Tarone and Gart, 1975; Page and Harris, 1975; Page et al, 1976) were omitted.

The water variable was handled differently by the investigators, i.e., population groups studied either obtained none of their water from the Mississippi River, or obtained some or all from the river. The results are in agreement with Page and Harris results that show a positive relationship between cancer mortality and drinking water for gastrointestinal cancer. The cancer mortality rates for southern parishes of Louisiana whose source of drinking water is the Mississippi River tend to be higher than the southern parishes whose source of drinking water is not Mississippi River water for the following: stomach, NWF; rectum, WM; large intestine, WF, NWF; cervix, NWF; lung, NWF; and total cancer, NWF.

The cancer mortality rates tend to be slightly higher for the southern parishes with river water than northern parishes for cancer of the urinary tract, gastrointestinal tract, and the lung.

In another set of analyses and comments, DeRouen and Diem (1975) discuss the problems associated with interpretation of regression coefficients as they relate to the Page and Harris report, particularly the problem of making inferences from indirect studies. They concluded that the inconsistencies in the data and failure to see the same relationships for other sex-race groups damages the credibility of the hypothesis.

An analysis was done by McCabe (1975) of EPA using 50 of the 80 cities from the NORS data. Only those cities with a 1950 population greater than 25,000 and 70% or more of the city's population receiving water comparable to that sampled by EPA were included in the study. The results showed a statistically significant correlation between the chloroform concentrations in the drinking water and the cancer mortality rate by city for total cancer combined.

In a second analysis done by McCabe (1977) using Region V data, correlations between $CHCl_3$ THM and total cancer mortality were not positive. When the same correlations were done using Region V plus NORS data for $CHCl_3$ and THM concentration levels, a positive statistically significant result was obtained.

Several epidemiological studies have been conducted in the Ohio River area. Buncher (1975) conducted a study of 88 counties bordering the Ohio River in which 14 of the counties used the Ohio River as a drinking water source. The results do not show a significant relationship between drinking water from the Ohio River and the higher cancer mortality rates. There was a weak positive correlation between the chloroform concentrations in 23 cities and the cancer mortality rate for all cancer sites in white males.

Similar results were found in 77 cities (59 surface water suppliers) between chloroform concentrations and pancreas cancer mortality in white males. For cities that accounted for more than 70% of the country population, there was a significant correlation between chloroform concentration and bladder cancer mortality rates for both white males and white females.

Another study by Kuzma et al (1977) considered the 88 counties of Ohio, which were classified as either ground water or surface water counties based on the source of the drinking water used by a majority of the county residents. A two-stage analysis was performed and no statistically significant results were shown between the drinking water from the Ohio River and cancer mortality rates. Mortality rates for stomach, bladder, and total cancers were slightly higher for white males and for stomach cancer for white females in counties served by surface water supplies than in counties served by ground water supplies.

Reiches et al (1976) treated the 88 counties of Ohio by using a different methodology. Correlations between the surface drinking water variable and cancer mortality rates of stomach cancer and total cancers for both white males and females were statistically significant. The correlations between the drinking water variable and cancer mortality rates of the pancreas, bladder, esophagus, gastrointestinal tract, and urinary organs were significant for white males only.

Although several studies defined the water quality parameter by the amount of chlorination or levels of chloroform, only one study has done an analysis of all trihalomethanes, both collectively and separately. Cantor et al (1976) studied the correlation of cancer mortality at sixteen anatomical sites with the presence of THM concentration levels in drinking water for whites. Counties were grouped according to the percent of the county population served by the sampled water supply. In both sexes, there was a gradient of increasing correlation between halomethane concentration and bladder cancer in going from the low to intermediate to high percent served county groups or strata.

The correlation was stronger for the brominated THM than with chloroform. There was a negative correlation in white females of stomach cancer with total THM levels. Kidney cancer in white males showed a weakly positive correlation with chloroform levels. Lung cancer in white females showed a positive correlation with THM levels. Among white males, non-Hodgkins' lymphoma showed a positive correlation with the brominated trihalomethanes. A gradient of increasing association was observed between brain cancer mortality in both sexes and chloroform, but the associations were not strong.

Alavanja et al (1976) conducted a retrospective, case-control study of female cancer mortality and its relationship to drinking water chlorination in seven selected New York counties. A statistically significant association was found between drinking from a chlorinated drinking water supply and combined gastrointestinal and urinary tract cancer mortality rates. Further, there was a higher mortality for the summed gastrointestinal and urinary cancer in urban areas served by chlorinated surface or ground drinking water supplies than in urban areas served by nonchlorinated supplies.

Kruse (1977) conducted a retrospective, case control study of white males and females in Washington County, Maryland. The relationship between mortality and morbidity from liver (including biliary passages) and kidney cancer in areas supplied by chlorinated public water supplies was analyzed. While there was a slightly higher incidence of liver cancer among the exposed group, i.e., the group which consumed chlorinated drinking water, the correlations were not statistically significant. It should be noted that the sample size was relatively small.

Salg (1977) also conducted a retrospective study of various cancer mortality rates and drinking water as defined by source of supply and type of treatment in 346 counties in seven states bordering the Ohio River Valley Basin. Salg looked at mortality rates for white and nonwhite males and females. With weighted regression analyses, surface water usage showed weak but statistically significant associations with the following: white males–esophagus, lung, larynx, trachea, large intestine, rectum, bladder, other urinary organs and lymphosarcoma and reticulosarcoma; white females–breast and rectum; and nonwhite females–esophagus and larynx.

Only rectal cancer showed positive correlations across all race-sex groups. It should be noted that the test of significance utilized for this study was $p>0.10$ which is less stringent than all other studies, except for Cantor's study, which used even less stringent criteria for some correlations.

Mah et al (1977) conducted a retrospective study in the Los Angeles County area. The relationship between cancer mortality and morbidity and the chlorinated drinking water supply was analyzed for white populations only. Results did not reveal any trends and were not significant either for mortality or morbidity cancer rates. The authors point out several methodological problems, including the diluting effect of migration in the highly mobile area covered by this study.

Hogan et al (1977) also utilized the chemical analysis of the NORS and Region V data sets and applied various statistical procedures to the data in order to determine the appropriateness of the statistical model. Thus, it is not surprising that results were similar to previous studies showing a positive correlation between rectal-intestinal and bladder cancer mortality rates and chloroform levels in drinking water when a weighted regression analysis was applied.

In summary, many but not all of the studies have found positive correlations between drinking water and various cancer mortality/morbidity rates. It is beyond the scope of this document to evaluate each study in depth, however, it is pertinent to consider the interpretations and conclusions that can be drawn from these retrospective epidemiological studies collectively.

It is also extremely important in the evaluation process to consider the results from other epidemiological studies as they develop hypotheses of potential causal associations between cancer mortality and other agents. For example, the confounding factors of diet, occupation, and smoking all have been suggested as potential causative agents of bladder cancer (Cole, 1977). Therefore, any epidemiological study that investigates the possible association between bladder cancer and drinking water should either control for the aforementioned variables or analyze to avoid the problems that result in confounding of the data.

None of the studies completed thus far have obtained data on or controlled for diet; several studies have attempted to control for occupational exposure: Page and Harris, (1974 and 1975), Cantor et al (1976); only one study by Kruse (1977) obtained smoking history data. Only a few studies considered four sex-race groups (the number of nonwhites is too small in some of the geographic areas) and of those studies only a few showed consistent patterns of association of specific cancer sites, i.e., Salg–rectum. Several studies which considered only white populations found positive correlation coefficients for both sexes: Buncher (1975)–bladder; Reiches (1976)–stomach; and Cantor (1976)–bladder.

A decreasing level of association from high to low levels of the water quality variable, i.e., chloroform or THM, with cancer mortality rates is an important criteria in evaluating the evidence. This pattern of association should be observed if the difference in mortality rates is due to the water variable. Only a few studies defined the water quality variable by the chloroform concentrations (McCabe, 1975; Buncher, 1975; Cantor et al, 1976; Hogan et al, 1977), and by the THM concentrations (Cantor et al, 1976).

Of particular interest are the correlations of liver and kidney cancer mortality rates with drinking water, since the animal exposure data indicate that hepatocellular carcinomas and hepatic modular hyperplasias have been observed in B6C3F1 strains of mice after lifetime exposure. Several of the preliminary studies grouped the cancer sites for the anatomical systems, i.e., gastrointestinal and urinary organs, in order to increase the sample size.

Only one of the studies (Cantor, 1976) which considered site-specific cancer mortality showed a positive association between drinking water and cancer of the kidney in white males. The absence of any positive association between drinking water and liver cancer mortality may be due in part to small sample sizes, very low incidence of the disease, or because the exposure levels of contaminants in trace amounts over a lifetime may be below the no-effect level (Weisburger, 1977).

Thus, the evidence is incomplete and the trends and patterns of association have not been fully developed. As stated previously, a causal relationship cannot be established, nor can it be disproven. When viewed collectively, the epidemiological studies completed thus far provide sufficient evidence for maintaining a hypothesis that there may be a health risk and that the positive correlations may be due to some association between drinking water and cancer mortality. Only when viewed in conjunction with animal studies, both acute and chronic toxicity studies, is the evidence evaluated in the appropriate context for policy decision making. Additional direct epidemiological studies may provide evidence regarding the strength of the associations and the possibility of a causal relationship between drinking water and cancer mortality.

RISK ASSESSMENT

The establishment of chloroform as an animal carcinogen, plus the epidemiological data and mutagenesis data on THM, show that a potential human risk exists from the consumption of trihalomethanes, but these data do not quantify the risk. Methods have been developed to estimate quantitatively the size of the risk under the assumption that there is no threshold level for a carcinogen. The state of the art at the present time is such that no experimental tools can accurately define, with any degree of certainty, the absolute numbers of excess cancer deaths attributable to chloroform in drinking water. Due to the biological variability and number of assumptions required, each of the risk estimates reports different absolute numbers with a wide degree of variability.

It is generally agreed that it is not possible to project with accuracy risk estimates to absolute numbers of cancers in human population from exposure to a given agent, using statistical extrapolation models with animal data. Given that caveat, it may be useful to apply one or more risk estimation procedures in an attempt to estimate a possible range of impact to affected populations both in the absence of the interim proposed standard and at some alternate standard levels.

The EPA Science Advisory Board (SAB), using the highest levels of chloroform then reported in drinking water by the NORS data (0.300 mg/l) and assuming a maximum daily intake of 4 liters of water for a 70-kg man, attempted to compute an estimated risk in 1975. The estimates were based on the Eschenbrenner and Miller animal data, which are highly speculative since the experimental protocol involved only 5 animals per sex per dose.

Using a linear extrapolation of the animal data over more than 2 orders of magnitude of dose from mice to humans at the 0.300 mg/l concentration level, the lifetime incidence of liver tumors in man were estimated in the range of 0 to 0.001 (95% confidence limits) or 0 to 100 x 10^{-5} in a lifetime. This rate may be compared with the lifetime incidence of 260 x 10^{-5} for malignancy of liver derived from data of the Third National Cancer Survey (1976). This estimate would range from zero to approximately 40% of the observed incidence of liver cancer in the United States that may be attributable to exposure to chloroform in drinking water at the 0.300 mg/l level.

It should be noted that this value is at the upper limit of the confidence interval and the linear nonthreshold dose-effect model allows an estimate of maximal risk where a risk has actually been observed. Other models would all yield lower estimates. The SAB, however, also stated that a more reasonable assumption would yield lower estimates of the risk.

Tardiff (1976), using four different models, calculated the maximum risk from chloroform ingestion via tap water. Using a margin of safety of 5,000 applied to the minimum effect animal dose, the safe level was calculated to be 0.02 mg/kg per day. Using the log probit model and the slope recommended by Mantel and Bryan, the conclusion reached was that at a maximum daily dose of 0.01 mg/kg the risk would be between 0.016 and 0.683 cancer per million exposed population per year. Using the identical data, but with the actual slope of the dose response curve as opposed to the slope of the one in the previous calculation, the conclusion reached was that a maximum daily dose of 0.01 mg/kg would produce less than one tumor per million population per lifetime.

Using the linear or one hit model, usually considered to be the most conservative, a risk estimate of between 0.42 and 0.84 cancer per million population per year was calculated to result from a maximum dosage level of 0.01 mg/kg/day. The two-step model produced an estimated maximum risk of between 0.267 and 0.283 cancer per million population per year at a maximum dose level of 0.01 mg/kg/day.

In the NAS report on drinking water and health, lifetime risks were estimated from the NCI animal data. For concentrations of 10 ppb exposure, the number of excess cases of cancer was computed as one for every 50,000 exposed persons, assuming a risk of 2×10^{-5} and 2 liters per day of water consumed. If the U.S. population using chlorinated water is assumed to be approximately 160 million people, this translates into 3,200 excess lifetime deaths from cancer or 45.7 cases per year.

For a concentration of chloroform at 1 μg/liter the estimated lifetime cancer risk would fall at approximately 3.7×10^{-7} at the upper 95% confidence limits. In evaluating the risk estimates, it is important to compare the calculated maximum risk with the current cancer mortality data. Both liver and kidney cancer are rare diseases in the U.S. The standardized mortality rates in the U.S. for white males and females combined are 52.5 per million per year for liver carcinoma and 29.2 per million per year for kidney carcinoma.

Based on his risk estimates, Tardiff (1976) calculated that the percent of the cancer mortality rates attributable to chloroform in drinking water would be 1.60 and 1.44% for liver and kidney cancer incidence per year respectively, assuming the maximum exposure levels. Applying these percentages to the actual cancer mortality rates, the number of cancer deaths per year would be 168 from liver carcinoma or 84 from kidney carcinoma; an estimated maximum of 252 cancer deaths per year attributable to chloroform in drinking water.

EPA's Carcinogen Assessment Group's (CAG) risk estimations for chloroform exposure are shown in Table 2.7. The risks were computed for several exposure levels and were extrapolated from data from the National Cancer Institute (NCI, 1976) bioassay with the male rat and female mouse. Human exposure from drinking water was computed using a weighted average of chloroform concentrations in drinking water for 160 million people whose drinking water supplies are chlorinated.

In the absence of a THM standard the CAG statistical risk model would predict from 23 to 207 total tumors per year in the exposed human population, depending which animal data (rat or mouse) is utilized as the base.

TABLE 2.7: RISK ASSESSMENT FOR CHLOROFORM

ACTION	TOTAL TUMORS PER YEAR	TOTAL TUMORS PER YEAR REDUCED BY ACTIONS
NONE	23.1*-207.0**	---------
CITIES GREATER THAN 50,000 REDUCE TO:		
100 ugm/l	18.6-166.6	4.5-40.4
50 ugm/l	15.6-140.1	7.5-66.9
10 ugm/l	9.6-85.6	13.5-121.4
CITIES GREATER THAN 75,000 REDUCE TO:		
100 ug/l	19.6-175.3	3.5-31.7
50 ug/l	17.3-155.4	5.8-51.6
10 ug/l	12.1-108.5	11.0-98.5

Note: Risks extrapolated from NCI bioassay data
*Male rat
**Female mouse

Computations of estimated human risk and risk reduction at various levels of control were made for the total population in cities larger than 75,000 which are affected by this regulation.

A standard of 100 μg/l would reduce the annual risk according to the statistical model to 19 to 175 total tumors; a standard of 50 μg/l would reduce the risk to 17 to 155 total tumors; a standard of 10 μg/l would reduce the risk to 12 to 108 tumors.

Given that it is not possible to project with certainty or accuracy from risk estimates based on animal data to absolute numbers of cancers in a human population, such extrapolations are useful in attempting to quantify a range of possible impacts of alternate standards.

It should be noted, however, that these average exposure levels which refer to chloroform alone and do not consider the risk from other contaminants in the impacted population are overestimtes of the risk in light of the facts that: (1) the computations are based upon lifetime exposure, while in actuality the proposed interim standard is a temporary, phased standard which will be reduced in the future and therefore, the lifetime exposure values would be less; and (2) the interim standard clearly calls for maximum reductions obtainable using available technology thus indicating a lower average exposure. They may be underestimated since the risk estimates are based upon toxicity exposure data from chloroform, which is only a portion of the total THM and other contaminants found in drinking water. Therefore the magnitude of the contribution to the risk of the other THM (bromohalomethanes), which in some cases consists of a substantial portion of the THM, and the many other possible contaminants is unknown.

SUMMARY

The occurrence of trihalomethanes in drinking water supplies of various communities across the United States has been documented. Chloroform was found at concentrations ranging from 0.001 to 0.540 mg/l and trihalomethanes potential concentrations as high as 0.784 mg/l have been detected. The concentration of THM increased on treatment of raw water supplies with chlorine in the process of disinfection and subsequent preparation of water for drinking purposes. The THM concentrations may also be indicative of the presence of other undefined chemicals that are produced in water during chlorination.

Besides the presence of chloroform in drinking water, humans are exposed to chloroform from air and food. An analysis of the relative contribution of chloroform in drinking water as compared with air and food exposures considered various relative levels of exposures. Depending upon the ranges of chloroform concentrations that have been detected in air, food and water (which are a function of location, urbanization and industrialization), drinking water may contribute from zero to more than 90% of the total dietary intake.

Chloroform has been shown to be rapidly absorbed on oral and intraperitoneal administration and subsequently metabolized to carbon dioxide and unidentified metabolites in urine. The metabolic profile of chloroform in animal species such as mice, rats and monkeys is indicated in Table 2.4 and is qualitatively similar to that in man.

Biological responses on exposure of chloroform to mammals include its effect on the central nervous system resulting in narcosis, hepatotoxicity, nephrotoxicity, teratogenicity, and carcinogenicity. These responses are discernible in mammals on exposure to high levels of chloroform ranging from 30 to 350 mg/kg; the intensity of response was dependent upon the dose. Although less toxicological information is available for brominated trihalomethanes, mutagenicity and carcinogenicity have been detected in some test systems. Physiological chemical activity should be greater for the brominated THM than for chloroform.

Exposure to the low levels of trihalomethanes presently found in drinking water supplies may not manifest detectable responses in populations. It is the prolonged human exposure to trihalomethanes that should be a matter of major concern. Prolonged administration of chloroform at relatively high dose levels (100 to 138 mg/kg) to animals, specifically rats and mice, manifested oncogenic effects. The oncogenic effect was not observed at a lower dose level (17 mg/kg). Assuming that methods do not exist to establish a nonthreshold level for long-term effects for carcinogenesis (NAS, 1977), the preceding data do not imply that a safe level of exposure can be established.

Epidemiological evidence is inconclusive, although positive correlations have been found in several studies. There have been 11 retrospective studies that have investigated some aspect of a relationship between cancer mortality or morbidity and use of drinking water. Due to various limitations in the epidemiological methods, in the water quality data, and problems with the individual studies, the present evidence cannot lead to a firm conclusion that there is an association between contaminants in drinking water and cancer mortality/morbidity.

Causal relationships cannot be proven on the basis of results from epidemiological studies. The evidence from these studies thus far is incomplete and the trends and patterns of association have not been fully developed. When viewed collectively, however, the epidemiological studies provide sufficient evidence for maintaining the hypothesis that there may be a potential health risk and that the positive correlations may be reflecting a causal association between constituents of drinking water and cancer mortality.

Preliminary risk assessments made by the Science Advisory Board (SAB), the National Academy of Sciences (NAS), Robert Tardiff of EPA, and the Carcinogen Assessment Group (CAG) using four different models have estimated the cancer risks associated with the exposure from chloroform in drinking water. The exposure to THM from air and food have not been included in these computations. The total cancer risk estimates associated with the maximum contaminant level (MCL) at the 0.1 mg/l level range from the NAS estimated lifetime risk of 4×10^{-5} to the CAG's estimate of 2×10^{-4} using somewhat different assumptions. These risks are similar to the lifetime estimated risks of other known carcinogenic standards, e.g., for vinyl chloride emissions (about 10^{-5}) and ionizing radiation exposure to the general public (about 10^{-5}).

On the basis of the available toxicological data summarized in the above reports, chloroform has been shown to be a carcinogen in rodents (mice and rats) at high dose levels. Since its metabolic pattern in animals is qualitatively similar to that in man, it may prove to be a human carcinogen. Epidemiological studies also imply a human risk. Therefore, because a potential human health risk does exist, levels of chloroform in drinking water should be reduced as much as is technologically and economically feasible using methods that will not compromise protection from waterborne infectious disease.

SELECTED MAXIMUM CONTAMINANT LEVELS (MCL)

Since it is evident from the foregoing that a risk to the public exists from exposure to the trihalomethanes in drinking water, the risk should be reduced as much as is technologically and economically feasible without increasing the risk of microbiological contamination. This can be accomplished by several means, and the Safe Drinking Water Act (P.L. 93-523) provides two major regulatory avenues: (1) the establishment of an MCL or (2) the institution of a treatment requirement.

EPA has determined that the establishment of an MCL through a phased approach, along with monitoring requirements, is the most effective and conservative approach to regulating the levels of trihalomethanes in drinking water. The administrator has determined that monitoring is both technically and economically feasible. Measures taken to reduce the THM concentrations will concurrently provide the additional benefit of reducing human exposure to the other undefined by-products and possibly other synthetic organic contaminants.

Since it is known that chlorination of water is primarily responsible for the relatively high levels of trihalomethanes in drinking water, modifications in the chlorination process, the substitution of other disinfectants, and the use of adsorbents to remove precursor chemicals are possible approaches for control. The optimal approach would be to reduce organic precursor concentrations by adsorbents or other means prior to addition of the disinfectant.

Use of a chlorine residual in a less active form such as combined chlorine or chloramine will significantly reduce trihalomethane formation, however, chloramines are much less potent disinfectants than free chlorine and therefore it would not always be appropriate to adopt this approach. The two chemicals most often mentioned as substitute disinfectants, ozone and chlorine dioxide, are both well known as effective disinfectants and chemical oxidants. The issues of the bioeffects and toxicology of these disinfectants and their by-products are being clarified by studies under way. In the meantime, the application of chlorine dioxide should be limited to 1 mg/l.

The National Organics Monitoring Survey found that the mean total trihalomethane concentrations in the drinking water systems evaluated were approximately 0.068, 0.117, 0.053 and 0.100 mg/l for Phase I, II, III (dechlorinated) and III (terminal) respectively with the highest levels of 0.784 mg/l in Phase II (refer to Table 2.1).

It is reasonable to assume that the calculated risk estimates for chloroform from various studies do indicate a potential risk to public health. It is possible that a percentage of the total number of liver and/or kidney cancers are attributable to exposure to chloroform in drinking water, although it is most likely that drinking water interacts with a number of other variables such as smoking and diet as effect modifiers in a multifactorial manner. It is also likely that the other trihalomethanes are a potential risk.

Thus, based upon a number of risk extrapolations, assuming various levels of exposure to chloroform in drinking water, it has been estimated that such exposures may cause an excess of cancers in the U.S. population (ranging from zero to several hundred). At higher levels of exposure of chloroform (>0.300 mg/l) the risk estimates would result in larger numbers of excess cancer cases.

The reduction of the total trihalomethanes to the MCL level of 0.10 mg/l would reduce the unnecessary and excessive exposure to these potential human carcinogens, mutagens, and chronic toxicants and may result in the reduction of excess cases of cancer. At the same time, measures taken to reduce THM levels (such as the use of adsorbents) will concurrently result in reduction of human exposure to other contaminants in drinking water.

Since it is economically and technologically feasible to reduce the THM levels in drinking water and there is a benefit achieved by reducing the health risks to exposure, EPA has decided to set the MCL at 0.10 mg/l as an initial step in a phased, regulatory approach. As more data becomes available from implementation experience, standards will become more restrictive in the future. In the meantime, EPA will take steps as necessary on a case by case basis to provide adequate protection for the delivery of safe drinking water to the public.

REFERENCES

Bellar, T.A., Lichtenberg, J.J., and Kroner, R.C., "The Occurrence of Organohalides in Chlorinated Drinking Water," *Journal AWWA,* 66:703 (1974).

Butterfield, C.T., and Wattie, E., "Relative Resistance of *E. coli* and *E. typhosa* to Chlorine and Chloramines," *Pub. Health Rpts.,* 59, 1661 (1944).

Butterfield, C.T., and Wattie, E., "Influence of pH and Temperature on the Survival of Coliforms and Enteric Pathogens When Exposed to Chloramine," *Pub. Health Rpts.,* 61, 157 (1946).

Butterfield, C.T., "Comparing the Relative Bactericidal Efficiencies of Free and Combined Available Chlorine," *J. AWWA,* 40, 1305 (1948).

Clarke, N.A., Berg, G., Kabler, P.W., and Chang, S.L., "Human Enteric Viruses in Water: Source, Survival and Removability," International Conf. Water Pollution Research, Pergamon Press, London (Sept., 1962).

Cole, P.C., Hoover, R., and Friedell, G.H., "Occupation and Cancer of the Lower Urinary Tract," *Cancer,* 29: 1250-60, (1972).

DeRouen, T.A. and Diem, J.E., "Ethnic, Geographical Difference in Cancer Mortality in Louisiana," Tulane University, School of Public Health and Tropical Medicine, Unpublished (1975).

DeRouen, T.A. and Diem, J.E., "The New Orleans Drinking Water Controversy: A Statistical Perspective," *American Journal of Public Health,* 65: (No. 10): 1060 (1975).

DeSalva, S., Volpe, A., Leigh, G., and Regan, T., "Long-Term Safety Studies of a Chloroform-Containing Dentrifice and Mouth Rinse in Man," *Food Cosmet. Toxicol,* 13: 529-532 (1975).

Environmental Protection Agency, Region VI, Dallas, Texas, "Industrial Pollution of the Lower Mississippi River in Louisiana," (April 1972).

Environmental Protection Agency, Region VI, Dallas, Texas, "Analytical Report, New Orleans Area Water Supply Study," (1974).

Environmental Protection Agency, Science Advisory Board, "A Report–Assessment of Health Risk from Organics in Drinking Water by an Ad Hoc Study Group to the Hazardous Materials Advisory Committee," Unpublished, (1975).

Environmental Protection Agency, Region V, Chicago, Illinois, "Region V Joint Federal/State Survey of Organics and Inorganics in Selected Drinking Water Supplies," (June 1975).

Environmental Protection Agency, Office of Pesticides Programs Criteria and Evaluation Division, "Risk Evaluation of Chloroform," Unpublished, (1976).

Environmental Protection Agency, Cancer Assessment Group, "Chloroform Risk Assessment in Drinking Water," Unpublished (1977).

Environmental Protection Agency, "Economic Impact Analysis of a Trihalomethane Regulation for Drinking Water," prepared by Temple, Barker, and Sloan, Inc. for EPA, Office of Water Supply, (August 1977).

Environmental Protection Agency, "Interim Treatment Guide for the Control of Chloroform and Other Trihalomethanes," EPA, Water Supply Research Division, MERL (June 1976).

Eschenbrenner, A.B. and Miller, E., "Induction of Hepatomas in Mice by Repeated Oral Administration of Chloroform with Observation on Sex Differences," *J. Natl. Cancer Inst.,* 5: 251-255, (1945).

Fair, G.M., Morris, J.C., and Chang, S.L. "The Dynamics of Water Chlorination," *S. NEWWA,* 61, 285 (1947).

Fridyland, S.A. and Kagan, G. "Experimental Validation of Standard for Residual Chlorine Dioxide in Drinking Water," *Hygiene and Sanitation,* No. 36, pp 18-21, (1971).

Fry, F.J., Taylor, T., and Hathaway, E.D., "Pulmonary Elimination of Chlorform and Its Metabolites in Man," *Arch. Int. Pharmacodyn,* 196: 98-111, (1972).

Heilbrunn, G., Liebert, E., and Szanto, P.B., "Chronic Chloroform Poisoning–Clinical and Pathological Report of a Case," *Arch. Neurol. Psych.,* 53: 68-72, (1945).

Hogan, M.D., Chi, P., Mitchell, T.J., and Hoel, D.G., "Association Between Chloroform Levels in Finished Drinking Water Supplies and Various Site-Specific Cancer Mortality Rates," National Institute of Environmental Health Sciences, Environmental Biometry Branch, Research Triangle Park, North Carolina, Draft Report (1977).

Ilett, K.F., Reid, W.P., Sipes, I.G. and Krishna, G., "Chloroform Toxicity in Mice: Correlation of Renal and Hepatic Necrosis with Covalent Binding of Metabolites to Tissue Macromolecules," *Exptl. Molec. Pathol.,* 19: 215-229, (1973).

Jones, W.M., Marguis, G. and Stephen, C.R., "Hepatotoxicity of Inhalation Anesthetic Drugs," *Anesthesiology,* 19: 715-23, (1958).

Kabler, P.W., "Relative Resistance of Coliform Organisms and Enteric Pathogens in the Disinfection of Water with Chlorine," *J. AWWA,* 43, 553 (1953).

Kelly, S.M., and Sanderson, W.W., "The Effect of Chlorine in Water on Enteric Viruses II., The Effect of Combined Chlorine on Poliomyelitis and Coxsackie Viruses," *Amer. J. Pub. Health,* 59, 14 (1960).

Kruse, C.W., "Chlorination of Public Water Supplies and Cancer–Washington County, Maryland Experience," A Preliminary Report from the Johns Hopkins University, School of Hygiene and Public Health to the Office of Research and Development, Health Effects Research Laboratory, Cincinnati, Ohio, Unpublished Draft, (1977).

Kubic, V.L., Anders, M.W., Engel, R.R., Barlow, C.H. and Caughey, W.S., "Metabolism of Dihalomethanes to Carbon Monoxide. In Vivo Studies," *Drug Metabolism and Disposition,* 2: 53-57, (1974).

Kuzma, R.J., Kuzma, C.J., and Buncher, R.C., "Ohio Drinking Water Source and Cancer Rates," *American Journal of Public Health,* Submitted for publication August or September, (1977).

Liu, O.C. and McGrowan, F., "Effect of Chlorination on Human Enteric Viruses in Partially Treated Water from the Potomac River Estuary," in *Virus In Water,* edited by G. Berg et al, APHA, Washington, D.C. (1976).

Musil, J. et al, "Toxicological Aspects of Chlorine Dioxide Application for the Treatment of Water Containing Phenol", *SF, VYS. Sk. Chem–Technol,* Vol 8, pp 327-345, (1963).

National Academy of Sciences, "Drinking Water and Health," Washington, D.C. (1977).

National Academy of Sciences, "Nonfluorinated Halomethanes in the Environment," Washington, D.C., (1977).

National Cancer Institute, "Report on Carcinogenesis Bioassay of Chloroform," (1976).

National Institute of Occupational Safety and Health, "Criteria for a Recommended Standard, Occupational Exposure to Chloroform," (1974).

Page, T. and Harris, R.H., "Relation Between Cancer Mortality and Drinking Water in Louisiana," Unpublished (1975).

Page, T., Harris, R.H., and Estein, S.S., "Drinking Water and Cancer Mortality in Louisiana," *Science,* 193: 55-57, (1976).

Page, T., Talbot, E., and Harris, R.H., "The Implications of Cancer Causing Substances in Mississippi River Water," A Report by the Environmental Defense Fund (1974).

Paul, B.B., and Rubinstein, D., "Metabolism of Carbon Tetrachloride and Chloroform by the Rat," *J. Pharm. Exp. Therap.,* 141: 141-148, (1963).

Plaa, G.L., Evans, E.A. and Hine, G.H., "Relative Hepatotoxicity of Seven Halogenated Hydrocarbons," *J. Pharmacol. Exp. Therap,* 123: 224-229, (1963).

Plaa, G.L. and Larson, R.E., "Relative Nephrotoxic Properties of Chlorinated Methane, Ethane and Ethylene Derivatives in Mice," *Toxicol. Appl. Pharmacol.,* 7: 37-44, (1965).

Poirier, L.A., Stoner, G.D., and Shimkin, M.B., "Bioassay of Alkyl Halides and Nucleotide Base Analogs by Pulmonary Tumor Response in Strain A Mice," *Cancer Research*, 35: 1411-1415, (1975).

Reiches, N.A., Page, T., Talbot, P., and Harris, R.H., "Carcinogenic Hazards of Organic Chemicals in Drinking Water," Unpublished, (1976).

Roe, F.J.C., "Preliminary Report of Long-Term Tests of Chloroform in Rats, Mice and Dogs," Hazleton Laboratories, Vienna, Virginia (1976).

Rook, J.J., "Production of Potable Water from a Highly Polluted River," *Water Treatment and Examination* 21, Part 3, 259-274, (1972).

Rook, J.J., "Formation of Haloforms During Chlorination of Natural Waters," *Water Treatment and Examination,* 23: 234-243, (1976).

Rook, J.J., "Chlorination Reactions of Fulvic Acids in Natural Waters," *Env. Sci. Technol.,* 11: 478-482, (1977).

Salg, J., "Cancer Mortality Rates and Drinking Water in 346 Counties of the Ohio River Valley Basin," Final Report from the University of North Carolina, Department of Epidemiology to the Office of Research and Development, Health Effects Research Laboratory, Cincinnati, Ohio, Unpublished, (1977).

Schwetz, B.A., Leong, B.K.J. and Gehring, P.J., "Embryo and Fetotoxicity of Inhaled Chloroform in Rats," *Toxicol. Appl. Pharmacol.* 28: 442-451, (1974).

Sontheimer, H., and Kuhn, W., The Engler-Bunte Institute, University of Karlsruhe, Karlsruhe, Germany, Personal Communication, (1977).

Storms, W.W., "Chloroform Parties," *JAMA,* 225:160, (1973).

Symons, J.M., Bellar, T.A., Carswell, J.K., Demarco, J., Kropp, K.L., Robeck, G.G., Seeger, D.R., Slocum, C.J., Smith, B.L. and Stevens, A.A., "National Organics Reconnaissance Survey for Halogenated Organics," *J. Am. Water Works Assoc.,* 67: 634-646, (1975).

Tardiff, R.G., "Health Effects of Organics: Risk and Hazard Assessment of Ingested Chloroform," The 96th Annual Conference of the American Water Works Association, New Orleans, (1976).

Tarone, R.E. and Gart, J.J., "The Implications of Cancer-Causing Substances in Mississippi Rain Water," An unpublished review of the study by R.H. Harris, (1975).

Taylor, D.C., Brown, D.M., Keeble, R. and Langley, P.F., "Metabolism of Chloroform II. A Sex Difference in the Metabolism of [C] Chloroform in Mice," *Xenobiotica* 4: 165-174, (1974).

Theiss, J.C., Stoner, G.D., Shimkin, M.B., and Weisburger, E.K., "Test for Carcinogenicity of Organic Contaminants of United States Drinking Waters by Pulmonary Tumor Response in Strain A Mice," *Cancer Research,* 37: 2717-2720, (1977).

United States Department of Health, Education and Welfare, National Institute of Occupational Safety and Health, "Toxic Substance List," (1973).

Von Oettingen, W.F., "The Halogenated Hydrocarbons: Toxicity and Potential Dangers," Public Health Service No. 414, Washington, D.C: U.S. Govenment Printing Office, (1955).

Wallace, C.J., "Hepatitis and Nephrosis Due to Cough Syrup Containing Chloroform," *Calif. Med.,* 73: 442, (1959).

Watanabe, P.G., McGowan, G.R., Madrid, E.O. and Gehring, P.J., "Fate of [C] Vinyl Chloride Following Inhalation Exposure in Rats," *Toxicol. Appl. Pharmacol.,* 37: 49-59, (1976).

Weisburger, J.H., "Social and Ethical Implications of Claims for Cancer Hazards," *Medical and Pediatric Oncology,* 3: 137-140, (1977).

APPENDIX I

The following material (Table 2.8) is taken from "National Organics Reconnaissance Survey for Halogenated Organics," by J.M. Symons, T.A. Bellar, J.K. Carswell, J. DeMarco, J.L. Kropp, G.C. Robeck, D.R. Seeger, C.J. Slocum, B.L. Smith and A.A. Stevens, in *J. Am. Water Works Assoc.* 67:634 (1975).

TABLE 2.8: NATIONAL ORGANICS RECONNAISSANCE SURVEY WATER QUALITY DATA (FROM 80 UTILITIES FOR FOUR TRIHALOMETHANES, CARBON TETRACHLORIDE, AND 1,2-DICHLOROETHANE)

Utility Name (Plant name when applicable)	Chloroform	Bromodichloromethane	Dibromochloromethane	Bromoform	1,2-Dichloroethane	Carbon Tetrachloride	Nonvolatile Total Organic Carbon
	....	....	....	μg/l	....	....	....
1. Lawrence Water Works	<0.1	<0.2	NF*	NF	NF	NF	3.7
	91	9	0.6	NF	NF	NF	1.6
2. Waterbury Bureau of Water	NF	NF	NF	NF	NF	NF	2.2
(Morris Treatment Station)	93	10	0.6	<1	<0.2	<2	2.9
3. Metropolitan District Commission	NF	NF	NF	NF	NF	NF	2.1
(Norumbego Treatment Station)	4	0.8	NF	NF	NF	NF	2.0
Boston, Massachusetts							
4. Newport Dept. of Water	NF	NF	NF	NF	NF	NF	4.6
(South Pond Reservoir Treatment Plant #1)	103	42	13	1	NF	NF	4.1
5. Department of Water Resources	NF	NF	NF	NF	NF	NF	3.0
New York, New York	22	7	0.9	NF	NF	NF	2.5
6. Puerto Rico Aqueduct and Sewer Authority	<0.2	NF	NF	NF	NF	NF	2.0
(Sergio Cuevas Water Treatment Plant)	47	29	16	2	NF	NF	2.0
7. Passaic Valley Water Commission	0.3	NF	NF	NF	<0.2	<2	3.6
Little Falls, New Jersey	59	16	2	NF	<0.2	<2	1.9
8. Toms River Water Company	0.4	NF	NF	NF	NF	NF	<0.05
	0.6	<0.8	3	NF	NF	NF	<0.05
9. Buffalo Water Department	NF	NF	NF	NF	NF	NF	2.6
	10	10	4	NF	<0.2	NF	1.7
10. Village of Rhinebeck Water Department	0.3	NF	NF	NF	3	NF	3.5
	49	11	1	NF	2	NF	1.6
11. Philadelphia Water Department	0.2	NF	NF	NF	3	NF	2.6
(Torresdale Plant)	86	9	5	NF	6	NF	1.7
12. Wilmington Suburban Water Corp.	0.3	<0.4	NF	NF	NF	NF	2.8
(Stanton Plant)	23	11	3	NF	<0.4	<2	1.8
Claymont, Delaware							
13. Artesian Water Company	0.2	NF	NF	NF	NF	NF	0.05
(Llangolen Well Field Plant)	0.5	0.5	1	<1	<0.2	NF	0.2
14. Washington Aqueduct	<0.2	NF	NF	NF	NF	NF	1.8
(Delacarlia Plant)	41	8	2	NF	<0.3	NF	1.2

(continued)

TABLE 2.8: (continued)

Utility Name (Plant name when applicable)	Chloro-form	Bromo-dichloro-methane	Dibromo-chloro-methane	Bromo-form	1,2-Dichloro-ethane	Carbon Tetra-chloride	Nonvolatile Total Organic Carbon
				μg/l			
15. Baltimore City Bureau of Water Supply	NF	NF	NF	NF	NF	NF	1.8
(Montebello Plant #1)	32	11	2	NF	NF	NF	1.2
16. Western Pennsylvania Water Company	0.3	NF	NF	NF	NF	NF	0.9
(Hays Mine Plant)	8	2	0.4	NF	NF	NF	0.8
17. Strasburg Borough Water System	NF	NF	NF	NF	NF	NF	0.2
	<0.1	NF	NF	NF	NF	NF	0.05
18. Fairfax County Water Authority	<0.2	<0.4	NF	NF	NF	NF	4.7
(New Lorton Plant)	67	6	<0.6	NF	NF	NF	2.7
19. Virginia-American Water Company	0.2	NF	NF	NF	NF	NF	4.2
Hopewell District	6	1	0.8	<2	NF	NF	0.2
20. Huntington Water Corp.	1	NF	NF	NF	<0.3	4	2.2
	23	16	5	NF	<0.4	3	1.0
21. Wheeling Water Department	0.2	NF	NF	NF	<0.3	NF	3.2
	72	28	17	NF	<0.4	NF	1.8
22. Miami-Dade Water and Sewer Authority	NF	NF	NF	NF	<0.2	<2	9.8
(Preston Plant)	311	78	35	3	<0.2	NF	5.4
23. Jacksonville Dept. of Public Works	NF	NF	NF	NF	NF	NF	2.40
(Highlands Pumping Station)	9	4	2	NF	NF	NF	2.3
24. Atlantic Waterworks	<0.2	NF	NF	NF	<0.3	NF	1.3
(Chattahoochee Plant)	36	10	2	NF	NF	NF	0.9
25. Owensboro Municipal Utilities	NF	NF	NF	NF	NF	NF	1.7
	13	20	17	3	NF	NF	2.0
26. Greenville Water Department	0.3	NF	NF	NF	NF	NF	3.3
	17	6	3	<1	<0.2	NF	4.0
27. Tennessee American Water Company	0.9	NF	NF	NF	NF	NF	1.1
Chattanooga, Tennessee	30	9	0.7	NF	<0.4	NF	0.6
28. Memphis Light, Gas and Water Division	<0.2	NF	NF	NF	NF	NF	0.2
(Malloy Plant)	0.9	2	1	NF	NF	NF	0.2
29. Metropolitan Water and Sewage Dept.	<0.1	NF	NF	NF	NF	NF	1.2
(Lawrence Plant)	16	5	<0.4	NF	NF	NF	0.8

(continued)

TABLE 2.8: (continued)

Utility Name (Plant name when applicable)	Chloro-form	Bromo-dichloro-methane	Dibromo-chloro-methane	Bromo-form	1,2-Dichloro-ethane	Carbon Tetra-chloride	Nonvolatile Total Organic Carbon
	....	....	....	...... μg/l	....	....	....
30. Commissioners of Public Works	<0.2	NF	NF	NF	NF	NF	11.4
(Stoney Plant)	195	9	0.8	0.8	NF	NF	4.1
Charleston, South Carolina							
31. Cincinnati Water Works	0.5	NF	NF	NF	NF	<2	2.3
	45	13	4	NF	<0.4	<2	1.1
32. Chicago Dept. of Water and Sewers	<0.2/0.4**	NF/<0.5	NF/NF	NF/NF	NF/NF	NF/NF	1.9/1.7
(South District Water Filt. Plant)	15	10	4	NF	<0.4	NF	1.5
33. Clinton Public Water Supply	<0.2	NF	NF	NF	NF	NF	7.7
	4	0.5	NF	NF	NF	NF	6.7
34. Indianapolis Water Company	0.1	NF	NF	NF	<0.3	NF	5.1
(White River Plant)	31	8	<2	NF	NF	2	2.6
35. Whiting Water Department	16***	11	3	NF	NF	NF	2.0
	0.1†	<0.8	NF	NF	NF	NF	1.9
	0.5	0.3	NF	NF	NF	NF	1.5
36. Detroit Metro Water Department	<0.2	NF	NF	NF	0.5	NF	2.6
(Water Park Plant)	12	9	3	NF	0.4	NF	1.2
37a. Mt. Clemens Water Purification	NF	NF	NF	NF	NF	NF	2.0
	11	6	<2	NF	<0.4	NF	1.4
37b. Mt. Clemens Water Purification	0.9	NF	NF	NF	<0.2	NF	6.7
(after replacement of granular act. carbon)	6	3	<2	NF	NF	NF	1.4
38. St. Paul Water Authority	<0.2	NF	NF	NF	NF	NF	7.9
	44	7	2	NF	NF	NF	4.4
39. Cleveland Division of Water	NF	NF	NF	NF	NF	NF	2.2
(Division Filtration Plant)	18	9	4	NF	NF	NF	1.8
40. City of Columbus	0.1	NF	NF	NF	NF	NF	6.8
(Dublin Road Plant)	134	8	<0.4	NF	NF	NF	2.3
41. Dayton Water Works	NF	NF	NF	NF	NF	NF	0.9
(Ottawa Plant)	8	8	11	4	<0.2	<2	0.7
42. Indiana Hill Water Supply	<0.2	NF	NF	NF	NF	NF	0.8
	5	7	11	NF	NF	NF	0.9
43. Piqua Water Supply	NF	NF	NF	NF	NF	NF	6.0
	131	13	3	NF	<0.2	NF	4.2

(continued)

TABLE 2.8: (continued)

Utility Name (Plant name when applicable)	Chloro-form	Bromo-dichloro-methane	Dibromo-chloro-methane	Bromo-form	1,2-Dichloro-ethane	Carbon Tetra-chloride	Nonvolatile Total Organic Carbon
	μg/l						
44. Mahoning Valley Sanitary District	NF	NF	NF	NF	NF	NF	4.7
Youngstown, Ohio	80	5	<1	NF	NF	NF	3.1
45. Milwaukee Water Works	<0.2	NF	NF	NF	NF	NF	2.4
(Howard Avenue Purification Plant)	9	7	3	NF	<0.2	NF	1.7
46. Oshkosh Water Utility	NF	NF	NF	NF	NF	NF	4.5
	26	4	<0.4	NF	<0.2	NF	3.3
47. Terrebonne Parish Waterworks	NF	NF	NF	NF	NF	NF	5.4
District #1, Houma, Louisiana	134	32	8	<1	0.2	NF	3.2
48. Camden Municipal Water Works	NF	NF	NF	NF	NF	NF	3.1
	40	19	7	NF	NF	NF	1.5
49. Town of Logansport System	0.7	NF	NF	NF	NF	NF	5.3
	28	39	24	3	NF	NF	3.5
50. City of Albuquerque	NF	NF	NF	NF	NF	NF	<0.05
	0.4	1	2	3	NF	NF	<0.05
51. Oklahoma City Water Department	NF	NF	NF	NF	NF	NF	3.6
(Hefner Plant)	44	28	20	6	<0.4	<2	2.8
52. Brownsville Public Utility Board	NF	NF	NF	NF	NF	NF	4.7
(Plant #2)	12	37	100	92	NF	NF	3.1
53. Dallas Water Utilities	<0.1	NF	NF	NF	NF	NF	3.4
(Bachman Plant)	18	4	<2	NF	NF	NF	2.9
54. San Antonio City Water	NF	NF	NF	NF	NF	NF	0.5
	0.2	0.9	3	3	NF	NF	0.5
55a. Ottumwa Waterworks	<0.2	NF	NF	NF	NF	NF	4.1
(2/17/75 sample)	0.8	NF	NF	NF	NF	NF	2.3
55b. Ottumwa Waterworks	NF	NF	NF	NF	NF	NF	4.9
(4/17/75 sample)	1	NF	NF	NF	NF	NF	2.4
56. Clarinda Iowa Water Works	<0.2	NF	NF	NF	NF	NF	3.5
	48	19	4	NF	NF	NF	3.0
57. Davenport Water Company	0.4	NF	NF	NF	NF	NF	6.5
	88	8	<0.6	NF	<0.4	NF	4.4

(continued)

TABLE 2.8: (continued)

Utility Name (Plant name when applicable)	Chloro-form	Bromo-dichloro-methane	Dibromo-chloro-methane	Bromo-form	1,2-Dichloro-ethane	Carbon Tetra-chloride	Nonvolatile Total Organic Carbon
	μg/l						
58. Topeka Public Water Supply	0.4	<0.8	NF	NF	NF	NF	3.4
(South Plant)	88	38	19	5	NF	3	2.2
59. Missouri Utility Company	0.2	NF	NF	NF	0.2	NF	4.5
Cape Girardeau, Missouri	116	21	2	NF	0.3	2	3.6
60. Kansas City Missouri Water Department	NF	NF	NF	NF	NF	NF	3.4
	24	8	2	NF	NF	NF	1.9
61. St. Louis County Water Company	NF	NF	NF	NF	0.3	NF	3.4
(Central Plant)	55	13	3	<1	0.4	NF	2.6
62. Lincoln Municipal Water Supply	NF	NF	NF	NF	NF	NF	1.4
	4	6	4	<2	NF	NF	1.4
63. City Water Department	NF	NF	NF	NF	NF	NF	9.2
Grand Forks, North Dakota	3	1	NF	NF	NF	NF	5.2
64. Denver Water Board	<0.2	NF	NF	NF	NF	NF	2.0
(Marston Plant)	14	10	3	NF	NF	NF	1.7
65. Pueblo Board of Waterworks	<0.2	NF	NF	NF	NF	NF	1.8
(Gardner Plant)	2	2	<2	NF	NF	NF	1.6
66. Huron Water Department	NF	NF	NF	NF	NF	NF	19.2
	309	116	49	8	NF	NF	12.2
67. Salt Lake Water Department	<0.2	NF	NF	NF	<0.4	NF	1.2
	20	14	8	NF	NF	NF	0.9
68. City of Tucson Water & Sewer Dept.	<0.1	NF	NF	NF	NF	NF	<0.05
(Plant #1)	<0.2	<0.8	2	13	NF	NF	<0.05
69. City of Phoenix Water & Sewer Dept.	<0.2	NF	NF	NF	NF	NF	1.0
(Verde Plant)	9	15	17	<4	NF	NF	1.0
70. Department of Supply & Purification	<0.2	NF	NF	NF	NF	NF	3.7
Coalinga, California	16	17	15	2	NF	NF	2.4
71. Contra Costa County Water Dept.	0.3	<0.3	NF	NF	NF	NF	3.4
(Bollman Plant, Concord, Calif.)	31	18	6	<1	NF	NF	1.9
72. City of Dos Palos Water Dept.	NF	NF	NF	NF	NF	NF	4.4
	61	53	34	7	NF	NF	2.9

(continued)

TABLE 2.8: (continued)

Utility Name (Plant name when applicable)	Chloro-form	Bromo-dichloro-methane	Dibromo-chloro-methane	Bromo-form	1,2-Dichloro-ethane	Carbon Tetra-chloride	Nonvolatile Total Organic Carbon
				μg/l			
73. Los Angeles Dept. of Water and Power	<0.1	NF	NF	NF	NF	NF	1.2
	32	6	3	NF	NF	NF	1.3
74. San Diego Water Utilities Dept.	NF	NF	NF	NF	NF	NF	2.9
(Miramar Plant)	52	30	19	3	NF	NF	2.8
75. San Francisco Water Department	NF	NF	NF	NF	NF	NF	1.3
(San Andreas Treatment Plant)	41	15	4	<0.8	NF	NF	1.6
76. Seattle Water Department	<0.2	NF	NF	NF	NF	NF	0.9
(End of Dist. System)	15	0.9	NF	NF	NF	NF	0.9
77. Douglas Water System	NF	NF	NF	NF	NF	NF	3.4
	40	0.8	<0.4	NF	NF	NF	2.8
78. Idaho Falls Water Department	<0.2	NF	NF	NF	NF	NF	0.5
	2	3	3	NF	NF	NF	0.3
79. City of Corvallis Utilities Div.	NF	NF	NF	NF	NF	NF	1.0
(Taylor Plant)	26	3	NF	NF	NF	NF	0.4
80. Ilwaco Municipal Water Department	0.1	NF	NF	NF	NF	NF	7.5
	167	35	5	NF	NF	NF	3.1
Range	NF-0.9	NF-0.8	NF	NF	NF-3	NF-4	<0.05-19.2
	<0.1-311	NF-116	NF-100	NF-92	NF-6	NF-3	<0.05-12.2

Note: The first line for each entry is raw water data and the second line is finished water data.

*NF is none found.
**Two samples.
***After pre Cl_2.
†Before pre Cl_2.

APPENDIX II

The material in this appendix is from "The National Organics Monitoring Survey: Samplings and Analyses for Purgeable Organic Compounds," by H.J. Brass, M.A. Feige, T. Halloran, J.W. Mello, D. Munch, and R.T. Thomas, in *Drinking Water Quality Enhancement Through Source Protection,* R.B. Pojasek, Ed., Ann Arbor Science (Ann Arbor, 1977).

The United States Environmental Protection Agency conducted the National Organics Monitoring Survey (NOMS) to determine the frequency of occurrence of specific contaminants in drinking water supplies and to provide data for the possible establishment of additional maximum contaminant levels of organic compounds or for a treatment requirement for the control of organic compounds in drinking water.

Included in the NOMS were 113 community water supplies, representing various types of sources and treatment processes, which were monitored three times (phases) during a twelve month period. The NOMS was designed as a study of finished water quality. In several cases, additional studies were made of raw water sources to obtain background information. The sampling dates for the first three phases are as follows: Phase I, from March to April of 1976; Phase II, from May to July of 1976; and Phase III, from November of 1976 to January of 1977. Data obtained would have reflected certain long-term and seasonal variations, if present.

A broad range of compounds may be present in a given water supply. The sources of these compounds are broadly classified as follows: naturally occurring organic matter (e.g., humus); contaminants introduced as a result of treatment (e.g., trihalomethanes); synthetic chemicals from point sources; and chemicals from nonpoint sources. Compounds selected for study are listed in Table 2.9.

TABLE 2.9: COMPOUNDS STUDIED IN THE NATIONAL ORGANICS MONITORING SURVEY

1.	Chloroform	12.	1,2,4-Trichlorobenzene
2.	1,2-Dichloroethane	13.	2,4-Dichlorophenol
3.	Carbon Tetrachloride	14.	Pentachlorophenol
4.	Bromodichloromethane	15.	Polychlorinated Biphenyls
5.	Trichloroethylene	16.	Fluoranthene
6.	Dibromochloromethane	17.	11,12-Benzofluoranthene
7.	Bromoform	18.	3,4-Benzofluoranthene
8.	Benzene	19.	1,12-Benzoperylene
9.	Vinyl Chloride	20.	3,4-Benzopyrene
10.	Bis[2-chloroethyl] ether	21.	Indeno[1,2,3-cd]Pyrene
11.	p-Dichlorobenzene		

They were chosen on the basis of possible occurrence, available toxicological data, and the existence of analytical methodology for their identification and quantification. In addition, several general organic class parameters (NPTOC, carbon chloroform extract, ultraviolet absorbance, and emission fluorescence scan) were studied in an attempt to determine a possible relationship between one or more of them and specific compound occurrences. Certain sampling and analytical modifications were instituted following Phases I and II to take into consideration the findings of the earlier phases.

Trihalomethanes (THM)

NOMS data for trihalomethanes (THM) are summarized in Tables 2.10 and 2.11. Clearly, the occurrences and concentrations of THM in finished water were greater than for any other specific compounds or classes of compounds. In Phase I, samples were collected and analyzed in a fashion similar to that employed in the National Organics Reconnaissance Survey (NORS). Samples were shipped from the utility and stored at 2° to 8°C for 1 to 2 weeks prior to analysis. In Phase II samples were allowed to stand at 20° to 25°C for 3 to 6 weeks prior to analyses.

Under these conditions THM formation was permitted to proceed to reaction end points (terminal values) generally corresponding to the consumption of free chlorine residual in the sampling vial. All supplies were sampled in Phase III with and without a chlorine reducing agent present at the time of sampling (quenched and terminal). In addition, samples with a reducing agent present were collected at five points in the distribution system in each of ten cities. A site at a distant point in most systems was sampled.

These experiments were incorporated in Phase III in order to more fully understand the distribution of total trihalomethane (TTHM) concentrations that can reach the consumer, since it is recognized that the extent of THM formation depends on the type and chemical composition of raw water, pH, water temperature, chlorine contact time, treatment method, and in certain cases season.

TABLE 2.10: NOMS MEAN AND MEDIAN CONCENTRATIONS OF TOTAL TRIHALOMETHANES

	Mean TTHM μg/l	Median TTHM μg/l
Phase I–111 Cities Iced*	68	45
Phase II–113 Cities Terminal**	120	87
Phase III–106 Cities Quenched**,***	52	38
Phase III–105 Cities Terminal**	100	78

*Samples shipped and stored at 2° to 8°C for 1 to 2 weeks prior to analyses.
**Samples stored at 20° to 25°C for 3 to 6 weeks prior to analyses.
***Sodium thiosulfate added

Phase II and Phase III terminal TTHM data (Table 2.10) show the mean and median values to be closer to each other than to quenched or iced values; these latter values also appear to be similar. The data can be interpreted as indicating that on a national basis, differences between values in Phases I to III were in a large measure due to the handling of samples prior to analyses. Generally, Phase III data show that THM levels in those distribution systems studied were between quenched and terminal values, provided water leaving the plant was the sole water source within the distribution system.

A seasonal variation in THM levels, though known to occur in specific water systems, was a secondary effect nationally. It should be noted that dichloroiodomethane was qualitatively identified in many samples, but not quantified. In most supplies, the contribution of dichloroiodomethane to TTHM was likely small. Up to an order-of-magnitude lower levels of TTHM were found in supplies using ground rather than surface water sources.

Raw Water Contaminants

Other than trihalomethanes, compounds studied in the NOMS are generally present in finished drinking water due to raw water contamination. An exception is 2,4-dichlorophenol which is produced on treatment by chlorination of phenol, itself a raw water contaminant. 1,1,1-Trichloroethane and tetrachloroethylene were quantified in Phase III due to their frequent qualitative identification in Phase II. ortho- and meta-Dichlorobenzene and bis(2-chloroisopropyl) ether were quantified in Phases II and III because the determination of related compounds, para-dichlorobenzene and bis(2-chloroethyl) ether, made their analyses convenient.

Samplings and analyses for polynuclear aromatic hydrocarbons and chlorinated phenols were of a limited nature after Phase I due to either their low frequency of occurrence and concentration levels and/or the lack of adequate analytical methodologies. Additional raw water contaminants were identified by broad spectrum gas chromatography/mass spectrometry (GC/MS) analyses in Phases II and III.

Raw water contaminants occurred on a frequent basis at low concentrations in many supplies. However, the regularity of occurrences was not nearly as great as for trihalomethanes. Supplies using certain raw water sources are more susceptible to contamination. For example, supplies located on rivers upstream from large numbers of industries were more susceptible to raw water contamination.

General parameters did not correlate with the occurrences of specific organic compounds. An exception was the relationship between TTHM and finished water NPTOC concentrations. In Phases I, II, and III, the correlation coefficients (Pearson r) between TTHM and NPTOC were 0.55, 0.70, and 0.60 respectively. While statistically significant, these data are in no way predictive.

TABLE 2.11: SUMMARY OF FINDINGS OF THE NATIONAL ORGANICS MONITORING SURVEY, MARCH 1976 THROUGH JANUARY 1977

Compound	Number of Positive Analyses per Number of Analyses			Mean Concentration, μg/l Positive Results Only		
	Phase 1	Phase II	Phase III	Phase I	Phase II	Phase III
Chloroform						
Quenched**	102/111*	18/18	98/106	47*	68	38
Terminal***	—	112/113	101/105	—	84	73
Bromodichloromethane						
Quenched**	88/111*	18/18	100/106	22*	16	9.2
Terminal***	—	109/113	103/105	—	18	17
Dibromochloromethane						
Quenched**	47/111*	15/18	83/106	17*	13	7.5
Terminal***	—	97/113	97/105	—	14	11
Bromoform						
Quenched**	3/111*	6/118	19/106	21*	28	13
Terminal***	—	38/113	30/105	—	12	13
Dichloroiodomethane						
Terminal***	—	85/111	50/105	—	†	†
1,2-Dichloroethane						
Quenched**	1/111*	—	1/106	2.0*	—	1.3
Terminal***	—	2/113	1/105	—	0.9	1.2
1,1,1-Trichloroethane						
Quenched**	—	—	4/105	—	—	1.5
Terminal***	—	19/111	4/104	—	†	1.0
Carbon tetrachloride						
Quenched**	3/111*	—	8/106	2.9*	—	6.4
Terminal***	—	10/110	11/105	—	2.4	4.3
1,1,2-Trichloroethylene						
Quenched**	4/112*	—	10/106	11*	—	2.4
Terminal***	—	28/113	19/105	—	2.1	1.3
Tetrachloroethylene						
Quenched**	—	—	8/106	—	—	1.1
Terminal***	—	48/111	9/105	—	†	0.81

(continued)

TABLE 2.11: (continued)

Compound	Number of Positive Analyses per Number of Analyses			Mean Concentration, µg/l Positive Results Only		
	Phase I	Phase II	Phase III	Phase I	Phase II	Phase III
Benzene	0/111	7/113	4/16	–	0.4	0.95
o-Dichlorobenzene	–	0/113	4/110	–	–	2.5
m-Dichlorobenzene	–	0/113	2/110	–	–	0.10
p-Dichlorobenzene	2/111	20/113	29/110	2.0	0.14	0.067
1,2,4-Trichlorobenzene	1/112	2/113	10/110	10	0.29	0.090
Vinyl chloride	–	2/113	–	–	0.14	–
Methylene chloride	15/109	–	–	6.1	–	–
2,4-Dichlorophenol	56/108	0/9	–	0.18	–	–
Pentachlorophenol	86/108	5/10	–	0.07	0.008	–
Bis(2-chloroethyl) ether	0/112	13/113	8/110	–	0.10	0.024
Bis(2-chloroisopropyl) ether	–	8/113	7/110	–	0.17	0.11
PCBs	2/110	4/113	2/110	0.76	0.13	0.15
Fluoranthene	17/110	0/9	–	0.02	–	–
3,4-Benzofluoranthene	0/110	0/9	–	–	–	–
1,12-Benzoperylene	0/110	0/9	–	–	–	–
3,4-Benzopyrene	0/110	0/9	–	–	–	–
Indeno[1,2,3-cd] pyrene	0/110	0/9	–	–	–	–
Nonpurgeable total organic carbon	108/112	108/113	107/110	2.1 mg/l	2.1 mg/l	2.4 mg/l
Chemical oxygen demand	80/80	–	–	7.0 mg/l	–	–
Carbon chloroform extract	111/111	107/107	109/109	0.55 mg/l	1.1 mg/l	1.2 mg/l

*Samples were shipped iced (4°C), stored at 2°-8°C for 1 to 2 weeks prior to analyses.

**Samples were preserved with sodium thiosulfate at time of sampling, shipped at ambient temperature, stored at 20°-25°C for 3 to 6 weeks prior to analyses.

***Samples were shipped at ambient air temperature, stored at 20°-25°C for 3 to 6 weeks prior to analyses.

†Not quantified.

(continued)

TABLE 2.11: (continued)

Additional Organic Compounds Detected in NOMS

Acetone	Benzyl cyanide
Aniline	Atrazine
Benzaldehyde	α-Bergamotine and three related compounds
Bromotrichloroethylene	Trimethylisocyanurate
n-Butyl acetate	Tris(chloroethyl) phosphate
Di-n-butyl ether	Hexachloroethane
Chlorobenzene	1,1,2-Trichloroethane
Chlorostyrene	Trichlorofluoromethane
Chlorotoluene	Styrene
Chloroxylene	Tetralin
Dibromodichloroethylene	1,1- and 1,2-Dibromoethane
Ethylbenzene	Methylstyrene
Isobutyl acetate	Chloropicrin
1,2-Dichloropropane	Iodomethane
Isopropyl ether	Bromochloromethane
Methylene bromide	1-Naphthyl-N-methyl carbamate
Methycyclohexane	Fluorobenzene
Methylmethacrylate	Bromoethane
4-Methyl-2-pentanone	Methyl bromide
Naphthalene (or azulene)	Dimethyl disulfide
Phenol	Phenol
cis-β-Terpineol	Terpinenes
Terpineol isomer	Nitrotoluene
1,1,2,2-Tetrachloroethane	Trichlorobutane
Tetrahydrofuran	Acetonitrile
Toluene	cis- and trans-Dichloroethylene
Tribromochloroethylene	Diethyl adipate
Trichlorobenzene isomer	Dioctyl adipate
Xylenes	Trichlorophenol
Heptadecane	2-Chloropropane
N-methylaniline	Ethyl ether
1,1-Dichloroethene	

Statement of Basis and Purpose for the Amendment on Organic Compounds

The material in this chapter was taken from the report of the same title which was excerpted from the book, *Drinking Water and Health,* prepared for the USEPA by the Safe Drinking Water Committee of the Advisory Center on Toxicology, National Research Council, and published by the National Academy of Sciences in June 1977.

Carcinogenic Risk Estimates were arrived at by a study of the available set of dose-response data for the compound under study as described in the above book in the chapter on the margin of safety. This set of data was used to estimate both the lifetime risk and an upper 95% confidence bound on the lifetime risk at the low-dose level. These estimates are of lifetime human risks, corrected for species conversion on a dose per surface area basis. Risk estimates are expressed as a probability of cancer after a lifetime consumption of 1 liter of water per day containing Q ppb of the compound of interest. A risk of 1×10^{-6} Q implies a lifetime probability of 2×10^{-5} of cancer if 2 liters per day were consumed and the carcinogen concentration was 10 ppb (i.e., Q = 10). Thus, at a concentration of 10 ppb during a lifetime of exposure this compound would be expected to produce one excess case of cancer for every 50,000 persons exposed. If the population of the United States is 220 million people, this translates into 4400 excess lifetime deaths from cancer from this substance, or 62.8 per year.

References for the Pesticides section are on p 139; the remainder are on p 152.

PESTICIDES

Benzene Hexachloride (BHC) and Lindane

Benzene hexachloride (BHC) is the common name used to designate the mixed isomers of 1,2,3,4,5,6-hexachlorocyclohexane. Hexachlorohexane is the proper term for this compound; however, because it is more customary, the trivial name, benzene hexachloride (or BHC), will be used here.

BHC (technical grade) is a mixture of the eight possible isomers that constitute the different spatial arrangements of the six chlorine atoms on the trans form of the ring. Its composition approximates 65% α-isomer, 11% β, 13 to 14% γ, 8 to 9% δ, and 1% ϵ. The lowest-melting-point (112.8°C) isomer, which is also the most reactive known, is that designated as the γ-isomer. The commercial insecticide lindane is defined as a product containing at least 99% γ-isomer (the remainder being other BHC isomers). Technical BHC is prepared by photochlorination of benzene. Production of the γ-isomer for lindane is achieved by selective crystallization (Melnikov, 1971).

The different isomers have different solubilities in water and different vapor pressures: α, 10 mg/liter and 0.06 torr; β, 5 mg/liter and 0.17 torr; γ, 10 mg/liter and 0.14 torr. The relatively high water solubility and vapor pressure of lindane cause it to have relatively low persistence in the environment. Lindane has been detected in the finished water of Streator, Illinois at 4 mg/liter (USEPA, 1975j). The EPA has set an interim standard for lindane in finished water of 0.004 mg per liter (USEPA, 1975i).

The insecticidal properties of BHC were discovered and developed for commercial use in pest control beginning in 1942. When it was found that virtually all the insecticidal activity of BHC resided in the γ-isomer, major development of that isomer as an insecticide itself was rapid. Lindane has been marketed under a large number of trade names as an insecticide. It has had major use in insect control in domestic and commercial settings, in numerous agricultural and silvicultural applications, and in dips, sprays, and dusts for livestock and pets. Recent U.S. production has been under 1 million pounds a year.

Health Aspects: Surveys of human tissue for organochlorine insecticide residues frequently show the presence of the most persistent, the β-isomer of BHC. In a study on the concentration of organochlorine residues in fat and liver of terminal patients, the only BHC isomer noted was the beta. Its concentration in cancer patients did not differ significantly from that in people dying from infectious or other diseases (Radomski et al, 1968). Chronic liver damage (cirrhosis and chronic hepatitis) has been found, in liver biopsy, in eight workers heavily exposed to BHC, DDT, or both for periods ranging from 5 to 13 years. As far as was feasible, other conditions, such as alcoholism, were excluded as the cause of the cirrhosis (Schuttmann, 1968).

Over 30 cases of exposure to BHC or lindane and 21 cases of exposure to BHC and DDT followed by the development of aplastic anemia have been reported in the literature (Loge, 1965; West, 1967; Woodliff et al, 1966). No satisfactory animal model of that condition has been found and, despite efforts to study the question, a firm causal relationship between lindane or technical BHC exposure and aplastic anemia cannot be stated. Development of leukemia after lindane exposure was reported for two cases (Jedlica, 1958). That causal relationship is also inconclusive in relation to insecticide exposure.

Carcinogenic Risk Estimates: α-, β-, and γ-BHC have produced dose-related liver tumors when given orally to mice and rats (Ito et al, 1973 and 1975; Thorpe and Walker, 1973). For each compound the available sets of dose response data were individually considered. For α-BHC at a concentration of 1 μg/liter (Q = 1) the upper 95% confidence estimate of risk for man would fall between 2.2 to 6.5 x 10^{-6} Q. For β-BHC at a concentration of 1 μg/liter (Q = 1) the estimated risk for man would be from 1.1 to 3.5 x 10^{-6} Q. The upper 95% confidence estimate of risk at the same concentration would be between 2.5 to 5.8 x 10^{-6} Q. For lindane at a concentration of 1 μg/liter (Q = 1), the estimated risk for man would be from 3.3 to 8.1 x 10^{-6} Q. The upper 95% confidence estimate of risk at the same concentration would be from 5.6 to 13 x 10^{-6} Q.

Conclusions and Recommendations: The chronic toxicity of the BHC isomers is clearly related to the tumorigenic effects so far observed only in rodents. The alpha-isomer is the most strongly implicated; its activity is sufficient to account for the degree of hepatoma formation observed with technical BHC administration in mice. Lindane is a weaker tumorigen in mice, and is so far a questionable tumorigen in the rats.

As of 1972, the FAO/WHO ADI for lindane was set at 0.0125 mg/kg/day. Later, that value was reduced to 0.001 mg/kg/day and held under temporary status because of the newer data concerning carcinogenicity. A full-scale reevaluation of the chronic toxicity of lindane is scheduled by the FAO/WHO. The EPA has announced plans to issue a presumptive notice that lindane is too hazardous for continued registered use, with the intention of reevaluating its administrative position on this insecticide.

In light of the above and taking into account the carcinogenic risk projections, it is suggested that very strict criteria be applied when limits for BHC isomers are established.

Cyclodienes: Aldrin, Dieldrin, Endrin, Chlordane, Heptachlor and Heptachlor Epoxide

The cyclodiene insecticides are all derivatives of hexachlorocyclopentadiene produced by the Diels-Alder, or diene, reaction. Their discovery and development date from the synthesis of chlordane by Julius Hyman in 1944 (U.S. Patents 2,509,160 and 2,606,910). Perhaps 600 million pounds of these highly chlorinated, cyclic organic compounds have been dispersed into the soil, air, water, and food of the United States during the last 30 years. Little is certain about the degradation and fate of these compounds; however, traces of them and their stable epoxide oxidation products are ubiquitous in the environment and are heavily bioconcentrated in the lipids of terrestrial and aquatic wildlife, humans, and foods, especially animal fats and milk.

The cyclodienes have been used principally as preemergence soil insecticides for the control of corn rootworms, wireworms, cutworms, etc.; as seed treatments; as soil poisons for control of termites and ants; and on cotton for the control of the boll weevil and bollworms.

Domestic production in 1971 estimates were: chlordane, 25 million pounds; aldrin, 10 million pounds; heptachlor, 6 million pounds; dieldrin, less than 1 million pounds; and endrin, less than 1 million pounds. Because of their environmental persistence and carcinogenic behavior in laboratory animals, aldrin and dieldrin were banned by the EPA on October 1, 1974, and chlordane and heptachlor registrations for agricultural crops were suspended on April 1, 1976.

Because of their persistence, the cyclodienes and their epoxides (viz, dieldrin, heptachlor epoxide, and probably oxychlordane) are found in surface waters virtually everywhere. In an extensive 1958-1965 survey of the rivers of the United States, Breidenbach et al (1967) found the following average concentrations of the cyclodienes:

	ppb
Aldrin	<0.001-0.006
Dieldrin	0.8-0.122
Endrin	0.008-0.214
Heptachlor	0.0-0.0031
Heptachlor epoxide	<0.001-0.008
DDT	0.008-0.144

The highest concentrations were generally found in the lower Mississippi basin. In 1964, 74% of the grab samples were positive for dieldrin, 46% for endrin, 10% for aldrin, 17% for heptachlor, and 25% for heptachlor epoxide. In comparison, 44% were positive for DDT and 39% for DDE.

More than 500 grab samples of finished drinking water and related raw water from the Mississippi and Missouri Rivers were analyzed by Schafer et al (1969). More than 40% of the finished water samples contained dieldrin at up to 0.25 ppb, more than 30% contained endrin, and 20% contained chlordane at up to 0.5 ppb. Aldrin and heptachlor were found occasionally.

An extensive investigation of surface, subsurface and finished water in Iowa (Richard et al, 1974) showed dieldrin present at the following concentrations:

	ppt
S. Skunk River	2-76
Indian Creek	4-30
Des Moines River	2-12
Raccoon River	1-12
Red Rock Reservoir	3-36
Rothbun Reservoir	2-22
Cedar Rapids (surface)	42
Iowa River	22
Mississippi River	<0.5-7
Davenport (finished)	2
Iowa City (finished)	5
Des Moines (finished)	<0.4-2

It was concluded that water treatment plants were not removing substantial amounts of pesticides from raw water, even by filtration through activated carbon beds.

The New Orleans water supply contained dieldrin at 0.05 to 0.07 ppb (50 to 70 ppt) and endrin at 0.004 ppb (USEPA, 1974a). The 10-city drinking-water survey (USEPA, 1975j) found dieldrin at 1 to 2 ppt in drinking water of Miami, Seattle, Ottumwa (Iowa), and Cincinnati. Other cyclodienes identified in U.S. drinking water included aldrin, chlordane, chlordene, endrin (80 ppt), heptachlor, and heptachlor epoxide.

Standards have been proposed for the maximal permissible concentrations of the cyclodienes in finished water (Schafer et al, 1969). Concentrations suggested in 1965-1966 and based on those of the Subcommittee on Toxicology in 1965 were: aldrin, 32.0 ppb; dieldrin, 18.0 ppb; endrin, 1.0 ppb; heptachlor, 78.0 ppb;

heptachlor epoxide, 18.0 ppb; and chlordane, 52 ppb. The suggested DDT–T (DDT, DDE, and DDD combined) concentration was 42 ppb. These were drastically lowered in a 1967 recommendation (Ettinger and Mount, 1967) based on maximal reasonable stream allowances to: aldrin, 0.25 ppb; dieldrin, 0.25 ppb; endrin, 0.1 ppb; heptachlor, 1.0 ppb; heptachlor epoxide, 1.0 ppb; and chlordane, 0.15 ppb. The DDT–T allowance was 0.5 ppb.

The U.S. Public Health Service Advisory Committee recommended the following drinking water standards in 1968 (Mrak, 1969): aldrin, 17 ppb; dieldrin, 17 ppb; endrin, 1 ppb; heptachlor, 18 ppb; and heptachlor epoxide, 18 ppb.

The EPA has set an interim standard for endrin in finished water for 0.002 milligrams per liter (USEPA, 1975i).

The persistent organochlorine cyclodienes are present everywhere in the environment, are readily biomagnified through food chains, and are common trace contaminants of human food. Market Basket Surveys of the U.S. diet, collected in five major U.S. cities and designed to simulate the diet of a 16- to 19-year-old male, have shown that, in a 6-year period (1965 through 1970), the average daily intake of aldrin and endrin was 0.001 mg; of heptachlor epoxide was 0.002 mg; and of dieldrin was 0.005 mg.

The combined residues for aldrin and dieldrin are very close to the FAO/WHO acceptable daily intake (ADI) which is 0.0001 mg/kg/day for aldrin and dieldrin vs 0.00013 mg/kg/day in 1966. For heptachlor and heptachlor epoxide, the FAO/WHO ADI is 0.0005 mg/kg/day vs 0.00005 mg/kg/day in 1966 (Mrak, 1969). It should be pointed out that recent studies showing increases in mouse hepatomas at the lowest dosages fed (i.e., 0.1 ppm) demonstrated that the no-adverse-effect level for aldrin and dieldrin has never been determined and that the ADI is therefore too high.

Residues of the cyclodienes in milk are particularly high because of the ingestion of these insecticides with forage. Dieldrin has the highest retention time of all pesticides in milk, approximately 100 days (Mrak, 1969). For example, analogues of 1971 milk samples from 12 grade B daily farms in an intensive grain-producing area of northwestern Illinois showed combined aldrin and dieldrin concentrations in butterfat of 0.1314 to 0.5560 ppm (average 0.2921). There was a definite correlation between the overall dieldrin soil residue on each farm (0.0104 to 0.3859 ppm) and the concentration in the milk (Moore et al, 1973). Consumption of this milk (composite dieldrin concentration 0.22 ppm) by a 5-kg infant at 500 g/day would provide a daily dieldrin intake of approximately 0.75 μg/kg, or 7.5 times the FAO/WHO ADI.

The cyclodiene insecticides (especially the epoxides dieldrin, heptachlor epoxide, and oxychlordane) are very stable, both environmentally and biologically, and have high lipid-water partition coefficients. Thus, they pass through food chains and undergo biologic magnification (Lu et al, 1975; Metcalf et al, 1973).

These factors account for the bioaccumulation of these persistent products in human adipose tissue. The values shown in Table 3.1 indicate the average amounts found in fiscal 1970-1974 in over 1,400 bioassays of U.S. human fatty tissues.

TABLE 3.1: PESTICIDES IN FATTY TISSUE

Insecticide	Concentration, ppm				
	1970	1971	1972	1973	1974
Dieldrin	0.27	0.29	0.24	0.24	0.20
Heptachlor epoxide	0.17	0.12	0.12	0.12	0.10
Oxychlordane	--	--	0.15	0.15	0.15
DDT-T	11.65	11.55	9.91	8.91	7.83

Dieldrin provides a graphic example of the propensity to persist in food chains and to accumulate in lipid tissues. Gannon et al (1959) demonstrated that dieldrin fed to chickens was stored in body fat to very much higher concentrations than when fed to steers, hogs, and lambs. For example, dieldrin fed at 0.1 ppm was stored at 4.1 ppm, and that fed at 0.75 ppm was stored at 35.7 ppm. However, it was not until 1974 that it was demonstrated that as many as 20 million chickens in Mississippi fed on waste food stocks of soybean oil that had been processed on soybeans grown in aldrin-treated soil containing illegal residues of dieldrin (generally at 0.01 to 0.04 ppm) contained dieldrin residues in their body fat greatly exceeding the FDA "safe limit" of 0.3 ppm and ranging up to 30 ppm (Moore, 1975). The chickens had to be destroyed as unfit for human consumption.

Health Aspects: The effects of the cyclodienes on animal and human health are very subtle and complex. These are the most hazardous of all pesticides, because of their persistence, fat storage, and central nervous system target site. Their effects can be reviewed only briefly here.

Human illness and death have been observed after poisoning during the manufacture, spraying, or accidental ingestion of the cyclodienes. Typical symptoms of poisoning result from stimulation of the central nervous system and include headache, blurred vision, dizziness, slight involuntary muscular movements, sweating, insomnia, bad dreams, nausea, and general malaise. More severe illness is characterized by jerking of muscles or groups of muscles and epileptiform convulsions, with loss of consciousness, involuntary incontinence of urine and feces, disorientation, personality changes, psychic disturbances, and loss of memory. Such seizures may recur for 2 to 4 months after cessation of exposure and are marked by abnormal encephalographic patterns.

These sysmptoms of severe poisoning have developed in 10 to 20% of spraymen working in WHO house-spraying programs (Hayes, 1957, 1959), and such poisoning has not been eliminated in any spray program.

Epidemics of endrin poisoning have occurred after the eating of bread made from flour accidentally contaminated with endrin; there were 59 illnesses in one episode (Davies and Lewis, 1956) and 874, with 26 deaths, in Saudi Arabia (Weeks, 1967). At least 97 cases of fatal endrin poisoning were recorded through 1965 (USEPA, 1973a). It appears that ingestion of endrin at 0.2 to 0.25 mg/kg can produce convulsions in humans (Hayes, 1963).

Workers in a plant manufacturing and formulating aldrin, dieldrin, isodrin, and endrin had epileptiform convulsions (3.3%) and had encephalograms suggesting brain stem injury (20.5%). The encephalograms usually returned to normal within 3 to 6 months after exposure ceased (Hoogendem et al, 1962).

Carcinogenic Risk Estimates: Dieldrin, chlordane and heptachlor have produced dose-related hepatomas when fed to mice (Walker et al, 1973 and NCI, 1975). For each compound the available sets of dose-response data were individually considered.

For dieldrin at a concentration of 1 μg/liter (Q = 1) the estimated risk for man would fall between 0.8 to 1.9 x 10^{-4} Q. The upper 95% confidence estimate of risk at the same concentration would be between 1.9 to 2.4 x 10^{-4} Q.

For chlordane at a concentration of 1 μg/liter (Q = 1) the upper 95% confidence estimate of risk for man would be between 0.96 to 1.8 x 10^{-5} Q.

For heptachlor at a concentration of 1 μg/liter (Q = 1) the upper 95% confidence estimate of risk for man would be from 3.5 to 4.8 x 10^{-5} Q.

Conclusions and Recommendations: The cyclodiene insecticides (particularly the persistent epoxides—dieldrin, endrin, heptachlor epoxide, and oxychlordane) present the greatest hazards of all residual pesticides in water. At low dosages, they are highly active hepatocarcinogens and have a dangerous effect on the central nervous system of man and higher animals, leading to apparently irreversible changes in encephalographic and behavioral patterns. They are highly persistent biologically and can accumulate in animal fats and exist in milk.

In light of the above and taking into account the carcinogenic risk projections, it is suggested that very strict criteria be applied when limits for dieldrin, heptachlor and chlordane in drinking water are established. Before limits for aldrin, endrin and heptachlor epoxide in drinking water can be established, more toxicological data must be gathered and evaluated.

DDT, DDD, and DDE

DDT, or 2,2-bis-(p-chlorophenyl)-1,1,1-trichloroethane, was patented as an insecticide in 1939 by Swiss chemist Paul Muller. After a period of extensive use for the control of malaria, typhus, and other insect-transmitted diseases during World War II, it became the prototype of the synthetic insecticides. At the height of its use in the United States, in 1963, production was 176 million pounds, and DDT was registered for use on 334 agricultural commodities. DDT has been used very extensively all over the world, both in malaria control and in agriculture, and it is estimated that more than 4.4 billion pounds have been used for insect control since 1940, about 80% in agriculture. Because of the extensive environmental problems resulting from its stability and high lipid-water partitioning, DDT was banned for all but essential public health use in the United States on January 1, 1973.

The persistence of DDT, DDE and DDD has made them ubiquitous contaminants of water. The total residues are commonly referred to as DDT-T. In an extensive 1958–1965 survey of the rivers of the United States, Breidenbach et al (1967) found DDT in every river surveyed at 0.008 to 0.144 ppb; DDE at 0.002 to 0.011 ppb, and DDD at 0.004 to 0.080 ppb. The highest concentrations were generally found in the West and South, where 44% of the samples were positive for DDT and 38% for DDE. More than 500 grab samples of finished drinking water and related raw water from the Mississippi and Missouri Rivers were analyzed by Schafer et al (1969); more than 33% of the finished water samples contained DDT-T.

Lake Michigan contains DDT-T at an average of 6 ppt. Most fishes from Lake Michigan contain DDT residues in excess of the 7 ppm FDA safe limit and the overall biomagnification from water to fish may exceed a factor of 3×10^6.

Water standards have been proposed for DDT in finished water (Schafer et al, 1969). The DDT-T concentration suggested in 1965–1966 and based on maximal acceptable concentrations of the Subcommittee on Toxicology was 42 ppb. This was drastically lowered to 0.5 ppb by Ettinger and Mount (1969) on the basis of a maximal reasonable stream allowance.

DDT and its breakdown products are ubiquitous and, because of biomagnification and persistence, are in virtually every food product. Market-basket surveys of the U.S. diet, collected in five major U.S. cities and designed to represent the diet of a 16- to 19-year-old male, show the daily dietary intake of DDT-T over a six-year average period to be 0.053 mg.

Health Aspects: There are no definite examples of human fatality due to ingestion of DDT, but a dosage of 10 mg/kg has produced illness in some (but not all) subjects, without convulsions. Convulsions have frequently occurred at 16 mg/kg or higher. Human volunteers have consumed 35 mg/day (about 0.5 mg/kg/day) for as long as 25 months without ill effects (Hayes, 1963). These subjects stored 101 to 466 ppm in their body fat after 12 months and 105 to 659 ppm after 21 months.

Carcinogenic Risk Estimates: Despite the positive results in mice, oral administration of DDT to rats has not provided convincing evidence of carcinogenicity. Feeding studies on dogs and monkeys have also not shown DDT to be carcinogenic, but these studies are of limited value due to small group size and short duration.

Studies on human workers occupationally exposed to DDT have not shown an increased incidence of cancer, but these studies are limited by time factors. Terminal cancer patients have been observed to have higher fat concentrations of DDT, but a causal relationship is difficult to prove (IARC, 1974).

Only the data from feeding studies in mice can be statistically treated to provide an estimate of risk for man. Several species of mice have developed hepatomas after oral exposure to DDT.

For DDT at a concentration of 1 μg/liter (Q = 1) there are several risk estimates depending on which feeding study is evaluated. Four studies (Innes et al, 1969; Tomatis et al, 1972; Walker et al, 1972; and Thorpe and Walker, 1973) provide a a risk to man of from 0.18 to 13.0×10^{-6} Q. The upper 95% confidence estimate of risk at the same concentration is from 0.65 to 20.0×10^{-6} Q.

Conclusions and Recommendations: DDT is of moderate acute toxicity to man and most other organisms. However, its extremely low solubility in water (0.0012 ppm) and high solubility in fat (100,000 ppm) result in great bioconcentration. Its principal breakdown product DDE has very similar properties. Both compounds are also highly persistent in living organisms, so the major concern about DDT toxicity is related to its chronic effects. In light of the above and taking into account the carcinogenic risk projections, it is suggested that very strict criteria be applied when limits for DDT and DDE in drinking water are established.

Hexachlorobenzene

Hexachlorobenzene (HCB, Anticarie, perchlorobenzene) was introduced as a cereal and seed treatment in 1945 and is registered in the United States as a fungicide for various cereal, vegetable, and other crops. U.S. production for pest control is estimated at only 700,000 pounds but HCB is encountered in much larger quantities as an intermediate or waste by-product in various chemical syntheses, including that of pentachlorophenol, and as a contaminant of other pesticides. HCB has also been identified as a prominent constituent of the air collected in the vicinity of a plant manufacturing perchloroethylene.

HCB is produced commercially by exhaustive chlorination of benzene in the presence of a catalyst. HCB has a very low solubility in water: only 6 ppb. HCB is extremely lipophilic and resistant toward degradation. It has been identified in the tissues of marine birds (Gilbertson and Reynolds, 1972), predatory birds (Cromartie et al, 1975; Vos et al, 1968), and starlings (Nickerson and Barbehann, 1975); surface water (Herzel, 1972); freshwater and marine fish (Zitko, 1971); and other aquatic organisms (Koeman et al, 1969). HCB residues in edible freshwater fish reached 0.34 ppm, and in one case carp contained 62 ppm (Johnson et al, 1974). Marine fish oils contained 0.06 to 0.38 ppm.

HCB residues have also been found in human adipose tissue and blood from different parts of the world (Abbott et al, 1972; Acker and Schutle, 1970; Brady and Siyali, 1972; Burns and Miller, 1975; Curley et al, 1973; Siyali, 1972); in human milk (Graca et al, 1974; Stacey and Thomas, 1975); and in food products (Manske and Corneliussen, 1974; Manske and Johnson, 1975; Smyth, 1972).

The EPA established interim HCB tolerances of 0.5 ppm in the fat of cattle and other domestic animals and 0.3 ppm in fat or milk and other dairy products. HCB has been detected at 0.010 ppb in raw and 0.006 ppb in finished U.S. drinking water (USEPA, 1975j).

Health Aspects: In the period 1955–1959, an outbreak of human poisoning occurred in Turkey as a result of the comsumption of HCB-treated wheat (Cam and Nigogosyan, 1963; DeMatteis et al, 1961; Schmid, 1960). Some deaths resulted, but the major syndrome was cutaneous porphyria, with skin lesions, porphyrinuria, and photosensitization. The estimated dosage was approximately 50 to 200 mg/day (0.71 to 2.9 mg/kg/day) for presumably long periods before toxic manifestations became apparent.

Conclusions and Recommendations: The acute toxicity of HCB is relatively low, but subchronic or chronic exposure of laboratory animals or humans to HCB results in the development of severe porphyria, especially in females. An ADI was calculated at 0.001 mg/kg/day based on a 10-month feeding study in rats. The toxicity data and calculations of ADI are summarized in Table 3.2. A conditional acceptable daily intake of 0.0006 mg/kg/day was derived by the FAO/WHO as the upper limit for residues. The FAO/WHO suggested extreme caution with the compound and indicated that available information is insufficient to establish a firm acceptable intake for HCB.

HCB can be readily determined by electron capture gas chromatography at concentrations as low as 0.0001 ppb.

TABLE 3.2: SUMMARY OF TOXICITY DATA FOR HCB

Species	Duration of Study	Dosage Level and Number of Animals per Group	Highest No-Adverse-Effect Level or Lowest Minimal-Effect Level	Effect Measured	Reference
Quail	90 days	0-80 ppm in diet, 15 animals	1 ppm (0.14 mg/kg/day)*	no toxic effect	Vos et al, 1968
Rat	9-10 mo	0-160 ppm in diet, 12 animals	20 ppm (1mg/kg/day)* **	sleeping time reduced	Grant et al, 1974
Rat	13 wk	0-625 ppm in diet	25 ppm (1.25 mg/kg/day)*	no toxic effect	EPA, 1973c
Rat	13 wk	0-200 mg/kg/day, orally, 5 animals	2 mg/kg/day	no toxic effect	EPA, 1973c
Rat	12 wk	0-32 mg/kg/day, orally, 140 animals	2 mg/kg/day	no toxic effect	Kuiper-Goodman et al, 1974
Rat	4 mo	0-1,000 ppm in diet, 30 animals	100 ppm (5 mg/kg/day)*	adrenal hyperplasia and hematologic effects	Kimbrough and Linder, 1974
Rat	30 days	0-100 mg/kg/day	10 mg/kg/day	no toxic effect	EPA, 1973c

Note: Using an uncertainty factor of 1,000, suggested no-adverse-effect level in drinking water is calculated as follows:

$$\frac{1}{1{,}000} = 0.001 \text{ mg/kg/day (ADI)}, \; 0.001 \times 70 \times 0.1 = 0.007 \text{ mg/liter}$$

This assumes the average weight of a human adult is 70 kg; and also assumes the average daily intake of water for man is 2 liters, and that 20% of this total intake is from water.

*Assumes weight of rat is 0.4 kg and average daily food consumption of rat is 0.02 kg.

**Test study from which to calculate suggested no-adverse-effect level.

There are a number of puzzling differences in the highest no-effect and lowest minimal-toxic-effect dosages found for HCB in rats (Table 3.2). These differences may be the results of using different rat strains or different HCB formulations in the various studies. They may also result from the use of HCB of uncertain purity. The source of the observed variations should be established. No subchronic- or chronic-toxicity studies have been conducted with HCB in mammalian species other than rats. It is especially important to conduct 2-year feeding experiments and carcinogenicity studies with HCB in two species, because HCB has been found to be extremely toxic on long-term exposure and is on the list of suspected carcinogens.

Kepone

Kepone is the trade name of chlordecone. Kepone was introduced in 1958. More recently it has been produced solely by the Life Science Products Company for Allied Chemical in Hopewell, Virginia, with the total output purchased by Allied Chemical. The Hopewell plant was closed in July 1975 when a number of its workers were found seriously ill. In August 1975, the EPA ordered that sales and use of Kepone be stopped and prohibited further manufacture. Kepone was registered for the control of root boorers on bananas with a residue tolerance of 0.01 ppm. This constituted the only food or feed use of Kepone. Nonfood uses included wireworm control in tobacco fields and bait to control ants and other insects in indoor and outdoor areas. U.S. production of Kepone for 1974 and 1975 was about 850,000 lb/yr. 99.2% was exported to Latin America, Europe and Africa, the remainder being used here for ant and roach traps or baits.

Metabolism: Kepone is very stable in the environment. No degradation products have been reported, although ultraviolet produced dechlorinated products in a laboratory study (Alley et al, 1974). No metabolic products have been reported; cows fed 5.0 ppm in the diet for 60 days excreted 90 ppb of Kepone in milk 35 days after cessation of treated feeding (Smith and Arant, 1967).

Health Aspects: Kepone came to public attention after Life Science Products Company, which produced Kepone, was closed down when many employees became seriously ill with such afflictions as tremors, nausea, dizziness, impaired vision, and impotence. The mean latency between start of employment and onset of symptoms was 6 weeks. Symptoms have persisted for as long as 6 months after employment was terminated (Public Health Service, 1976).

Companies using chlordecone have received OSHA notices that this substance is hazardous and that its use should be strictly controlled. The OSHA suggested that worker exposure to Kepone be kept below 100 $\mu g/m^3$ of air for up to 10 hr/day or 40 hr/wk, over a working lifetime. There is no specific OSHA standard for this chemical.

According to a report in the March 1976 issue of the *Occupational Health Letter,* Kepone is the cause of sterility in exposed human males, in addition to harmful effects on the nervous system and liver. Philip S. Guzelin, spokesman for a team of researchers at the Medical College of Virginia, reported on studies of 23 former industrial workers heavily exposed to Kepone who exhibited overt signs and symptoms of toxicity. In their investigations, the concentration of Kepone in whole blood has been in parts per million, or thousands of times greater than the minimal detectable concentration, and concentrations measured in biopsies of liver, muscle, and adipose tissue have been several times those in whole blood. Effective therapy for Kepone-associated toxicity is unknown. In the absence of data on the pharmacokinetics of Kepone and the mechanisms of its toxicity, rational treatment of exposed or afflicted people is impossible.

Carcinogenic Risk Estimates: Kepone has produced dose-related hepatomas when fed to mice and rats (NCI, 1976). The available sets of dose-response data were individually considered.

For Kepone at a concentration of 1 μg/liter (Q = 1) the estimated risk for man would be between 2.2 to 6.0 x 10^{-5} Q. The upper 95% confidence estimate of risk at the same concentration would be from 1.4 to 8.0 x 10^{-5} Q.

Conclusions and Recommendations: Kepone is a very toxic compound and is persistent in the environment. Test results clearly suggest that liver lesions, including cancer, were induced in both sexes of rats and mice fed chlordecone. In addition, the time to detection of the first hepatocellular carcinoma observed at death was shorter for treated than for control mice and, in both sexes and both species, it appeared inversely related to the dose.

In light of the above and taking into account the carcinogenic risk projections, it is suggested that very strict criteria be applied when limits for Kepone in drinking water are established.

Apparently, little is known about the pharmacokinetics of Kepone and its mechanisms of toxicity. There is a pressing need for systematic investigation of the absorption, distribution, biotransformation, and excretion of Kepone in humans and experimental animals, to gain an understanding of its toxicity and to provide a basis for rational therapy.

There is also very little information on the environmental transport mechanisms of Kepone and its degradation products, its persistence, and its degradation in soil. Kepone residues have been found in food crops grown in rotation with Kepone-treated tobacco.

Toxaphene

Toxaphene, a complex mixture of largely uncharacterized chlorinated camphene derivatives, is the most heavily used and least understood organochlorine insecticide. The major reason that so little is known about either its structure or its metabolism is its complex nature: it is a mixture of at least 175 compounds, of which the structures of fewer than 10 are known.

Toxaphene is widely used as a foliage insecticide on a variety of food, feed, and fiber crops (USEPA, 1974c). A tolerance of 7 ppm was established in 1950 for a variety of crops. More recently, tolerances of 5, 3, and 2 ppm have been established for small grains, for cottonseed, and for bananas and soybeans, respectively. In addition, there is a temporary tolerance of 7 ppm for residues in or on sugar beets and sunflower seeds. Similar foreign tolerances have also been established. A tolerance of 7 ppm for residues of toxaphene in the fat of meat has also been established. Other regulations have established interim tolerances for toxaphene residues in or on alfalfa at 1 ppm and in milk at 0.05 ppm.

Like many organochlorine insecticides, toxaphene is known to be somewhat persistent in the environment, particularly in soil. The EPA has set an interim standard for toxaphene in finished water of 0.005 mg/liter.

Health Aspects: Although some cases have been reported, acute toxaphene poisoning in humans is rare. When toxaphene was introduced, four cases of poisoning by ingestion in children under 4 years old were reported (McGee et al, 1952). The same study contained a description of severe toxaphene poisoning in adults after its misuse in agriculture. The authors estimated that three patients ingested toxaphene at 9.5 to 47 mg/kg. Human volunteers have participated in toxaphene toxicity studies. In one study, 50 volunteers inhaled mist containing toxaphene at 0.0004 mg/liter for 10 min/day for 15 days; there were no subjective or objective results (USEPA, 1974c). In another study a mist containing toxaphene at 0.25 mg/liter of air was inhaled by 25 people for 30 min/day for 13 days, with no evidence of local or systemic toxicity (USEPA, 1974c).

Conclusions and Recommendations: Toxaphene is a widely used organochlorine insecticide that apparently has not caused a great deal of environmental harm, although it has been used in agriculture for many years. Because it is a complex mixture of uncharacterized camphene derivatives, very little is known about its metabolism in plants or other higher organisms. Considerable information is available, however, on its toxicity in laboratory animals and various aquatic organisms. An ADI at 0.00125 mg/kg/day was calculated on the basis of the chronic toxicity data.

A summary of the results of examination of over 100,000 samples of raw agricultural commodities by the FDA between 1963 and 1969 (Duggan et al, 1971) shows that toxaphene residues are seldom present. Thus, the possibility that large quantities of toxaphene residues could be found in drinking water is not great.

Triazines: Atrazine, Simazine, Propazine, Cyanazine

These four herbicides are all derivatives of cyanuric chlorides and are closely related in environmental properties. They are used largely in preemergence applications for corn, sorghum, and sugar cane, with minor use on pineapple, macadamia orchards, and turf grasses, especially atrazine. Simazine is also used in citrus, deciduous fruits, pineapple, turf grasses, ornamentals, and nursery plantings.

United States production is estimated as: atrazine, 90 million pounds; simazine, 5 million pounds; propazine, 4 million pounds; and, cyanazine, 1 million pounds. Atrazine is the pesticide most heavily used in the United States.

Health Aspects: No case of poisoning in man from simazine, atrazine, propazine, or cyanazine has been reported, although exposure to simazine has caused acute and subacute dermatitis in the U.S.S.R., characterized by erythema, slight edema, moderate pruritus, and burning lasting 4 to 5 days (Elizarov, 1972).

Conclusions and Recommendations: Atrazine, propazine and simazine all appear to have low chronic toxicity. The only good carcinogenicity feeding study done on these compounds did not reveal a significant increase in cancer incidence over controls. On the basis of these chronic studies an ADI was calculated for each of these compounds. The ADI for atrazine is 0.0215 mg/kg/day, for propazine 0.0464 mg/kg/day, and for simazine 0.215 mg/kg/day.

References

Abbott, D.C., G.B. Collins, and R. Goulding. 1972. Organochlorine pesticide residues in human fat in the United Kingdom 1969-1971. *Br. Med. J.* 2:553-556.

Acker, L., and E. Schulte. 1970. Appearance of chlorinated biphenyls and hexachlorobenzene along with insecticides in human milk and fat tissues. *Naturwissenschaften* 57:497.

Alley, E.G., B.R. Layton, and J.P. Minyard, Jr. 1974. Identification of the photoproducts of the insecticides Mirex and Kepone. *J. Agric. Food Chem.* 22:442-445.

Brady, M.N., and D.S. Siyali. 1972. Hexachlorobenzene in human body fat. *Med. J. Aust.* 1:158-161.

Breidenbach, A.W., C.G. Gunterson, F.K. Kawahara, J.J. Lichtenberg, and R.S. Green. 1967. Chlorinated hydrocarbon pesticides in major river basins, 1957-65. *Public Health Rep.* 82:139-156.

Burns, J.E., and F.M. Miller. 1975. Hexachlorobenzene contamination: Its effects in a Louisiana population. *Arch. Environ. Health* 30:44-48.

Cam, C., and G. Nigogosyan. 1963. Acquired toxic porphyria cutanea tarda due to hexachlorobenzene. Report of 348 cases caused by this fungicide. *J. Am. Med. Assoc.* 183:88-91.

Cromartie, E., W.L. Reichel, L.N. Locke, A.A. Belisle, T.E. Kaiser, T.G. Lamont, B.M. Mulhern, R.M. Prouty, and D.M. Swineford. 1975. Residues of organochlorine pesticides and polychlorinated biphenyls and autopsy data for bald eagles, 1971-72. *Pestic. Monit. J.* 9:11-14.

Curley, A., V.W. Burse, R.W. Jennings, E.C. Villanueva, L. Tomatis, and K. Akazaki. 1973. Chlorinated hydrocarbon pesticides and related compounds in adipose tissue from people of Japan. *Nature* 242:338-340.

Davies, G.M., and I. Lewis. 1956. Outbreak of food-poisoning from bread made of chemically contaminated flour. *Br. Med. J.* 2:393-398.

De Matteis, F., B.E. Prior, and C. Rimington. 1961. Nervous and biochemical disturbances following hexachlorobenzene intoxication. *Nature* 191:363-366.

Duggan, R.E., G.Q. Lipscomb, E.L. Cox, R.E. Heatwole, and R.C. Kling. 1971. Pesticide residue levels in foods in the United States from July 1, 1963 to June 30, 1969. *Pestic. Monit. J.* 5:73-212.

Elizarov, G.P. Vestn. 1972. Professional'nye zabole vaniia kozhi, vyzvannye simazinom i protazinom. *Dermatol. Venerol.* 46:27-9.

Ettinger, M.B. and D.I. Mount. 1967. A wild fish should be safe to eat. *Environ. Sci. Technol.* 1:203-205.

Gannon, N., R.P. Link, and G.C. Decker. 1959. Storage of dieldrin in tissues of steers, hogs, lambs, and poultry fed dieldrin in their diets. *J. Agric. Food Chem.* 7:826-828.

Gilbertson, M., and L.M. Reynolds. 1972. Hexachlorobenzene (HCB) in the eggs of common terns in Hamilton Harbour, Ontario. *Bull. Environ. Contam. Toxicol.* 7:371-373.

Graca, I., A.M.S. Silva Fernandes, and H.C. Mourao. 1974. Organochlorine insecticide residues in human milk in Portugal. *Pestic. Monit. J.* 8:148-156.

Grant, D.L., F. Iverson, G.V. Hatina, and D.C. Villeneuve. 1974. Effects of hexachlorobenzene on liver porphyrin levels and microsomal enzymes in the rat. *Environ. Physiol. Biochem.* 4:159-165.

Grant, D.L., F. Iverson, G.V. Hatina, and D.C. Villeneuve. 1974. Effects of hexachlorobenzene on liver prophyrin levels and microsomal enzymes in the rat. *Toxicol. Appl. Pharmacol.* 29:101. Abstr. no. 66.

Hayes, W.J. Jr. 1963. *Clinical Handbook of Economic Poisons.* U.S. Department of Health, Education, and Welfare, Public Health Service Communicable Disease Center, Atlanta, Ga. Public Health Service Publication no. 476.

Hayes, W.J. Jr. 1957. Dieldrin poisoning in man. *Public Health Rep.* 72:1087-1091.

Hayes, W.J. Jr. 1959. The toxicity of dieldrin to man. Report on a survey. *Bull. WHO* 20:891-912.

Herzel, F. 1972. Organochlorine insecticides in surface waters in Germany—1970 and 1971. *Pestic. Monit. J.* 6:179-187.

Hoogendam, I.J., P.J. Versteeg, and M. DeVlieger. 1962. Electroencephalograms in insecticide toxicity. *Arch. Environ. Health* 4:86-94.

Innes, J.R.M., B.M. Ulland, M.G. Valerio, L. Petrucelli, L. Fishbein, E.R. Hart, A.J. Pallotta, R.R. Bates, H.L. Falk, J.J. Gart, M. Klein, I. Mitchell, and J. Peters. 1969. Bioassay of pesticides and industrial chemicals for tumorigenicity in mice. A preliminary note. *J. Nat. Cancer Inst.* 42:1101–1114.

International Agency for Research on Cancer. 1974. *IARC Monographs on the Evaluation of Carcinogenic Risk of Chemicals to Man. Vol. 5: Some Organochlorine Pesticides.* Lyon, France.

Ito, N., H. Nagasaki, M. Arai, A. Masayuki, S. Makiura, S. Sugihara, and K. Hirao. 1973. Histopathologic studies on liver tumorigenesis induced in mice by technical polychlorinated biphenyls and its promoting effect on liver tumors induced by benzene hexachloride. *J. Nat. Cancer Inst.* 51:1637–1646.

Ito, N., H. Nagasaki, H. Aoe, S. Sugihara, Y. Miyata, M. Arai, and T. Shirai. 1975. Development of hepatocellular carcinomas in rats treated with benzene hexachloride. *J. Nat. Cancer Inst.* 54:801–805.

Jedlica, V.L., Z. Hermankaska, I. Smida, and A. Kouba. 1958. Paramyeloblastic leukemia appearing simultaneously in two blood cousins after simultaneous contact with gammexane (hexachlorocyclohexane). *Acta. Med. Scand.* 161:447–451.

Johnson, J.L., D.L. Stalling, and J.W. Hogan. 1974. Hexachlorobenzene (HCB) residues in fish. *Bull. Environ. Contam. Toxicol.* 11:393–398.

Kimbrough, R.D., and R.E. Linder. 1974. The toxicity of technical hexachlorobenzene in the Sherman strain rat. A preliminary study. *Res. Commun. Chem. Pathol. Pharmacol.* 8:653–664.

Koeman, J.H., M.C. ten Noever de Brauw, and R.H. DeVos. 1969. Chlorinated biphenyls in fish, mussels and birds from the river Rhine and the Netherlands coastal area. *Nature* 221:1126–1128.

Kuiper-Goodman, T., D. Grant, G. Korsrud, C.A. Moodie, and I.C. Munro. 1974. Toxic effects of hexachlorobenzene in the rat: Correlations of electron microscopy with other toxic parameters. *Toxicol. Appl. Pharmacol.* 29:101, Abstr. no. 67.

Loge, J.P. 1965. Aplastic anemia following exposure to benzene hexachloride (lindane). *J. Am. Med. Assoc.* 193:110–114.

Lu, P.-Y., R.L. Metcalf, A.S. Hirwe, and J.W. Williams. 1975. Evaluation of environmental distribution and fate of hexachlorocyclopentadiene, chlordane, heptachlor, and heptachlor epoxide in a laboratory model ecosystem. *J. Agric. Food Chem.* 23:967–973.

Manske, D.D., and P.E. Corneliussen. 1974. Pesticide residues in total diet samples (VII). *Pestic. Monit. J.* 8:110–124.

Manske, D.D., and R.D. Johnson. 1975. Pesticide residues in total diet samples (VIII). *Pestic. Monit. J.* 9:94–105.

McGee, L.C., H.L. Reed, and J.P. Fleming. 1952. Accidental poisoning by toxaphene. *J. Am. Med. Assoc.* 149:1124–1126.

Melnikov, N.N. 1971. Chemistry of pesticides. Translated as *Residue Reviews* vol. 36. Springer-Verlag. pp. 371-372.

Metcalf, R.L. 1973. A century of DDT. *J. Agric. Food Chem.* 21:511–519.

Moore, S., III, W.N. Bruce, D.E. Kuhlman, and R. Randell. 1973. A study of the sources of insecticide residues in milk on dairy farms in Illinois—1971. *Pestic. Monit. J.* 6:233–237.

Mrak, E.M., Chairman. 1969. Report of the Secretary's Commission on Pesticides and Their Relationship to Environmental Health. U.S. Department of Health, Education, and Welfare. Washington, D.C.

National Cancer Institute. 1975. Preliminary report on the carcinogenesis bioassay of chlordane and heptachlor. Draft document. Carcinogenesis Program, Division of Cancer Cause and Prevention, National Institute of Health, Bethesda, Md.

Nickerson, P.R., and K.R. Barbehenn. 1975. Organochlorine residues in starlings, 1972. *Pestic. Monit. J.* 8:247–254.

Radomski, J.L., W.B. Deichmann, and E.E. Clizer. 1968. Pesticide concentrations in the liver, brain, and adipose tissue of terminal hospital patients. *Food Cosmet. Toxicol.* 6:209–220.

Richard, J.J., G.A. Junk, M.J. Avery, N.L. Nehring, J.S. Fritz, and H. Svec. 1975. Analysis of various Iowa waters for selected pesticides: atrazine, DDE, and dieldrin—1974. *Pestic. Monit. J.* 9:117–123.

Schafer, M.L., J.T. Peeler, W.S. Gardner, and J.E. Campbell. 1969. Pesticides in drinking water. Waters from the Mississippi and Missouri Rivers. *Environ. Sci. Technol.* 3:1261-1269.

Schmid, R. 1960. Cutaneous porphyria in Turkey. *New Engl. J. Med.* 263:397-398.

Schuttmann, W. 1968. Chronische Leberer-krankungen nach beruflicher Einwirkung von Dichlorodiphenyltrichlorothan (DDT) and Hexachlorocyclohexan (HCH). *Int. Arch. Gewerbepathol. Gewerbehyg.* 24:193-210.

Siyali, D.S. 1972. Hexachlorobenzene and other organochlorine pesticides in human blood. *Med. J. Aust.* 2:1063-1066.

Smith, J.C., and F.S. Arant. 1967. Residues of Kepone in milk from cows receiving treated feed. *J. Econ. Entomol.* 60:925-927.

Smyth, R.J. 1972. Detection of hexachlorobenzene residues in dairy products, meat fat, and eggs. *J. Assoc. Offic. Anal. Chem.* 55:806-808.

Spencer, E.Y. 1973. *Guide to the Chemicals Used in Crop Protection.* Publication no. 1093. Research Branch, Agriculture Canada, Ottawa, Canada. 542 pp.

Stacey, C.I., and B.W. Thomas. 1975. Organochlorine pesticide residues in human milk, Western Australia–1970-71. *Pestic. Monit. J.* 9:64-66.

Thorpe, E., and A.I.T. Walker. 1973. Comparative long-term oral toxicity studies in mice with dieldrin, DDT, phenobarbitone, β-BHC and γ-BHC. *Food Cosmet. Toxicol.* 11:433-42.

Tomatis, L., V. Turusov, N. Day, and R.T. Charles. 1972. Effect of long-term exposure to DDT on CF-1 mice. *Int. J. Cancer* 10:489-506.

U.S. Environmental Protection Agency. 1973a. Aspects of pesticidal use of endrin on man and the environment. Office of Toxic Substances, Washington, D.C.

U.S. Environmental Protection Agency. 1973b. Unpublished Report. BHC-Lindane. Criteria and Evaluation Division, Office of Pesticide Programs, Washington, D.C.

U.S. Environmental Protection Agency. 1973c. Environmental contamination from hexachlorobenzene. Office of Pesticide Programs, Washington, D.C.

U.S. Environmental Protection Agency. 1973d. The toxicology and environmental hazards of ethylenebisdithiocarbamate fungicides and ethylenethiourea. Report of the Special Pesticide Review Group. (Privileged Document)

U.S. Environmental Protection Agency. 1974a. Draft Analytical Report New Orleans Area Water Supply Study. Prepared and submitted by Lower Mississippi River Facility, Slidell, La. Surveillance and Analysis Division, Region VI, Dallas, Texas. EPA-906/10-74-002.

U.S. Environmental Protection Agency. 1974b. The herbicide 2,4-D. Office of Pesticides Programs, Washington, D.C.

U.S. Environmental Protection Agency. 1974c. Report: Aspects of pesticidal use of toxaphene and strobane on man and the environment.

U.S. Environmental Protection Agency. 1974d. Respondent's first proposed findings of fact on environmental effects of aldrin and dieldrin. Office of Pesticide Programs, Washington, D.C.

U.S. Environmental Protection Agency. 1975i. National interim primary drinking water regulations. *Fed. Reg. Vol. 40,* no. 248. 59566-59588. December 24, 1975.

U.S. Environmental Protection Agency. 1975j. Preliminary assessment of suspected carcinogens in drinking water. Report to Congress. Office of Toxic Substances, Washington, D.C. EPA-560/4-75-005. PB-25096.

U.S. Public Health Service. 1976. Kepone poisoning–Virginia. An internal report. Atlanta, Ga. Communicable Disease Center. (Unpublished, for administrative use only.)

Vos, J.G., H.A. Breeman, and H. Benschop. 1968. The occurrence of the fungicide hexachlorobenzene in wild birds and its toxicological importance: A preliminary communication. *Meded. Rijksfac. Landbouwwet. Gent* 33:1263-1269.

Walker, A.I.T., E. Thorpe, and D.E. Stevenson. 1973. The toxicology of dieldrin (HEOD). I. Long-term oral toxicity studies in mice. *Food Cosmet. Toxicol.* 11:415-432.

Walker, E.M., Jr., R.H. Gadsden, L.M. Atkins, and G.R. Gale. 1972. Some effects of 2,4-D and 2,4,5-T on Ehrlich ascites tumor cells in vivo and in vitro. *Ind. Ed. Surg.* 41:22-27.

Weeks, D.E. 1967. Endrin food-poisoning: A report on four outbreaks caused by two separate shipments of endrin-contaminated flour. *Bull. WHO* 37:499-512.

West, I. 1967. Lindane and hematologic reactions. *Arch. Environ. Health* 15:97-101.

Woodliff, H.J., P.M. Connor, and J. Scopa. 1966. Aplastic anemia associated with insecticides. *Med. J. Aust.* 1:628-629.

Zitko, V. 1971. Polychlorinated biphenyls and organochlorine pesticides in some freshwater and marine fishes. *Bull. Environ. Contam. Toxicol.* 6:464-470.

OTHER ORGANIC CONSTITUENTS

Benzene

Benzene is produced by petroleum refining, coal tar distillation, coal processing, and coal coking. The United States production of benzene in 1973 was over 10 billion pounds. It is used primarily as a chemical intermediate in the manufacture of styrene, cyclohexane, detergents, and pesticides.

Motor gasoline usually contains less than 5% benzene; the concentrations of benzene in the ambient air of gas stations were found to be from 0.001 to 0.008 milligrams per liter.

Benzene is slightly soluble in water (0.8 ppm by weight at 20°C). Four of the ten water supplies surveyed by the EPA contained benzene (EPA, 1975a,d) at levels between 0.1 and 0.3 μg/liter. The highest concentration of benzene reported in finished water was 10 μg/liter.

Metabolism: Benzene is excreted rapidly. In human retention studies, Nomiyama and Nomiyama (1974) reported 30% retention in man when exposed to 52 to 62 ppm for 4 hours in air; Hunter and Blair (1972) noted that humans retained 230 mg after exposure to 80 to 100 μg/liter for 6 hours. Benzene metabolism has been shown to be inhibited by 3-amino-1,2,4-triazole.

Health Aspects: Acute Effects – Single exposures to benzene at 20,000 ppm have proved to be fatal in man. Industrial air concentrations of benzene have been reported to give rise to nausea, giddiness, headache, unconsciousness, convulsions, and paralysis (Browning, 1965; Eckhardt, 1973).

Chronic Effects – The chronic exposure of humans to benzene has been reported to produce thrombocytopenia, leukopenia, myelocytic anemia, and leukemia. Despite negative animal toxicity data, the evidence that benzene is a leukemogen for man is convincing. The pathogenesis of leukemia is usually preceded by many observed effects on the hematopoietic system (Snyder and Kocsis, 1975; Mallory et al, 1939; Browning, 1965; Gerarde, 1960). The NIOSH (1974) recommended the occupational exposure to benzene at not in excess of 10 ppm, determined as a time-weighted average exposure for up to a 10-hour workday. A review on the health effects of benzene (NAS, 1976) concludes that:

> Most cases of severe benzene intoxication have been reported in workers exposed to rather high concentrations of benzene under somewhat unhygienic working conditions. It is probable that all cases reported as "leukemia associated with benzene exposure" have resulted from exposure to rather high concentrations of benzene and other chemicals.
>
> It has been suggested in the literature that "benzene-induced leukemia" may occur only in individuals who are highly sensitive because of genetic constitution or because of synergistic action of other chemical or physical environmental agents. A coleukemogenic role for benzene would explain the failure to induce leukemia in benzene-exposed animals.

The state of the benzene literature makes it very difficult or impossible to reach a firm conclusion on the dose-response relationship in chronic exposure of humans to benzene. The details of the extent of exposures are either inadequate or absent. Even in cases where some concentrations of benzene are reported, the stated concentrations were based on occasional measurements of short durations. The role of benzene metabolism in its toxicity and the significance of benzene-induced chromosome aberrations are currently unclear. It appears that a metabolite of benzene may be responsible for its myelotoxic effects.

Based on available literature, it can be concluded that benzene may be associated with leukemia; therefore, benzene must be considered as a suspect leukemogen. More definitive data are required for an accurate assessment of the myelotoxic, leukemogenic, and chromosome-damaging effects of benzene.

Carcinogenic Risk Estimates: On review of the original data from the carcinogenicity study by Ward et al (1975), it is concluded that the observed increased occurrence of granulocytic leukemia in benzene-treated animals is not statistically significant, even when time to response is incorporated into the analysis. Therefore, statistical extrapolation from these data would be unwarranted. An additional difficulty in extrapolation is posed by the fact that the experimental route of exposure was subcutaneous injection, whereas man's exposure would be through ingestion of drinking water.

Occupational studies on human exposure (Aksoy et al, 1972; 1974a,b; Ishimaru et al, 1971; Thorpe, 1974) do not contain adequate information on degree of exposure or size of population at risk. In addition, because the workers in benzene-related occupations were probably exposed to other chemicals, extrapolation of benzene-induced cancer risk from such data as these would be tenuous.

Conclusions and Recommendations: The acute effects of benzene cover a wide range of signs and symptoms. The effects are transitory but may lead to more lasting chronic effects such as anemia; if exposure is continuous and great enough, leukemia is a strong possibility for susceptible members of the population. There are no dose-response data on animals and the data on humans are inadequate to calculate a risk estimate for benzene with mathematical models.

In summary, there is no adequate source of data (animal or human) on which to base a statistical extrapolation from high to low exposure. More data are needed on the mutagenicity and teratogenicity of benzene. The cocarcinogenic effect of benzene should be further explored. If data are available on industrial benzene exposure, then systematic monitoring should be started with a view to following the population groups at risk.

Before limits for benzene in drinking water can be established more extensive toxicological data must be gathered and evaluated.

Benzo[a] pyrene

Benzo[a] pyrene is a ubiquitous polycyclic aromatic hydrocarbon that is produced largely, if not exclusively, in the pyrolysis of naturally occurring hydrocarbons.

It was isolated early in pure form from coal tar, one of the so-called industrial carcinogens. Benzo[a]pyrene is found as a constituent in coal, petroleum, shale, and kerosene. It has been reported that it is present in the combustion products of fuels and cigarette smoke.

Benzo[a]pyrene is very persistent in water and is soluble at 0.004 mg/liter at 27°C. It has been detected in finished water (EPA, 1976).

Health Aspects: There is no firm evidence that benzo[a]pyrene alone produces toxicity, including teratogenicity, mutagenicity, or carcinogenicity in humans. On the other hand, mixtures of compounds which contain benzo[a]pyrene as a constituent have been associated with cancer in man. In such cases the exact role of benzo[a]pyrene is difficult to assess.

Carcinogenic Risk Estimates: Numerous carcinogenesis studies have been conducted in rodents with oral administration of benzo[a]pyrene. Stomach tumors in mice have been observed in several studies, as well as leukemia and lung adenomas. In rats, mammary tumors and papillomas in the esophagus and forestomach have been found. In hamsters, tumors were found in the forestomach, esophagus, and intestine. No satisfactory human data are available.

In the above studies, the oral administration of benzo[a]pyrene showed evidence of dose-response relationships. However, it was generally fed for less than six months; this is not adequate for estimating lifetime effects. Thus, without specific biologic assumptions that relate short-term to lifetime effects, no reasonable risk estimates can be attempted.

Conclusions and Recommendations: The occurrence of upper gastrointestinal tract tumors in animals fed benzo[a]pyrene, skin tumors at sites painted with it, and subcutaneous sarcomas at sites where it was injected demonstrates that benzo[a]pyrene is a potent contact carcinogen. In light of the above, it is suggested that strict criteria be applied when limits for benzo[a]pyrene in drinking water are established.

Bis(2-Chloroethyl) Ether

Bis(2-chloroethyl)ether (dichloroethyl ether) is used as a soil fumigant; as an insecticide and acaricide; as a solvent for fats, waxes, greases, and cellulose esters; as a scouring agent for textiles; in paints, varnishes, and lacquers; as a paint remover; in dry cleaning; and in the manufacture of medicinals and pharmaceuticals (EPA, 1975d).

Bis(2-chloroethyl)ether is moderately persistent and insoluble in water. It can be formed during chlorination of water treatment when ethyl ether is present. Bis(2-chloroethyl)ether has been identified in finished water at 0.5 μg/liter in Philadelphia and 0.44 μg/liter in New Orleans (EPA, 1975a,c).

Health Aspects: Schrenk et al (1933) exposed humans to bis(2-chloroethyl)-ether and found that a concentration of 550 ppm was intolerable and caused irritation of the eyes and nasal passages. Concentrations of 100 and 260 ppm were irritating, but tolerable; at 35 ppm, there was no irritation, but a nauseous odor persisted.

Carcinogenic Risk Estimates: Bis(2-chloroethyl)ether has produced dose-related hepatomas when given orally to mice (Innes et al, 1969). The female mice of one strain did not develop any hepatomas. The available sets of dose-response data were individually considered. For bis(2-chloroethyl)ether at a concentration of 1 μg/liter (Q = 1) the estimated risk for both sexes is 8.1×10^{-7} Q. The upper 95% confidence estimate is 1.2×10^{-6} Q.

Conclusions and Recommendations: Even though the acute and chronic effects of bis(2-chloroethyl)ether have not been fully established, it has been shown to produce dose-related tumors when given orally to mice.

In view of this potential in humans and taking into account the risk estimates, it is suggested that very strict criteria be applied when limits for bis(2-chloroethyl)ether in drinking water are established.

Carbon Disulfide

Carbon disulfide is produced in petroleum and coal tar refining. Its principal uses are in the manufacture of rayon, rubber, chemicals, solvents, and pesticides. U.S. production in 1973 was over 77 million pounds. It is soluble in water (0.294% at 20°C). Carbon disulfide was detected in 5 of the 10 water supplies surveyed by the EPA (1975a).

Metabolism: Absorption and elimination of carbon disulfide differ with species and route of administration. Most of the work on absorption and elimination has been related to inhalation toxicity. The little work with cutaneous exposure has shown a change in blood proteins and a change in zinc content (Brieger and Teisinger, 1967). Inhalation-toxicity studies have shown that 18% of the carbon disulfide inhaled is exhaled unchanged. Of the remaining inhaled dose, 70% is excreted as free or bound carbon disulfide and urinary sulfates, and 30% is stored in the body and slowly excreted as carbon disulfide and its metabolites.

Large concentrations of both free and bound carbon disulfide are found in the brain (guinea pig studies) and peripheral nerves (rat studies) of exposed animals. The ratio of bound to free carbon disulfide in the brain is 3:1. Blood and fatty tissues contain mainly bound carbon disulfide, while the liver contains mainly free. Carbon disulfide can also chelate trace metals, especially copper and zinc.

Health Aspects: The effects of carbon disulfide inhalation have been examined in the light of their central nervous system and vascular effects on man. The lowest lethal concentration has been reported as 4,000 ppm in 30 minutes; in the same study, a person subjected to a concentration of 50 mg/m^3 for 7 years had central nervous system effects (Registry of Toxic Effects of Chemical Substances, 1975). Moderate chronic exposure at less than 65 mg/m^3 for several years has been reported by Cooper (1976) to cause polyneuropathy. In a study by Baranowska et al (1966), humans have been shown to absorb 8.8 to 37.2 mg of carbon disulfide from an aqueous solution containing 0.33 to 1.67 g/liter. This was over a period of 1 hour of hand-soaking.

The U.S. occupational standards recommend a maximum of 20 ppm. The standards allow a peak concentration of 100 ppm for 30 minutes in an 8-hour period. U.S.S.R. has set a standard of 4 ppm as the safe maximum, and the Czechoslovakians maintain 10 ppm as the safe limit (Hamilton and Hardy, 1974).

Conclusions and Recommendations: Carbon disulfide has been demonstrated to produce disturbances in reproduction as well as teratogenic effects in animals when inhaled. There are no data available on teratogenicity following oral exposure. In view of the relative paucity of data on the mutagenicity, carcinogenicity, and long-term oral toxicity of carbon disulfide, estimates of the effects of chronic oral exposure at low levels cannot be made with any confidence. It is recommended that studies to produce such information be conducted before limits in drinking water are established.

Carbon Tetrachloride

Over 1 billion pounds of carbon tetrachloride (tetrachloromethane) was produced in the United States in 1973. It is used mostly in the manufacture of chlorofluoromethanes, but also in grain fumigants, fire extinguishers, solvents, and cleaning agents.

Carbon tetrachloride is highly persistent and insoluble in water. Carbon tetrachloride was identified in District of Columbia drinking water at 5 μg/liter (Scheiman et al, 1974). The EPA's 80-city survey showed that it was detected in 10 locations in trace amounts, at 3 μg/liter or less (EPA, 1975a). The survey of 10 water utilities showed that it was present in 8 supplies. The Region V survey of 83 water supplies indicated that carbon tetrachloride is not formed during chlorination water treatment (EPA, 1975b).

Metabolism: Carbon tetrachloride in rats and humans is rapidly absorbed and distributed and is excreted primarily through the lungs. The excretion products are 85% parent compound, 10% carbon dioxide, and smaller quantities of other products, including chloroform. In animal studies, chloroform, hexachloroethane, and two unidentified metabolites were found in rabbits.

Health Aspects: Carbon tetrachloride is readily absorbed from the gastrointestinal tract and by inhalation through the lungs. A fatal dose for children has been reported as low as 3 ml, but there is great variation in individual susceptibility. Intestinal absorption is enhanced by fats, oils and alcohol (Masoud et al, 1973).

Some persons exposed to carbon tetrachloride will develop severe liver damage with little or no evidence of renal involvement, while others will suffer renal shutdown and no hepatic disease. The reasons for this are not known (Eckhardt, 1965).

In eight instances of either acute or subacute carbon tetrachloride poisoning, seven patients suffered renal insufficiency and six of these required dialysis. Two died from heart failure. None of the six survivors showed any evidence of liver damage (Dume et al, 1969).

Following a large acute exposure the primary finding is of centrilobular necrosis. If exposure is not too massive, repair may begin after 3 or 4 days and be complete in three weeks. Chronic exposure usually results chiefly in symptoms of gastrointestinal upset such as nausea and vomiting, and nervous system symptoms such as headache, drowsiness, and excessive fatique. It is rare to find jaundice following either acute or chronic exposure (Browning, 1961).

Gastric symptoms have been reported following chronic inhalation of from 45 to 100 ppm carbon tetrachloride. When exposure is from 100 to 300 ppm the symptoms, in addition to gastrointestinal upset, include apathy or mental confusion and weight loss (Lewis, 1961).

Carcinogenic Risk Estimate: Carbon tetrachloride has been given orally in a number of studies with mice, rats, hamsters, and dogs. It has also been used as a positive control in cancer bioassays (NCI, 1976). In the trichloroethylene study, the carbon tetrachloride positive control produced much higher incidences of hepatocellular carcinomas than did trichloroethylene. The available sets of dose-response data from the NCI trichloroethylene bioassay were individually considered.

For carbon tetrachoride at a concentration of 1 μg/liter (Q = 1) the projected risk for man would fall between 4.5 and 5.4 x 10^{-8} Q. The upper 95% confidence estimate of risk at the same concentration would be from 1.1 to 1.8 x 10^{-7} Q.

Conclusions and Recommendations: The acute toxicity effects of carbon tetrachloride are best characterized as hepatic nodular hyperplasia and cirrhosis and renal dysfunction in both experimental animals and man. It had no mutagenic potential in in vitro and in vivo test systems. Its teratogenic potential has not been firmly established. Carcinogenic bioassays have produced hepatomas in mice, rats, and hamsters, associated in most cases with regenerative nodular hyperplasia or postnecrotic cirrhosis.

In light of the above and taking into account the risk projections, it is suggested that very strict criteria be applied when limits for carbon tetrachloride in drinking water are established.

Di-n-Butyl Phthalate

Di-n-butylphthalate (DBP) is used in the manufacture of plasticizers, plastics, and in cosmetics, explosives, solid rocket propellants, textile lubricating agents, safety glass, insecticides, printing inks, paper coatings, and adhesive (EPA, 1975d). The United States production of di-n-butylphthalate in 1973 was about 38 million pounds.

This compound is insoluble in water and very persistent in the environment. Six of the ten water supplies surveyed by the EPA (1975a) contained di-n-butylphthalate; the highest concentration was 5 μg/liter in Miami.

Metabolism: Williams (1959) suggested that di-n-butylphthalate is one of the phthalate esters likely to be hydrolyzed in vivo, yielding phthalic acid and an alcohol. No accumulation of dibutylphthalate or monobutylphthalate was found by Williams and Blanchfield (1975) in tissues of rats fed di-n-butylphthalate at 1 g/kg of feed for 12 weeks. These authors reported that 80 to 90% of a dose of [^{14}C] di-n-butylphthalate was metabolized and excreted in the urine in 48 hours. Phthalic acid, monobutylphthalate, mono(3-hydroxybutyl)phthalic acid, and mono(4-hydroxybutyl)phthalate were found in the urine.

Smith (1953) found that DBP was hydrolyzed in vitro by pancreatic lipases; this suggested that it is metabolized like fat that is normally in the diet.

Health Aspects: Fairhall (1957) reviewed a report of an accidental ingestion of 10 g of DBP. Hours after ingestion, the patient was nauseous and giddy. His eyes were inflamed and painful, with photophobia and lacrimation. Toxic nephritis occurred, and his urine contained red and white cells and albumin. The patient recovered fully.

Conclusions and Recommendations: The major effect of di-n-butylphthalate in animals involves disturbances in reproduction and teratogenicity. There are no data available on mutagenicity and the chronic feeding studies did not show any carcinogenesis.

An ADI was calculated on the basis of the chronic toxicity data. Based on these data, the ADI is 0.11 mg/kg/day.

Pentachlorophenol

Pentachlorophenol (PCP) has been used since 1936 for wood preservation. Domestic production of PCP is estimated at 46 million pounds a year.

Health Aspects: Menon (1958) reported loss of appetite, respiratory difficulties, anesthesia, hyperpyrexia, sweating, dyspnea, and rapidly progressive coma in humans exposed to PCP. A number of cases of human poisoning by PCP are reviewed by Armstrong et al (1969). The minimum lethal dose for humans is estimated to be 29 mg/kg (The Toxic Substances List, 1974). Work done in the U.S.S.R. has established a maximum permissible concentration of 40 ng/m^3 in the air (Tabakova, 1969).

Conclusions and Recommendations: There are substantial disagreements in the results of several of the subacute and chronic toxicity experiments with PCP, perhaps because of the use of inadequately characterized PCP preparations in these studies. In addition, two-year chronic toxicity experiments in one or more species have not yet been conducted with this extensively used chemical. High doses (>5 mg/kg/day) of PCP have been shown to be teratogenic in rats and hamsters administered during susceptible days of gestation. There is also need for an adequate determination of the carcinogenic potential of this chemical. On the basis of the available chronic toxicity data, an ADI for pentachlorophenol has been calculated to be 0.003 mg/kg/day.

Polychlorinated Biphenyls

Polychlorinated biphenyls (PCBs) are mixtures of chlorinated biphenyls produced commercially by the chlorination of biphenyl. PCBs are used in the production of capacitors and transformers.

PCBs are highly persistent and can accumulate in the environment. They are soluble in water at 0.04 to 0.2 ppm. Trichlorinated, tetrachlorinated, and pentachlorinated biphenyls have been detected in a few water supplies: at 3.0 μg/liter in Winnebago, Illinois and 0.1 μg/liter in Sellersberg, Indiana (EPA, 1975b).

Metabolism: The degradation and elimination of PCB congeners appear to take place via the hepatic microsomal enzyme system. Two possible mechanisms for biotransformation have been suggested by Ecobichon (1976). The first and most rapid mechanism involves the formation of an arene oxide intermediate and requires the presence of unsubstituted adjacent carbon atoms in the nucleus. The second and much slower uses a different hydroxylation system for isolated unsubstituted positions, as are found in highly chlorinated biphenyls. Two adjacent unsubstituted carbon atoms appear to be important in metabolism, in that their presence facilitates the formation of arene oxides by the hepatic mixed-function oxidases.

Polychlorinated biphenyls are known to be strong inducers of hepatic mixed-function oxidase enzymes. The potency increases with increasing chlorination of the biphenyl rings. The PCBs are somewhat unique in this regard in that they induce both type I and type II P-450, but this may be due in part to contaminants in the PCBs tested.

These effects raise a concern that such induction may increase metabolism of birth control hormones with possible harmful effects. Also to be considered is the possibility that environmental carcinogens may be activated at a faster rate.

Health Aspects: PCBs are liver toxins and cause chloracne and possibly peripheral neuropathy in man (Murai and Kuroiwa, 1971).

There is no information on acute adverse effects on humans. Although Yusho disease (an acute outbreak of disease which occurred in Japan) has generally been ascribed to the ingestion of PCBs in rice oil (0.07 mg/kg/day for at least 50 days), recent evidence provided by Kuratsune et al (1976) suggests that the rice oil was contaminated with polychlorinated dibenzofurans (PCDFs). These compounds may have played a significant role in the observed toxicity. Initial measurements of the concentrations of PCDFs in the PCBs suggested that these materials contributed at least as much to the toxicity as the PCBs themselves.

Symptoms included excessive fatigue, headache, phymata in articular regions, fever, cough, digestive disturbances, numbness, and menstrual disorders. Physical signs consisted primarily of cutaneomucosal abnormalities, such as acneiform eruptions and black comedones on face, buttocks, and intertriginous sites; increased pigmentation of face, palpebral conjunctiva, gingiva, and nails; ocular signs consisting of swelling and hypersecretion of the meibomian gland and palpebral edema.

Recent surveys have indicated that PCB can be found in the milk of nursing mothers. The highest reported level was 10.6 ppm with a mean of 1.8 ppm for all samples.

This information is especially meaningful in light of Allen's recent work (1975). Female rhesus monkeys orally exposed to 2.5 and 5.0 ppm of PCB (Aroclor 1248) developed facial acne, erythema, subcutaneous edema conjunctivitis and loss of eyelashes. All infants born to PCB-exposed mothers had PCBs in their tissues at birth. These infants also developed skin lesions as a result of nursing PCB-contaminated milk. Fifty percent of the infants died within four months.

Carcinogenic Risk Estimates: Only the study by Kimbrough et al (1975) is of sufficient duration to permit a statistical extrapolation of risk to man. For PCB Aroclor 1260 at a concentration of 1 μg/liter (Q = 1) the projected risk for man is 2.2×10^{-6} Q. The upper 95% confidence estimate of risk at the same concentration is 3.1×10^{-6} Q.

It should be noted that this extrapolation is based on a one-hit mathematical model which may be invalid for this chemical. In addition, the extrapolation pertains to Aroclor 1260 and not to any of the polychlorinated biphenyls which make up this compound. It would be impossible from this limited study to single out any one of the PCBs as the carcinogenic agent and it should be kept in mind that they may be acting synergistically.

Conclusions and Recommendations: Although there are considerable data on toxicity of mixtures of PCBs, there is a paucity of data on the pure congeners present in these mixtures. Whether chronic toxicity is related to the metabolism of the PCBs and their intermediates or to the highly chlorinated stored PCBs remains to be determined. Considerably more attention must be directed to the detection of impurities in PCBs at very low concentrations.

Polychlorodibenzofuran may constitute only one of several significant contaminating compounds responsible for PCB toxicity. Populations at special risk–both the industrially exposed and those heavily exposed by the ingestion of contaminated foods–should be carefully evaluated.

Despite the current lack of evidence in the United States that dietary PCBs have any deleterious effects on health, there is a growing concern with long-range effects of the contamination of our ecosystem with these chemicals. There is an urgent need for epidemiologic studies of exposed populations, more precise identification of all sources of PCB contamination, and efforts directed at the control of disposal of PCBs. Because of the demonstrated carcinogenic potential, studies on individual congeners, both those metabolized and those stored by man, are urgent.

Styrene

Styrene monomer is synthesized from ethylene and benzene. It is used in the manufacture of polystyrene plastics, resins, insulators, synthetic rubber, and protective coatings. The United States production of styrene in 1973 was over 2.56 billion pounds. Styrene is insoluble in water. It has been detected in the finished water of three of the ten water supplies surveyed by the EPA (1975a).

Health Aspects: Wolf et al (1956), in the course of conducting inhalation studies on animals, exposed human subjects to various concentrations of styrene monomer. They reported very strong odor with eye and nasal irritation at 600 ppm, detectable odor with no irritation at 60 ppm, and no detectable odor at 10 ppm and less.

Stewart et al (1968) exposed human volunteers to styrene vapors at approximately 50, 100, 200, and 375 ppm for periods of 1 to 7 hours. Only at 375 ppm did the subjects experience subjective symptoms and objective signs of transient neurologic impairment. The vapors irritated the eyes and nose and in one subject produced a burning sensation of the facial skin. Neurologic effects were also manifested. Stewart showed that the amount of styrene exhaled after exposure indicated the extent of exposure. Urine hippuric acid content, however, was not a sensitive indicator.

Conclusions and Recommendations: Styrene at high concentrations is mainly an irritant whether ingested or exhaled. Humans have tolerated acute exposures of 200 ppm with no ill effects. An ADI was calculated on the basis of the available chronic toxicity data. Based on these data the ADI is 0.133 mg/kg/day.

Vinyl Chloride

Vinyl chloride (chloroethene) monomer is not known to occur in nature. It is commercially synthesized by the halogenation of ethylene. In the United States, the vinyl chloride monomer is used primarily in the production of polyvinyl chloride resins for the building and construction industries. Because it has been confirmed that vinyl chloride monomer is a human and animal carcinogen, the sale of propellents and all aerosols containing it was banned in 1974.

The occupational standard for atmospheric vinyl chloride in the United States is 1 ppm (2.56 mg/m^3) or less for 8 hr/day and 5 days/wk. The United States production of vinyl chloride in 1973 was 5.35 billion pounds.

Vinyl chloride monomer is slightly soluble in water ($<$0.11% by weight at 25°C). Results of the ten-city survey (EPA, 1975a) indicate that vinyl chloride was present in the finished water of Miami at 5.6 μg/liter, and Philadelphia at 0.27 μg/liter.

Health Aspects: In studies of the acute effects of vinyl chloride in man, it has been shown to produce central nervous system dysfunction, sympathetic-sensory polyneuritis, and organic disorders of the brain (Smirnova and Granik, 1970). In a series of studies on workers in the polyvinyl chloride industry, scleroma-like skin alterations, Raynaud's syndrome, and acroosteolysis were observed (Juehe and Lange, 1972; Juehe et al, 1974).

To date, 48 cases of hepatic angiosarcoma have been diagnosed in industrial vinyl chloride workers around the world. All authenticated cases were found in workers engaged in closed-in plants handling very large quantities of liquefied vinyl chloride under pressure (Makk et al, 1976). Exposure concentrations were high, probably ranging from 1,000 to several thousand ppm. Lesions of the skin, bones, liver, spleen, and lungs have also been reported after chronic exposure to the compound.

Carcinogenic Risk Estimates: In a recent study by Maltoni et al (1975), rats were given vinyl chloride in olive oil by gavage 4 or 5 times per week for 52 weeks and held for their life span. The available set of dose-response data was considered. For vinyl chloride at a concentration of 1 μg/liter (Q = 1) the estimated risk for man is 3.0×10^{-7} Q. The upper 95% confidence estimate at the same concentration is 5.1×10^{-7} Q.

Conclusions and Recommendations: Vinyl chloride has acute and chronic toxic effects in animals and man. In addition to its chronic toxicity, it has been clearly shown to be a carcinogen in animals and man, with dose- and time-related properties, and to be a mutagen in in vitro systems. Its carcinogenic property has been demonstrated by oral administration. This route is more effective (efficient) in producing the characteristic tumor, angiosarcoma, than is loading the atmosphere with similar amounts of vapor.

In animal studies, vinyl chloride has induced a wide variety of tumors in addition to the characteristic and otherwise rare angiosarcoma.

References

Aksoy, M., K. Dincol, S. Erdem, and G. Dincol. 1972. Acute leukemia due to chronic exposure to benzene. *Am. J. Med.* 52:160-166.

Aksoy, M., S. Erdem and G. Dincol. 1974a. Leukemia in shoe-workers exposed chronically to benzene. *Blood* 44:837-841.

Aksoy, M., S. Erdem, K. Dincol, T. Hepyuksel, and G. Dincol. 1974b. Chronic exposure to benzene as a possible contributory etiologic factor in Hodgkin's disease. *Blut* 28:293-298.

Allen, J.R. 1975. Response of the nonhuman primate to polychlorinated biphenyl exposure. *Fed. Proc.* 34:1675-1679.

American Industrial Hygiene Association. 1956. 1,2-Dichloroethane (ethylene dichloride), *Hygienic Guide Series. Am. Ind. Hyg. Assoc. J.* 17:447-448.

American Industrial Hygiene Association. 1965. 1,2-Dichloroethane (ethylene chloride, ethylene dichloride, glycol dichloride). *Hygienic Guide Series. Am. Ind. Hyg. Assoc. J.* 26:435-438.

American National Standards Institute. 1970. American national standard: Acceptable concentrations of ethylene dichloride: Approved September 5, 1969 (ANSI Z37.21-1969).

Armstrong, R.W., E.R. Eichner, D.E. Klein, W.F. Barthel, J.V. Bennet, V. Jonsson, H. Bruce, and L.E. Loveless. 1969. Pentachlorophenol poisoning in a nursery for newborn infants. II. Epidemiologic and toxicologic studies. *J. Pediatr.* 75:317-325.

Aulerich, R.J., R.K. Ringer, and S. Iwamoto. 1973. Reproductive failure and mortality in mink fed on Great Lakes fish. *J. Reprod. Fertil.* Suppl. 19:365-376.

Baranowska and Dutkiewicz. 1966. Influence of concentration and temperature of carbon disulfide solutions on its absorption through human skin. *Ann. Acad. Med. Lodz.* 8:169-74 (in Polish). *Chem. Abs.* 70:31443w, Feb. 24, 1969).

Bariliak, I.R., I.A. Vasil'eva, and L.P. Kalinovskaia. 1975. Effect of small concentrations of carbon disulfide and hydrogen disulfide on intrauterine developments in rats. *Arkh. Anat. Gistol. Embriol.* 68(5):77-81 (in Russian).

Blake, D.A., and G.W. Mergner. 1974. Inhalation studies on the biotransformation and elimination of [^{14}C] trichlorofluoromethane and [^{14}C] dichlorodifluoromethane in beagles. *Toxicol. Appl. Pharmacol.* 30:396-407.

Brem, H., A.B. Stein, and H.S. Rosenkranz. 1974. The mutagenicity and DNA-modifying effect of haloalkanes. *Cancer Res.* 34:2576-2579.

Brieger, H., and J. Teisinger, eds. 1967. Toxicology of Carbon Disulphide. *Proceedings of the 1st International Symposium.* Excerpta Medica Foundation, Amsterdam. pp 271.

Browning, E. 1965. *Toxicity and Metabolism of Industrial Solvents.* Elsevier Publishing Company, Amsterdam.

Browning, E. 1961. Toxicology of organic compounds of industrial importance. *Ann. Rev. Pharmacol.* 1:397-430.

Buhler, D.R., M.E. Rasmusson, and H.S. Nakaue. 1973. Occurrence of hexachlorophene and pentachlorophenol in sewage and water. *Environ. Sci. Tech.* 7:929-934.

Cooper, P. 1976. Carbon disulphide toxicology: the present picture. *Food Cosmet. Toxicol.* 14:57-59.

Dalvi, R., L. Hunter, and A. Neal. 1975. Toxicological implications of the mixed-function oxidase catalyzed metabolism of carbon disulfide. *Chem. Biol. Interact.* 10(5):347-361.

Dume, T., W. Herms, E. Schroeder, and E. Wetzels. 1969. Clinical aspects and therapy of carbon tetrachloride poisoning. *Dtsch. Med. Wochenschr.* 94(33):1646-1651.

Eckhardt, R.E. 1965. Toxic effects of carbon tetrachloride. *J. Am. Med. Assoc.* 194:1337-1338.

Eckhardt, R.E. 1973. Recent developments in industrial carcinogens. *J. Occup. Med.* 15:-904-907.

Ecobichon, D.J. 1976. Enzymatic and other biochemical responses to selected PCPs. In *Conference Proceedings, National Conference on Polychlorinated Biphenyls* (November 1975, Chicago, Illinois). Environmental Protection Agency, Office of Toxic Substances, Washington, D.C. 57-66.

El Masri, A.M., J.N. Smith, and R.T. Williams. 1958. *Biochem. J.* 68:199-204.

Fairhall, L.T. 1957. *Industrial Toxicology,* 2nd ed. Williams & Wilkins, Baltimore.

Feldman, R.G., R.M. Mayer, and A. Taub, 1970. Evidence for peripheral neurotoxic effect of trichloroethylene. *Neurology* 20:599-606.

Gerarde, H.W. 1960. *Toxicology and Biochemistry of Aromatic Hydrocarbons.* Elsevier Publishing Company, New York.

Gut, R. 1969. Sex differences in sensitivity to some chemicals. *Prac. Lek.* 21(10):453-458.

Hake, C.L., and V.K. Rowe. 1963. Ethers. F.A. Patty, ed. *Industrial Hygiene and Toxicology,* vol II, 2nd ed. 1655-1718. Interscience, New York.

Hamilton, A., and L. Hardy. 1974. *Industrial Toxicology,* 3rd ed. Publishing Sciences Group. Acton, Massachusetts.

Hunter, C.G., and D. Blair. 1972. Benzene: Pharmacokinetic studies in man. *Ann. Occup. Hyg.* 15:193-199.

Ikeda, M., and T. Imamura. 1973. Biological half-life of trichloroethylene and tetrachloroethylene in human subjects. *Int. Arch. Arbeitsmed.* 31:209-224.

Innes, J.R.M., B.M. Ulland, M.G. Valerio, L. Petrucelli, L. Fishbein, E.R. Hart, A.J. Pallotta, R.R. Bates, H.L. Falk, J.J. Gart, M. Klein, I. Mitchell, and J. Peters. 1969. Bioassay of Pesticides and industrial chemicals for tumorigenicity in mice: A preliminary note. *J. Nat. Cancer Inst.* 42:1101-1114.

Ishimaru, T., H. Okada, T. Tomiyasu, T. Tsuchimoto, T. Hoshino, and M. Ichimaru. 1971. Occupational factors in the epidemiology of leukemia in Hiroshima and Nagasaki. *Am. J. Epidemiol.* 93:157-165.

Johnson, R.L., P.J. Gehring, P.J. Kociba, and B.A. Schwetz. 1973. Chlorinated dibenzodioxins and pentachlorophenol. *Environ. Health Perspect.* 5:171-175.

Juehe, S., and C.E. Lange. 1972. Scleroderma-like skin alternations, Raynaud's syndrome, and acroosteolysis in workers in the polyvinyl chloride industry. *Dtsch. Med. Wochenschr.* 97:1922-1923.

Juehe, S., C.E. Lange, G. Stein, and G. Veltman. 1974. On the so-called vinyl chloride disease. A new occupational illness. *Berufs-Dermatosen.* 22,4-22.

Kimbrough, R.D., R.A. Squire, R.E. Linder, J.D. Strandberg, R.J. Montali, and V.W. Burse. 1975. Induction of liver tumors in Sherman strain female rats by polychlorinated biphenyl Aroclor 1260. *J. Nat. Canc. Inst.* 55:1453-1459.

Kleinfeld, M., and Tabershaw, I. 1954. Trichloroethylene toxicity. *Arch. Ind. Hyg. Occup. Med.* 10:134-41.

Kuratsune, M., Y. Masuda, and J. Nagayama. 1976. Some of the recent findings concerning Yusho. In *Conference Proceedings. National Conference on Polychlorinated Biphenyls* (November 1975, Chicago, Illinois). 14–29.

Lewis, C.E. 1961. The toxicology of carbon tetrachloride. *J. Occup. Med.* 3:82-86.

Makk, L., F. Delmore, J.L. Creech, Jr., L.L. Ogden II, E.H. Fadell, C.L. Songster, J. Clanton, M.N. Johnson, and W.M. Christopherson. 1976. Clinical and morphologic features of hepatic angiosarcoma in vinyl chloride workers. *Cancer* 37(1):149-163.

Mallory, T.B., E.A. Gall, and W.J. Brickley. 1939. Chronic exposure to benzene (benzol). III. The pathologic results. *J. Ind. Hyg. Toxicol.* 21:355-393.

Maltoni, C., A. Ciliberti, L. Gianni, and P. Chieco. 1975. The oncogenic effects of vinyl chloride administered by oral route in the rat. *Gli Ospedali della Vita* 2(6):65-66.

Masoud, A.N., and J.T. Elder, A.L. Czerwinski. 1973. Chemistry and pharmacology of common acute poisoning in children. *Paediatrician* 2:2-37.

Menon, J.S. 1958. Tropical hazards associated with the use of pentachlorophenol. *Br. Med. J.* 2:1156-1158.

Morgan, A., A. Black, and D.R. Belcher. 1970. The excretion in breath of some aliphatic halogenated hydrocarbons following administration by inhalation. *Ann. Occup. Hyg.* 13: 219-233.

Murai, Y., and Y. Kuroiwa. 1971. Peripheral neuropathy in chlorobiphenyl poisoning. *Neurology* 21:1173-1176.

National Academy of Sciences. National Research Council. 1975. Commission on Natural Resources. Environmental Studies Board. Pest Control: An Assessment of Present and Alternate Technologies. Vol I: Contemporary Pest Control Practices and Prospects: The Report of the Executive Committee. National Academy of Sciences, Washington, D.C. 506.

National Cancer Institute. 1976. Report on carcinogenesis bioassay of chloroform.

National Institute for Occupational Safety and Health. 1974. *Criteria for a Recommended Standard...Occupational Exposure to Chloroform.* HEW Publication no. (NIOSH)75-114. U.S. Government Printing Office. Washington, D.C.

Nomiyama, D., and H. Nomiyama. 1974. Respiratory retention, uptake and excretion of organic solvents in man. Benzene, toluene, n-hexane, trichloroethylene, acetone, ethyl acetate and ethyl alcohol. *Int. Arch. Arbeitsmed.* 32:75-83.

Norton, T.R. 1975. Metabolism of toxic substances. L.J. Casarett and J. Doull, eds. *Toxicology: The Basic Science of Poisons.* Macmillan, New York.

Plaa, G.L., and R.E. Larson. 1965. Relative nephrotoxic properties of chlorinated methane, ethane, and ethylene derivatives in mice. *Toxicol. Appl. Pharmacol.* 7:37-44.

Reid, W.D. 1973. Mechanism of renal necrosis induced by bromobenzene or chlorobenzene. *Exp. Mol. Pathol.* 19:197-214.

Ringer, R.K., R.J. Aulerich, and M. Zabik. 1972. Effect of dietary polychlorinated biphenyls on growth and reproduction of mink. *Proceedings American Chemical Society Meeting* (164th A.C.S. Meeting, New York), vol 12. 149-154.

Sagawa, K., H. Nishitani, H. Kawai, Y. Kuge, and M. Ikeda. 1973. Transverse lesion of spinal cord after accidental exposure to trichloroethylene. *Int. Arch. Arbeitsmed.* 31:257-264.

Scheiman, M.A., R.A. Saunders, and F.E. Saafeld. 1974. Organic contaminants in the District of Columbia water supply. *J. Biomed. Mass. Spectrom.* 1:209-211.

Schrenk, H.H., F.A. Patty, and W.P. Yant. 1933. Acute response of guinea pigs to vapors of some new commercial organic compounds. VII. Dichloroethyl ether. Public Health Report 48:1389-1398.

Schulz, C.O., and R.J. Rubin. 1973. Distribution, metabolism and excretion of di-2-ethylhexylphthalate in the rat. *Environ. Health Perspect.* 3:123-129.

Schwetz, B.A., P.A. Keeler, and P.J. Gehring. 1974. The effect of purified and commercial grade pentachlorophenol on rat embryonal and fetal development. *Toxicol. Appl. Pharmacol.* 28:151-161.

Shaffer, C.B., C.P. Carpenter, and H.F. Smyth, Jr. 1945. Acute and subacute toxicity of di-2-ethylhexylphthalate with note upon its metabolism. *J. Ind. Hyg. Toxicol.* 27:130-135.

Sipes, I.G., P.L. Gigon, and G. Krishna. 1974. Biliary excretion of metabolites of bromobenzene. *Biochem. Pharmacol.* 23(2):451-455.

Smirnova, N.A., and N.P. Granik. 1970. Long-term side effects of acute occupational poisoning by certain hydrocarbons and their derivatives. *Gig. Tr. Prof. Zabol.* 14(5):50-51.

Smith, C.C. 1953. Toxicity of butyl stearate, dibutyl sebacate, butyl phthalate and methoxyethyl oleate. *Arch. Ind. Hyg. Occup. Med.* 7:310-318.

Snyder, R., and J.J. Kocsis. 1975. Current concepts of chronic benzene toxicity. *CRC Crit. Rev. Toxicol.* 3:265-288.

Stewart, R.D., H.C. Dodd, E.D. Baretta, and A.W. Schaffer. 1968. Human exposure to styrene vapor. *Arch. Environ. Health* 16:656-662.

Stewart, R.D., E.A. Boettner, R.R. Southworth, and J.C. Cerny. 1963. Acute carbon tetrachloride intoxication. *J. Am. Med. Assoc.* 183:994-997.

Stofen, D. 1973. The maximum permissible concentrations in the U.S.S.R. for harmful substances in drinking water. *Toxicology* 1:187-195.

Tabakova, S.A. 1969. Experimental data for hygienic standards for dipterex. *Hyg. Sanit.* 34(10-12):187-193.

Teisinger, J. 1974. New advances in the toxicology of carbon disulfide. *Am. Ind. Hyg. Assoc. J.* 35(2):55-61.

Thiemes, C.H., and T.J. Haley. 1972. *Clinical Toxicology.* Lea and Febiger, Philadelphia.

Thorpe, J.J. 1974. Epidemiologic survey of leukemia in persons potentially exposed to benzene. *J. Occup. Med.* 16:375-382.

Toxic Substances List. 1974. Christensen, H.E., T.T. Luginbyhl, and B.S. Carroll, eds. National Institute for Occupational Safety and Health. HEW Publication no. (NIOSH)74-134. U.S. Government Printing Office, Washington, D.C.

USEPA. 1972. Region VI Surveillance and Analysis Division. Industrial pollution of the lower Mississippi River in Louisiana.

USEPA. 1974. Draft Analytical Report. New Orleans Area Water Supply Study. Surveillance and Analysis Division. Region VI. Dallas, Texas. EPA-906/10-74-002.

USEPA Advisory Committee. 1975. Report: Assessment of health risk from organics in drinking water.

USEPA. 1975a. Preliminary assessment of suspected carcinogens in drinking water. Report to Congress. Washington, D.C. EPA-56014-75-005 PB. 260961.

USEPA. 1975b. Region V. Joint Federal/State Survey of Organics and Inorganics in Selected Drinking Water Supplies.

USEPA. 1975c. Analysis of carbon and resin extracts. New Orleans Area Water Supply Study. Analysis of Carbon and Resin Extracts.

USEPA. 1975d. Identification of Organic Compounds in Effluents from Industrial Sources.

USEPA. 1976. Organic Compounds Identified in Drinking Water in the U.S. Health Effects Research Laboratory, Cincinnati, Ohio.

USHEW. 1973. PHS. *National Institute for Occupational Safety and Health Criteria for Recommended Standard, Occupational Exposure to Toluene.* ITS Publ. HSM 73-11023.

Ward, J.M., J.H. Weisburger, R.S. Yamamoto, T. Benjamin, C.A. Brown, and E.K. Weisburger. 1975. Long-term effect of benzene in C57BL/6N mice. *Arch. Environ. Health* 30:22-25.

Williams, D.T., and B.J. Blanchfield. 1975. The retention, distribution, excretion and metabolism of dibutyl phthalate-7-^{14}C in the rat. *J. Agric. Food Chem.* 23:854-858.

Williams, R.T. 1959. Detoxication Mechanisms. John Wiley & Sons, Inc. New York.

Wirtschafter, Z.T., and E.D. Schwartz. 1939. Acute ethylene dichloride poisoning. *J. Ind. Hyg. Toxicol.* 21:126-131.

Wolf, M.A., V.K. Rowe, D.D. McCollister, R.L. Hollingsworth, and F. Oyen. 1956. Toxicological studies of certain alkylated benzenes and benzene: Experiments on laboratory animals. *Arch. Ind. Health* 14:387-398.

World Health Organization. 1973. International Agency for Research on Cancer. *IARC Monographs on the Evaluation of Carcinogenic Risk of Chemicals to Man,* vol 3. International Agency for Research on Cancer, Lyon.

World Health Organization. 1972. *IARC Monograph on the Evaluation of Carcinogenic Risks of Chemicals to Man,* vol 1. 53-65.

Yllner, S. 1971d. Metabolism of 1,2-dichloroethane-^{14}C in the mouse. *Acta. Pharmacol. Toxicol.* 30:257-265.

Yodaiken, R.E., and J.R. Babcock. 1973. 1,2-Dichloroethane poisoning. *Arch. Environ. Health* 26:281-284.

ORGANIC COMPOUNDS FOR WHICH NO LIMITS WERE RECOMMENDED

Data were also collected on the following compounds:

bromobenzene	tetrachloroethane
chlorobenzene	tetrachloroethylene
1,2-dichloroethane	toluene
2,4-dichlorophenol	trichlorobenzene
hexachloroethane	trichloroethane
phenylacetic acid	trichloroethylene
phthalic anhydride	trichlorofluoromethane
propylbenzene	xylenes

In view of the relative paucity of data on the mutagenicity, carcinogenicity, teratogenicity, and long-term oral toxicity of these compounds, estimates of the effects of chronic oral exposure at low levels could not be made with any confidence. It was recommended that studies to produce such information be conducted before limits in drinking water are established for these compounds.

Review of Detoxification Processes

The material in this chapter is taken from the *Interim Treatment Guide for the Control of Chloroform and Other Trihalomethanes* and the *Interim Treatment Guide for Controlling Organic Contaminants in Drinking Water Using Granular Carbon* (see bibliography).

INTRODUCTION

Historically, the problem of organic contamination of drinking water was related to taste- and odor-producing compounds. Although of only indirect health significance, these compounds are still important as they are the ones that make drinking water disagreeable and cause some consumers to use poorly controlled sources of drinking water, bottled water or water treated by home treatment devices. In the past 5 years, through advances in organic analytic techniques and understanding of organic reactions, three other classes of organics, although undetected by consumers, have been shown to be important in drinking water.

The second of the four classes of concern is contaminants found in source waters that are of industrial origin. These are contained in industrial wastes, municipal wastes, and urban and rural runoff and have a potential adverse effect. The third of the four classes of organic compounds is that with which disinfectants react to produce disinfection by-products (precursors). These are largely, but not entirely, of natural origin, arising from urban and rural runoff and municipal wastes and have a potential secondary health effect, creating carcinogens and toxicants when reacting with disinfectants.

The fourth class of organic chemicals is the disinfection by-products themselves, specifically chloroform, bromodichloromethane, dibromochloromethane, and bromoform, of which chloroform appears most frequently and in the highest concentrations. These disinfection by-products come either from upstream disinfection practices or are created in the water treatment process itself. These too have a direct potential adverse health effect (see Chapter 3). A fifth class, of little concern includes the organics of natural origin that are not disinfection by-product precursors.

REMOVAL OF THMs AND THM PRECURSORS

Chlorine was first used as a disinfectant for municipal water supplies in the U.S. in 1908 to disinfect continuously the water supply of Jersey City, New Jersey. In many water supplies, chlorination is often the only treatment process used. When other treatment methods are used, disinfecting chlorine may be added to the raw water (prechlorination), the partially treated water, or the finished water (postchlorination). Where the distribution system contains open reservoirs, the treated water may be rechlorinated in the distribution system.

Because the formation of trihalomethanes is associated with chlorination, several options may be considered. Four approaches to chloroform concentration control would seem to be available: (1) a change of the point of chlorination; (2) a change of disinfectant; (3) treatment to reduce the precursor concentration prior to chlorination; and (4) treatment to reduce the chloroform concentration after formation.

Moving the Point of Chlorination

The EPA laboratory and pilot plant experimental results encouraged the water utility of Cincinnati, Ohio to attempt to reduce the chloroform content in their finished water by moving the point of chlorination so that clarified water was chlorinated rather than raw water. Figure 4.1 is a schematic diagram of the Cincinnati Waterworks.

FIGURE 4.1: SCHEMATIC OF CINCINNATI WATERWORKS

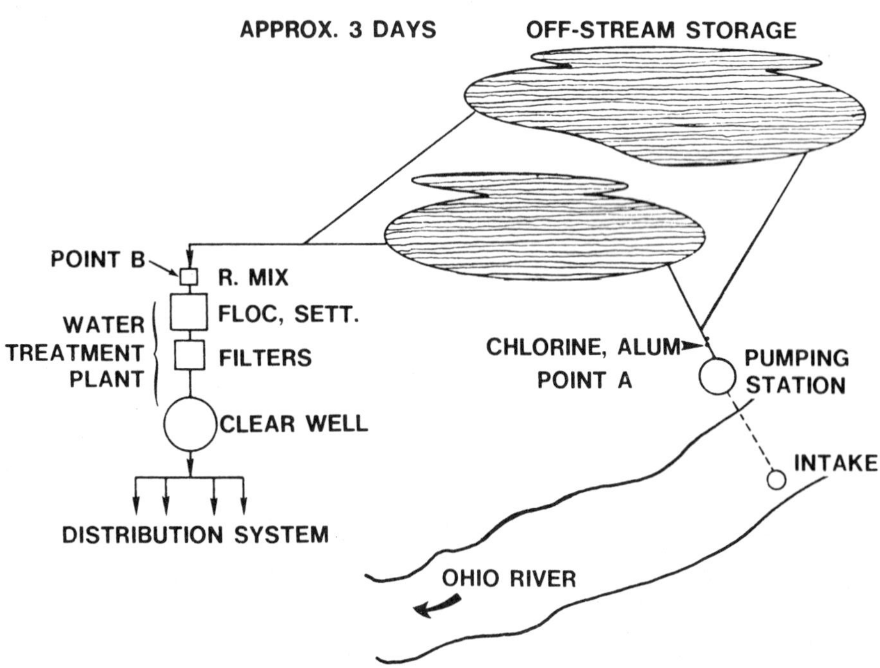

The city of Cincinnati pumps its water from the Ohio River into two large uncovered reservoirs with approximately three days' retention time. For the past several years, the practice has been to add alum to the water going to these reservoirs and sufficient chlorine to carry a free residual through the reservoirs and the treatment plant to the extremities of the distribution system. A consultant was hired to conduct the study and much effort went into collecting additional samples to insure the bacteriological integrity of the system during the experimentation.

In mid-July 1975, the point of chlorination was moved from point A to the headworks of the treatment plant, point B. Because of the addition of a coagulant prior to entering the off-stream storage reservoirs (point A) at the time of the study, the raw water turbidity was reduced from approximately 11 to 2 Turbidity Units (TU) as the water entered the treatment plant.

The sharp decline in chloroform concentrations in the tap water following moving of the point of chlorine application from point A to point B in mid-July is attributed to the change in chlorination practice (see Figure 4.2).

FIGURE 4.2: TRIHALOMETHANES IN CINCINNATI, OHIO TAP WATER

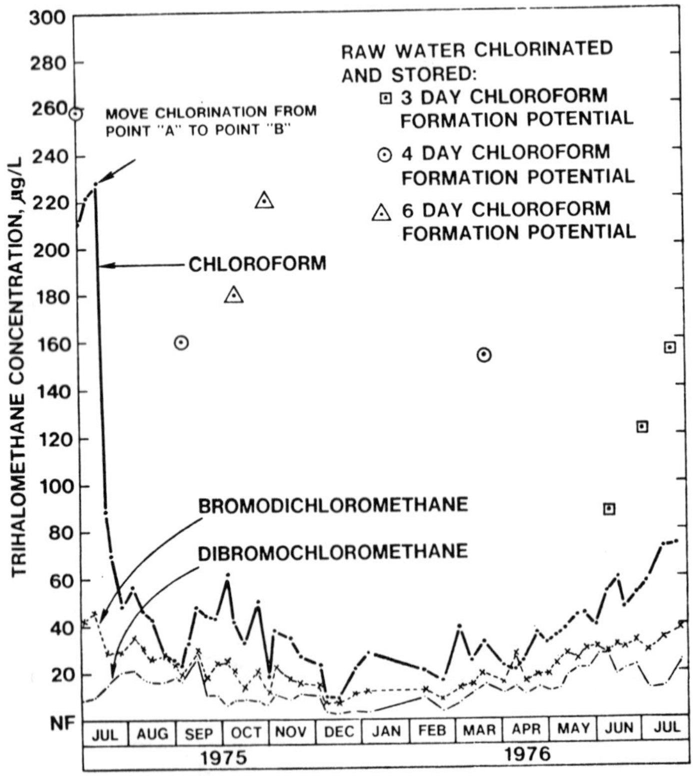

The change had little effect on the bromine-containing trihalomethanes. During the summer and early fall of 1975, other changes, such as the addition of small quantities of powdered activated carbon, variations in coagulants, and moving the point of chlorination to follow filtration, were made, but none of them had a significant observable effect on the chloroform concentration.

Other water utilities have also made in-plant modifications in treatment and have reported substantial reductions in the terminal chloroform concentration when chlorine was added after recarbonation instead of at the headworks prior to the high pH of lime softening. The importance of the pH control because of its effect on trihalomethane formation has also been reported.

Alternate Disinfectants

The benefits as well as the problems associated with changing to some alternative method of disinfectant must be weighed. For example, ozonation, while not creating chloroform, also does not provide any residual disinfectant in the distribution system. Moreover, it is conceivable that the by-products resulting from disinfecting with chlorine dioxide or ozone might be more harmful than chloroform itself. Mass spectrometric determination of these compounds is beginning and it is hoped that health effects studies of the identified species will follow.

Chlorine Dioxide: Chlorine dioxide may be a promising alternative to chlorine because a disinfectant residual can be maintained in the distribution system. This residual is reportedly comparable to chlorine as a biocide and, if the concentration of excess chlorine used to generate the chlorine dioxide can be kept low, the trihalomethane concentrations in the finished water will also be low.

Some question exists concerning the health effects of ingesting chlorite. Chlorite converts hemoglobin to methemoglobin, and the Norwegian Health Authority has recommended the absence of chlorite in drinking water. As of 1976, no restrictions were imposed on the use of chlorine dioxide for treating drinking water in the United States.

Other studies have begun to determine the inorganic end-products when chlorine dioxide is used in water treatment. For example, when 1.3 to 4.0 mg/l chlorine dioxide was applied to dual media filtered pilot plant effluent and stored for 3 days at 25°C, on a mg/l as Cl basis, 50% of the chlorine dioxide remained, 25% had become chlorite (ClO_2^-), 9% had become chlorate (ClO_3^-) and 14% was reduced to chloride. The unaccounted-for 2% was assumed to be analytical error. These preliminary results indicate that chlorite would be ingested by the consumer at concentrations depending on the original chlorine dioxide dosage. Further studies on chlorine dioxide end-products are under way.

Another problem with chlorine dioxide is its generation. The reaction of sodium chlorite ($NaClO_2$) and sulfuric or hydrochloric acid will produce chlorine dioxide without chlorine being present, but this reaction is inefficient. Therefore, chlorine dioxide is usually generated by reacting sodium chlorite with chlorine as in the following equation:

$$2NaClO_2 + Cl_2 \longrightarrow 2ClO_2 + 2NaCl$$

Because this reaction proceeds better at low pH, excess chlorine is usually added to reduce the pH. This produces a chlorine dioxide and chlorine mixture. The quantity of excess chlorine in chlorine dioxide can be reduced by adjusting the pH with acid and carefully controlling the ratio of chlorine to sodium chlorite. Although data indicate that the resultant chloroform concentration will be lower when chlorine and chlorine dioxide are used together when compared to the use of chlorine alone, trihalomethanes will not be absent if excess chlorine contacts the water.

Ozone: Ozone is another effective disinfectant that does not produce trihalomethanes, but it fails to provide a residual disinfectant in the distribution system and, like the other alternatives to chlorine, little is known about its organic by-products. Unless some other means of providing the residual disinfectant throughout the distribution system can be found, ozonation for disinfection followed by chlorination to provide a residual will not eliminate the trihalomethane concentrations reaching the consumer.

Chlorine plus Ammonia: Chloramines do not form trihalomethanes to any great extent and one alternate possibility would be to add chlorine and ammonia to provide a combined chlorine residual. The microbiological quality of the finished product is the foremost consideration in treating drinking water. Any disinfectant, therefore, must be an effective barrier against problem organisms.

Since combined chlorine is not as reactive as free chlorine for the formation of chloroform, if a utility should add ammonia in conjunction with chlorine addition or shortly thereafter such that no free chlorine residual ever existed for very long, chloroform formation should be low. The mere presence, however, of a combined chlorine residual in the finished water does not assure that free chlorine was not present sometime earlier during the treatment of the water. For example, many water utilities in the National Organics Reconnaissance Survey that had a finished water with only combined residual did have substantial quantities of chloroform in their water.

In these cases, free chlorine residual must have been reacting with chloroform precursor at some time during the treatment of the water. In spite of this, combined chlorination as a primary disinfectant is not discussed in detail here because not enough is known about its disinfecting power to judge its value at this time. Combined chlorine may, however, have a potential as a secondary disinfectant to provide a residual in the distribution system following ozonation.

Cost of Disinfection

In an effort to compare the cost of alternate disinfectants, calculations were made assuming no disinfection facilities at a water treatment plant. Therefore, costs of disinfection equipment, plus the cost of a disinfectant contact chamber is included for chlorine, ozone and chlorine dioxide.

The summary of these data, Table 4.1, shows that the costs of all three disinfectants are similar. Of course, in an existing plant, a change to ozone might mean the abandoning of some of the existing chlorination facilities so, on that basis, the costs in Table 4.1 might not be applicable. On the other hand, a water treatment plant now using chlorine could add chlorine dioxide capability at a small incremental cost.

TABLE 4.1: ESTIMATED COST OF DISINFECTION*

Design capacity	1 mgd	10 mgd	100 mgd	150 mgd
Average daily flow	0.7 mgd	7 mgd	70 mgd	105 mgd
Chlorine 15¢/lb**	4	1	0.7	0.6
O_3 generated by air***	6	2	0.9	0.8
O_3 generated by oxygen***	8	2	1	0.8
Chlorine 15¢/lb†	4	2	1	1

*All costs in cents/1,000 gallons. These costs will vary at different locations, so should be considered approximate; 1976 figures.

**2 mg/l chlorination, 30 minute contact time.

***1 mg/l ozone, 20 minute contact time.

†1 mg/l chlorine dioxide generated from $NaClO_2$; sodium chlorite cost equals $0.70/lb. Because sodium chlorate ($NaClO_3$) only costs $0.20/lb, chlorine dioxide generated from it might be less expensive than that generated from $NaClO_2$. Processes for this have yet to be adapted to water treatment practice, however.

In this cost analysis, a lower ozone and chlorine dioxide dose was compared to the chlorination dose on the basis that both of these disinfectants are more effective than chlorine and, therefore, less disinfectant would be required. During one experiment, for example, 0.5 mg/l of either ozone or chlorine dioxide was sufficient to adequately disinfect the effluent from the dual-media filter in the pilot plant. In a later experiment, 0.3 mg/l of ClO_2 reduced the standard plate count population to less than 1 per 1 ml. A dose of 1.3 mg/l was required for chlorine to accomplish the same disinfection, although the pilot plant was not performing too well at this time, so this might not be the minimum chlorine dose. Finally, using a lower disinfectant dose should cause the formation of less nontrihalomethane organic by-products, although this has not yet been demonstrated experimentally.

To determine the cost of disinfection to a typical household, a family of four was assumed to use 200 gallons of water per day. If this rate of consumption was steady throughout a calendar year, the annual usage would be 73,000 gallons. Therefore, multiplying any disinfection cost in cents per 1,000 gallons times 73 would produce an estimate of the annual cost to a typical household for a given treatment process. Using the costs in Table 4.1, chlorination contributes from about $0.40 to $2 per year to the water bill of an average household, depending on treatment plant size. On the same basis, changing to ozone would make these figures about $0.50 to $4 per year, while the use of chlorine dioxide would cost from about $0.70 to $2 per year.

Recommendations for Specific Treatment Changes

Although research on these and other alternate disinfectants is still in progress, enough is known about their effectiveness that any utility currently chlorinating that desires to lower its trihalomethane concentration should consider these alternate disinfectants. Care should be taken, however, during the transition period to insure the microbiological quality of the finished and distributed water. Utilities currently using chlorine dioxide for taste and odor control should consider using it both for taste and odor control and disinfection. They should attempt to generate the chlorine dioxide with as little chlorine as possible in it for the maximum reduction in trihalomethane concentration. Utilities currently adding ammonia to create chloramines should review their practice and reduce the elapsed time between the addition of chlorine and ammonia to the minimum compatible with good disinfection. Shortening the contact time of free chlorine residual should reduce the formation of trihalomethanes and, if mixing is relatively good, disinfection should be adequate.

Trihalomethane Precursor Removal

The following are definitions of three terms used in the following material:

1. Instantaneous trihalomethane concentration—the concentration of trihalomethanes in the water at the moment of sampling;
2. Terminal trihalomethane concentration—the concentration of trihalomethanes that occurs when a sample of water is stored for a specified time at a specified pH and temperature; and
3. Trihalomethane formation potential—the difference between the terminal and instantaneous trihalomethane concentrations.

Because no direct measurement exists for trihalomethane precursors, the degree of precursor removal was judged by comparing trihalomethane concentration upon chlorination of an untreated control (called the trihalomethane formation potential) to similar data collected on a treated water after similar chlorination. For example, if the effluent from a sand filter that was chlorinated and stored for 2 days yielded 50 μg/l chloroform and the same effluent passed through an adsorbent then chlorinated and stored under similar conditions produced 25 μg/l chloroform, the adsorbent was said to be 50% effective in removing chloroform formation potential. This example assumes that no instantaneous trihalomethanes are present in the filter effluents.

Aeration: Aeration was evaluated for reducing trihalomethane formation potential in Ohio River water. A countercurrent-flow aerator was fabricated out of 3.7 cm (1.5 in) diameter glass tubing with a fritted glass diffusor and used to aerate the river water at varying air-to-water ratios. The water was then chlorinated and stored at 25°C for 2 days in sealed vessels. The chlorine solution used in all the precursor removal studies was made by bubbling gaseous chlorine through nitrogen-purged distilled water and the residuals were determined amperometrically.

The contact time of 2 days was selected as a matter of convenience for maintaining free chlorine residuals and because experience had shown that the rate of trihalomethane formation for this water was fastest during the first few hours and then greatly reduced after 30 to 40 hours contact time. A companion river water sample was chlorinated and stored without being aerated. Table 4.2 shows the influence of aeration on trihalomethane formation potential to be insignificant even at an air-to-water ratio of 20 to 1.

Adsorption: Powdered Activated Carbon (PAC) – To determine the effectiveness of PAC on removal of trihalomethane formation potential, coagulated and settled water from the pilot plant (described below) was dosed with varying quantities of PAC, mixed at 100 rpm for 2 minutes, then centrifuged for 20 minutes at 1,500 rpm (480 *g*). The supernatant liquor was then decanted and chlorinated, rapidly mixed for 2 minutes, then stored for 2 days at 25°C. PAC had to be added in large doses before it had much effect on reducing the trihalomethane formation potential. For example, at the impractical PAC dose of 100 mg/l, 49% of the total trihalomethane formation potential remained after treatment (see Table 4.3).

TABLE 4.2: EFFECT OF AERATION ON REDUCING TRIHALOMETHANE FORMATION POTENTIAL

	Air:Water Ratio	. Concentration (2 day contact time). .			
		$CHCl_3$	$CHBrCl_2$	$CHBr_2Cl$	$CHBr_3$
		. . . . $\mu g/l$	. . . .		
Ohio River Water	–	NF*	NF	NF	NF
Ohio River Water + 13 mg/l Cl_2	–	66	28.0	8.0	<0.1
Aerated** Ohio River Water + 13 mg/l Cl_2	1:1	66	27.8	8.0	<0.1
Water as above	4:1	64	26.8	6.6	<0.1
Water as above	6:1	62	25.8	7.6	<0.1
Water as above	8:1	62	26.8	7.8	<0.1
Water as above	10:1	59	25.6	7.7	<0.1
Water as above	20:1	61	26.0	8.0	<0.1

*None found.
**Activated carbon-filtered compressed air, 10 minute contact time

TABLE 4.3: EFFECT OF POWDERED ACTIVATED CARBON (PAC) ON TRIHALOMETHANE FORMATION POTENTIAL

	PAC Dose mg/l	. . . Concentration** (2 day contact time). . . .				TTHM*** $\mu mol/l$	% Removal of TTHM
		$CHCl_3$	$CHBrCl_2$	$CHBr_2Cl$	$CHBr_3$		
		. . . . $\mu g/l$	. . . .				
Settled Water*	–	NF	NF	NF	NF	NF	–
Settled Water + Cl_2	–	27	15.2	10.4	<0.1	0.37	0
Settled Water + PAC + Cl_2	2	22	15.1	8.0	<0.1	0.31	16
Water as above	4	25	16.4	10.2	<0.1	0.36	2
Water as above	8	20	15.8	9.4	<0.1	0.32	14
Water as above	20	16	16.9	12.2	<0.1	0.29	22
Water as above	50	11	13.0	10.0	<0.1	0.22	41
Water as above	100	9	9.5	8.8	<0.1	0.18	51

*Alum coagulation + settling.
**Average of five experiments using three different commercial brands PAC.
***Total trihalomethane concentration.

Granular Activated Carbon – A pilot water treatment plant was fabricated to provide a continuous supply of treated, not yet chlorinated water for precursor removal studies. To minimize contamination from structural materials, the plant was built almost entirely of stainless steel, Teflon and glass. Through the assistance and cooperation of the Cincinnati Water Works, Ohio River water was provided as a source of raw water.

The pilot plant employed conventional alum coagulation, flocculation, sedimentation, and originally had three parallel filtration schemes: 76 cm (30 in) dual media (anthracite and sand); 76 cm (30 in) of granular activated carbon (GAC) that acted as both a filtering and adsorption media (which was termed filtration/adsorption); and a dual media column followed by 76 cm (30 in) of GAC. The reason for this latter arrangement (which was termed postadsorption) was to see if the life of a GAC bed could be extended if prefiltered water was applied. The nominal filter flow rates were 100 ml/min, which equaled a loading of 2 gpm/ft^2 with an apparent contact time within the GAC beds of slightly over 4 minutes. A simplified flow pattern is shown in Figure 4.3. The disinfection step (chlorine, chlorine dioxide, ozone or combinations of these) followed filtration so that trihalomethane formation potential removal could be monitored by measuring subsequent trihalomethane formation after an appropriate contact time.

FIGURE 4.3: SCHEMATIC OF PILOT PLANT FOR REDUCING THE TRIHALOMETHANE FORMATION POTENTIAL

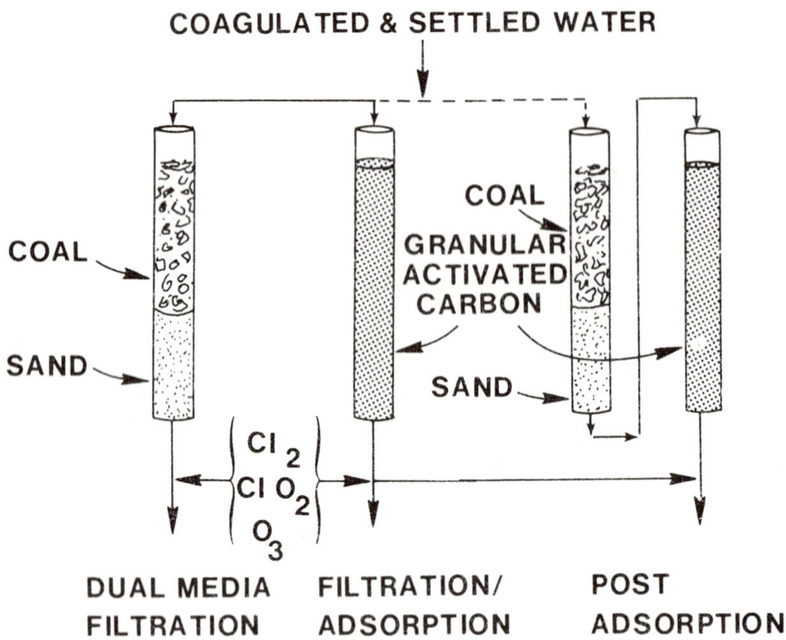

After 10 weeks of study, the conclusion was reached that little was gained in reducing the trihalomethane formation potential by prefiltering the water before exposing it to GAC. Although little difference appears in performance between the two modes, there are other considerations (such as GAC handling, flexibility in bed depths, etc.) that affect the overall economics of the system that must be included in any comparative evaluation.

The relative effectiveness of the GAC to prevent the formation of trihalomethanes was highest for chloroform and lowest for bromoform. Sufficient precursor was being passed through the GAC after one week to produce a measurable amount of trihalomethane during a 4-day chlorine contact time. A 4-day contact time was selected to approximate what would likely be a maximum retention time in a distribution system. The 76 cm deep GAC column was exhausted (i.e., the trihalomethanes formed upon chlorination of the GAC effluent equaled those levels formed in the chlorinated dual media effluent) in about 13 weeks for chloroform, 8 weeks for bromodichloromethane, 5 weeks for dibromochloromethane, and probably less than 2 weeks for bromoform. The reason for this might be that the GAC does not remove bromide effectively so bromide in the effluent, plus the first traces of precursors will form brominated trihalomethanes upon chlorination because the oxidation of bromide to bromine by chlorine followed by bromination occurs faster than the chlorination reaction. As the GAC ages and more precursors break through (bromide still being present) chloroform is produced.

Field data to support these general findings on the removal of precursor materials are lacking because the water utilities using GAC in the United States chlorinate prior to the filters and thus the GAC is exposed to some instantaneous trihalomethane levels. The use of granular activated carbon will be discussed in more detail in the next two chapters.

Oxidation: Experiments examining the prevention of trihalomethane formation were conducted with a pilot water treatment plant having conventional coagulation and sedimentation. The settled water was then divided between a dual media and two different depth granular activated carbon (GAC) filters. Following filtration, the effluent was disinfected with either chlorine, ozone, chlorine dioxide, or combinations of these oxidants. Bench-scale studies with powdered activated carbon (PAC) and aeration were also conducted. Table 4.4 lists the effectiveness of these various unit processes for reducing the chloroform formation potential.

TABLE 4.4: EFFECTIVENESS OF VARIOUS UNIT PROCESSES FOR REDUCING CHLOROFORM FORMATION POTENTIAL

Process	Chloroform Formation Potential μg/ℓ	Chloroform Formed μg/ℓ	Remarks
Aeration followed by chlorination	66[a]	66[a]	Diffused-air aeration with air to water ratios up to 20:1 did not reduce chloroform formation potential (10 min. Contact Time)
Coagulation, Sedimentation and Dual-Media Filtration followed by Chlorination	48[b]	13[b]	
Coagulation, Sedimentation Filtration/Adsorption by Granular Activated Carbon (5 min. contact time) followed by chlorination	48	≤ 1	GAC would be effective for 3 weeks
		≤ 10	GAC would be effective for 8 weeks
Powdered Activated Carbon added after Coagulation and Settling followed by Chlorination	27[a]	20[a]	at PAC dosage = 8 mg/ℓ
		9	at PAC dosage = 100 mg/ℓ PAC contact time = 2-20 min.
Ozone only	48	None found	O_3 neither forms trihalomethanes, nor removes precursors at disinfection doses

(continued)

TABLE 4.4: (continued)

Process	Chloroform Formation Potential µg/ℓ	Chloroform Formed µg/ℓ	Remarks
Ozone followed by chlorination	48	48	Disinfection doses (≤ 1 mg/ℓ)
Chlorine Dioxide only	74	< 1	ClO_2 does not form trihalomethanes
Coagulation, Sedimentation and Dual Media Filtration followed by:			
1. Chlorination	-	17[c]	1.3 mg/ℓ ClO_2 and 1.5 mg/ℓ Cl_2
2. Chlorine dioxide with chlorine	-	3[d]	Trihalomethane formation decreases as the ratio of ClO_2 to Cl_2 increases

All tests performed on Ohio River water. Chloroform Formation Potential is the amount of chloroform formed when raw water is chlorinated past break-point and stored at 25°C for a specified contact time.

a - chlorine contact time = 48 hours
b - chlorine contact time = 96 hours
c - chlorine contact time = 22 hours
d - contact time for combination of chlorine dioxide with chlorine = 22 hours

Removing Trihalomethanes After Formation

Aeration: Chloroform is lost to the atmosphere when water is held in open vessels or from a flowing stream accidentally contaminated by a chloroform spill. In beakers standing open at room temperature, almost all of the chloroform was lost from Cincinnati tap water on 3 days' standing. At one European utility, over 90% of the chloroform is lost during 3 weeks' standing in a deep holding reservoir just prior to water treatment. These data indicate that chloroform is volatile and will be lost from the water at any air-water interface. Therefore, studying aeration as a unit process for removing or reducing trihalomethanes after formation seemed logical.

A countercurrent-flow aerator was fabricated out of 3.7 cm (1.5 in) diameter glass tubing with a fritted glass diffusor. At an air to water (volume to volume) ratio common to water treatment aerator design for controlling taste and odor problems (1:1), the chloroform concentration in the tap water was not significantly changed. An increased air to water ratio of 8:1 yielded a 58% chloroform reduction and a further increase to 20:1 showed an 83% decrease.

In situations where a free chlorine residual persists through a water treatment plant, chloroform concentrations increase in spite of some loss to the atmosphere because of a continuing rapid reaction of chlorine with precursor. At the 20:1 air to water ratio, the chloroform reduction immediately after aeration was about 85%, yet the net reduction based on a 2-day storage time was only 50 to 55%. Treatment for reducing the precursor (trihalomethane formation potential) has already been covered.

Other Processes for Removal of Trihalomethanes After Formation: Studies were conducted on removing trihalomethanes from drinking water by the use of granular and powdered activated carbon adsorption, ozonation, and treatment with chlorine dioxide. Table 4.5 shows what conditions were found necessary to effect reductions in chloroform concentrations. The bromine-containing trihalomethanes were more effectively reduced in concentration than was chloroform by GAC. However, in all the processes tested, it was shown that trihalomethanes are not easily removed from water.

TABLE 4.5: EFFECTIVENESS OF VARIOUS UNIT PROCESSES FOR REDUCING CHLOROFORM IN DRINKING WATER

Process	Initial Chloroform Concentration, μg/ℓ	50%	25%	10%
Aeration				
Air to Water Ratios for diffused-air aeration: 10 min. contact time	100	6:1	15:1	25:1
Granular Activated Carbon				
Expected life for 5 min. contact time	55	7 weeks	5 weeks	4 weeks
Powdered Activated Carbon				
Dosage, mg/ℓ applied to:				
a. Chlorinated Raw Water	64	95 mg/ℓ	> 105 mg/ℓ	> 105 mg/ℓ
b. Chlorinated, Coagulated and Settled Water	44	27 mg/ℓ	90 mg/ℓ	105 mg/ℓ
Ozonation				
4 min. contact time	Up to 25 mg/ℓ O_3 had no effect on the chloroform concentration.			
Chlorine Dioxide				
Up to 48 hr. contact time	Up to 10 mg/ℓ ClO_2 had no effect on the chloroform concentration.			

REMOVAL OF OTHER TOXIC ORGANIC COMPOUNDS

The advantages and disadvantages of each of the four general types of organic removal unit processes will be briefly reviewed.

Precipitation

The addition of a metal coagulant for the removal of particulates or raising the pH for removal of calcium and magnesium results in the removal of some organic matter. This can be measured both by the reduction in the organic carbon concentration (see Figure 4.4) and color.

Although partially effective for the removal of disinfection by-product precursors, this process does little for the reduction of the other classes of organic compounds that are in true solution.

FIGURE 4.4: RELATIVE NONPURGEABLE ORGANIC CARBON REMOVAL DURING WATER TREATMENT IN THE PILOT PLANT

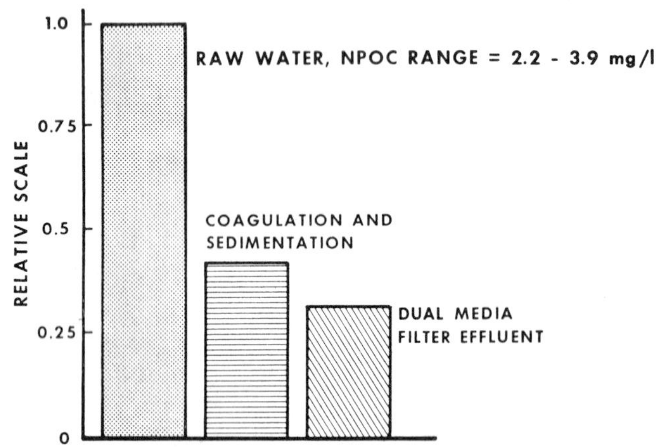

Oxidation

A second possibility for organic removal is to treat the organics with an oxidizing agent. If oxidation could be complete, this would be an attractive approach as the major end-products would be carbon dioxide and water. Although some organic oxidation has been demonstrated when oxidants, such as ozone, chlorine dioxide, chlorine, and potassium permanganate are added to water, doses must be uneconomically high to insure nearly complete oxidation (see Figure 4.5).

If oxidation is incomplete, the organic compounds will not be oxidized to carbon dioxide and water, but only to some intermediates of unknown toxicologic hazard. Thus far, even the addition of ultraviolet radiation to an ozonation process has not been demonstrated to cause nearly complete oxidation for many compounds.

Aeration

Because many of the organic compounds of concern that have recently been shown to occur in drinking water were determined by a procedure that employs a concentration step based on purging the compounds from the sample with diffused gas, aeration, either diffused air or spray, was considered as a possible organic removal process.

The advantage of such a process would be that the organic compounds are truly removed from the water without the formation of by-products or intermediates. The disadvantages are at least twofold: (1) the removed materials are transferred from the water phase to the air phase, possibly causing another problem, and (2) only organic substances with certain characteristics are removable even with excessive quantities of air (see Figure 4.6). Therefore, this approach is self-limiting.

FIGURE 4.5: REDUCTION IN NONPURGEABLE ORGANIC CARBON CONCENTRATION BY OZONATION

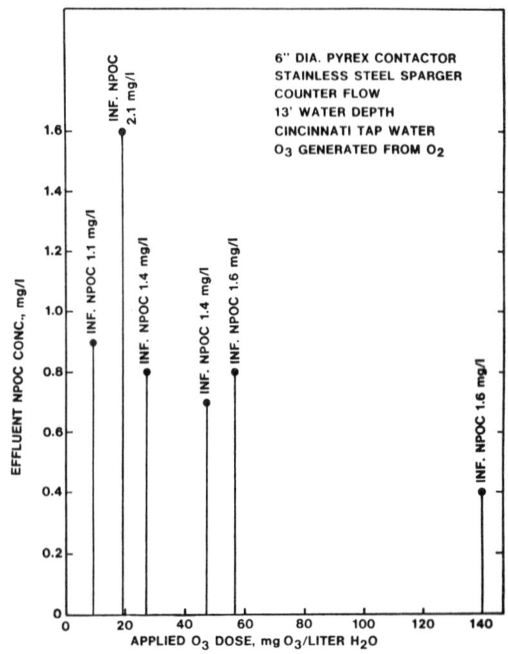

FIGURE 4.6: INFLUENCE OF AERATION ON TRIHALOMETHANE AND NONPURGEABLE ORGANIC CARBON CONCENTRATIONS

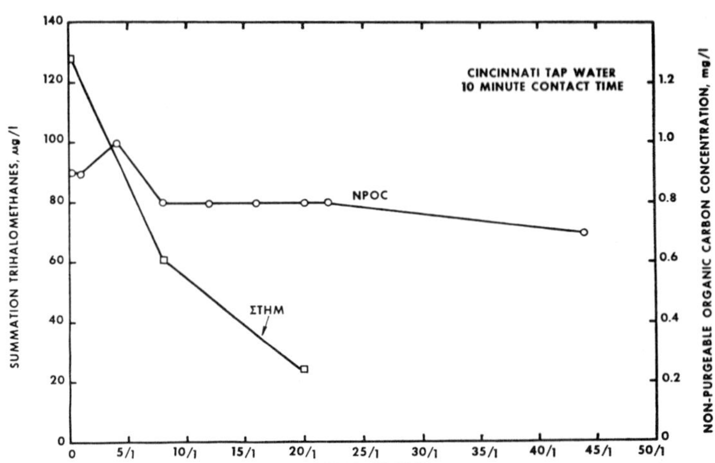

Adsorption

A fourth possible approach to control of organic contaminants is to remove them from the water by adsorption onto a surface. Two types of adsorbents are commercially available in the United States. These are: (1) synthetic resins and (2) activated carbon, either in the powder or granular form.

Synthetic Resins: Synthetic resins of the Amberlite XAD type (a class of macroreticular divinylbenzene-styrene or acrylic copolymers) have a very limited capacity for adsorption, although Ambersorb XE-340 (a carbonaceous adsorbent, chemically intermediate between the XAD type and activated carbon and manufactured by Rohm and Haas) has a high adsorption capacity for organic compounds such as chloroform and carbon tetrachloride, but does not have much capacity for the removal of a wide range of organic compounds as measured by organic carbon concentrations.

Activated Carbon: Activated carbon is a form of carbon that is activated by a carefully controlled oxidation process to develop a porous carbon structure with a large surface area. This large surface area gives the activated carbon a high capacity to adsorb dissolved organic materials from water. The major raw materials used in the manufacture of granular activated carbons are petroleum coke, bituminous coal and lignite.

After preliminary processing, these materials are heated to a high temperature and reacted with steam to develop the extensive internal pore structure required for adsorption. The substance is then crushed, graded, screened, and packaged. The raw material and activating conditions used affect adsorption and physical properties of the carbons produced.

Powdered activated carbon is a good adsorbent with a high surface area per gram of material, but as it is usually added to the water to be treated in a single dose, an equilibrium is established between the organic compounds and the adsorbent, thereby limiting the extent of removal possible. High doses of powdered activated carbon were unable to reduce the Threshold Odor Number (TON) to acceptable levels (three or less) in a test at a Nitro, W.Va. water plant. Further, as already shown, high doses of powdered activated carbon were required to greatly reduce the concentration of chloroform in tap water, as well as the concentrations of the disinfection by-product precursors as measured by the trihalomethane formation potential test. Because of the problem of interference of residual powdered activated carbon with the total organic carbon test, evaluating the ability of powdered activated carbon to remove organic carbon is difficult.

In contrast, when activated carbon is used in a column, it acts like a series of powdered activated carbon doses being applied to water of continually improving quality as the water passes through the column. This plate effect permits an activated carbon column to produce water containing much lower concentrations of organic matter when contrasted to reasonable additions (10 mg/l or less) of powdered activated carbon. Granular activated carbon, as the removal agent of preference, will be discussed at length in the next two chapters.

Granulated Activated Carbon for Removal of Organic Compounds

The material in this chapter was taken from various sections of the *Interim Treatment Guide for Controlling Organic Contaminants in Drinking Water Using Granular Activated Carbon,* compiled and edited by James M. Symons of the Water Supply Research Division, Municipal Environmental Research Laboratory, Office of Research and Development, Cincinnati, Ohio, and printed in January 1978 (see bibliography).

OVERVIEW OF GAC PERFORMANCE

Class I Compounds: An example of excellent performance of granular activated carbon for removing Class I type organic compounds, taste and odor producing, is shown in Figure 5.1. In this installation, the granular activated carbon replaced the sand in existing filter boxes and under these circumstances controlled tastes and odors for more than one year.

Class II Compounds: An early example demonstrating the ability of granular activated carbon beds to adsorb Class II type organic compounds, source water contaminants, showed that seven compounds present in the Kanawha River after aeration could be reduced to below detectable concentrations by fresh (two-day old) granular activated carbon beds. These compounds were bis(2-chloroethyl) ether, 2-ethylhexanol, bis(2-chloroisopropyl) ether, α-methylbenzyl alcohol, acetophenone, isophorone, and tetralin. A recent review of the literature has expanded this list considerably.

Current research efforts by the Water Supply Research Division continue to demonstrate the ability of granular activated carbon to adsorb a variety of synthetic organic contaminants. Figure 5.2 is an example of these data, showing the adsorption of carbon tetrachloride by granular activated carbon. The data in Table 5.1 indicate the adsorbability of other low molecular weight halogenated organic compounds by granular activated carbon.

FIGURE 5.1: THRESHOLD ODOR NUMBER BREAKTHROUGH PATTERN FOR NEW ACTIVATED CARBON FILTER, LAWRENCE, MASS.

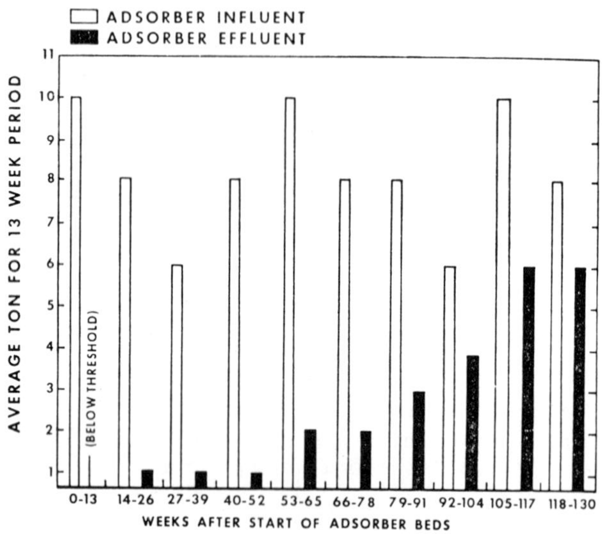

FIGURE 5.2: PERFORMANCE OF GRANULAR ACTIVATED CARBON BED FOR CARBON TETRACHLORIDE REMOVAL

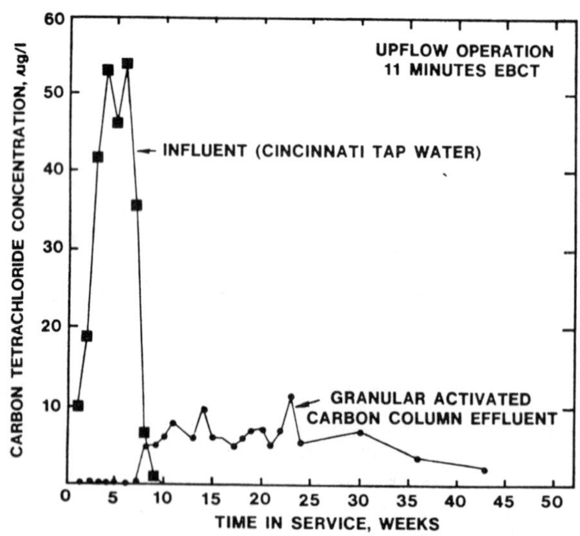

TABLE 5.1: USE OF GRANULAR ACTIVATED CARBON BEDS TO REMOVE ORGANIC CONTAMINANTS FROM A GROUND WATER SUPPLY

Contaminant	Influent Concentration Range, μg/l	.. . Average Effluent Concentration, μg/l. . 0-4 weeks	4-8 weeks	8-12 weeks
1,1,1-Trichloroethane	10.9-36.7	NF*	NF*	0.4
Trichloroethylene	1.9-7.7	NF*	<0.1	<0.1
Tetrachloroethylene	0.1-16.9	NF*	<0.1	<0.1

*NF = None found

NOTE: After six weeks of testing, two new contaminants, 1,1-dichloroethane and cis-1,2-dichloroethylene began appearing in the influent to the granular activated carbon column. No cis-1,2-dichloroethylene has yet been detected in the treated water, but the average concentration of 1,1-dichloroethane in the effluent for the 4 to 8 week period after first appearing was 0.7 μg/l.

The data in Figure 5.3 demonstrate that with Ohio River water, granular activated carbon adsorption treatment can initially produce an effluent containing only approximately 0.1 mg/l of organic carbon. This shows that under these circumstances, when fresh, the granular activated carbon adsorbed over one milligram per liter of organic carbon. Because most of the specific organic contaminants found in drinking water occur at μg/l concentrations, 1000-fold less, this further indicates the ability of granular activated carbon to adsorb a wide variety of organic contaminants.

FIGURE 5.3: ORGANIC CARBON BREAKTHROUGH CURVE FOR PILOT WATER TREATMENT PLANT

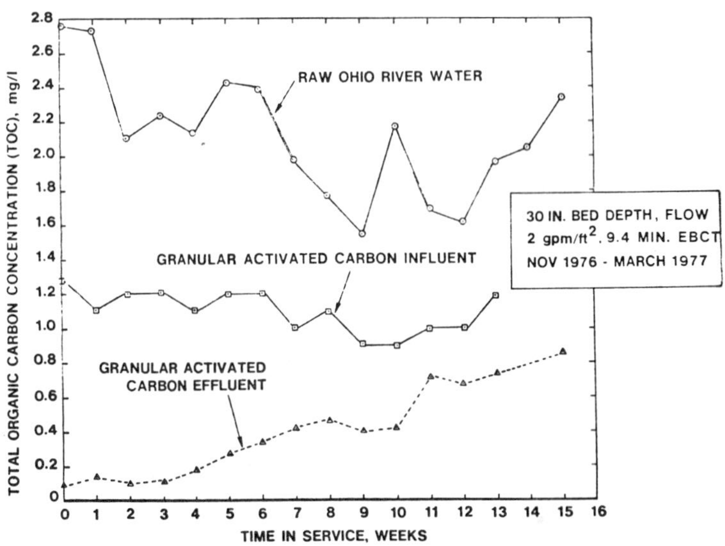

Finally, the data in Figure 5.4 further demonstrate this point, as indicated by the absence in the granular activated carbon bed effluent of many of the organic compounds that show up as peaks on the capillary gas chromatogram of the influent to the granular activated carbon column.

FIGURE 5.4: COMPARISON OF ORGANIC CONTENT OF A GRANULAR ACTIVATED CARBON EFFLUENT BY SOLVENT EXTRACTION AND CAPILLARY GAS CHROMOTOGRAPHY

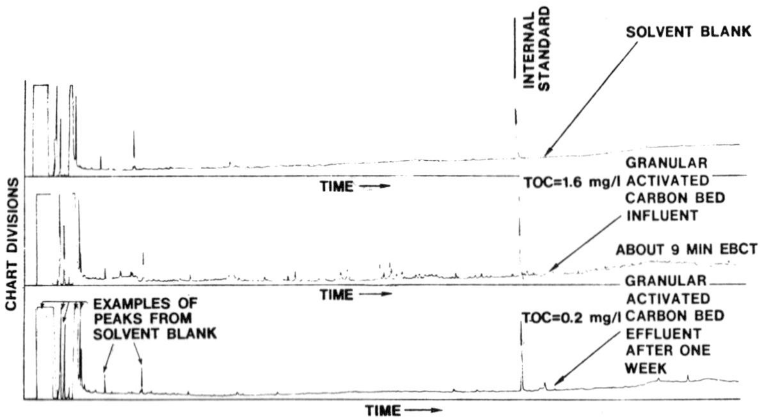

Class III Compounds: Figure 5.5 is an example of data collected by the Water Supply Research Division at one of its large-scale field research projects. These data show the ability of granular activated carbon to adsorb Class III type organic compounds, disinfection by-product precursors.

Class IV Compounds: The data in Figures 5.6A and 5.6B show that, when fresh, granular activated carbon has the ability to adsorb Class IV type organic compounds, disinfection by-products. The bromine-containing trihalomethanes are well adsorbed, chloroform breaking through the granular activated carbon treatment systems first.

Summary

Granular activated carbon has been selected as the best broad-spectrum process for the removal of organic compounds, without the production of by-products, currently available at this time (Fall 1977).

The rest of this chapter contains the detailed analysis of the literature and the Water Supply Research Division research findings that support this conclusion, plus a brief review of a variation of the usual granular activated carbon adsorption process called biological activated carbon employing ozonation preceding adsorption.

FIGURE 5.5: BREAKTHROUGH PATTERN OF CLASS III ORGANIC CONTAMINANTS, DISINFECTION BY-PRODUCT PRECURSORS

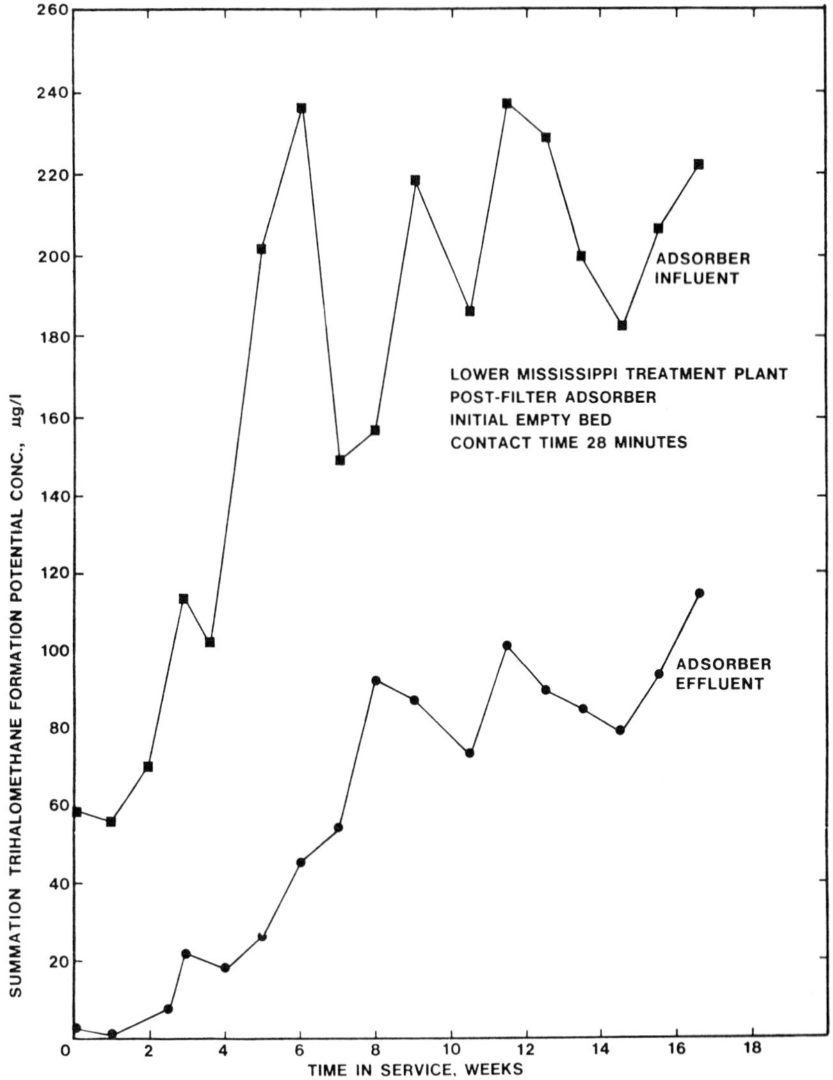

FIGURE 5.6: REMOVAL OF TRIHALOMETHANES FROM CINCINNATI, OHIO TAP WATER

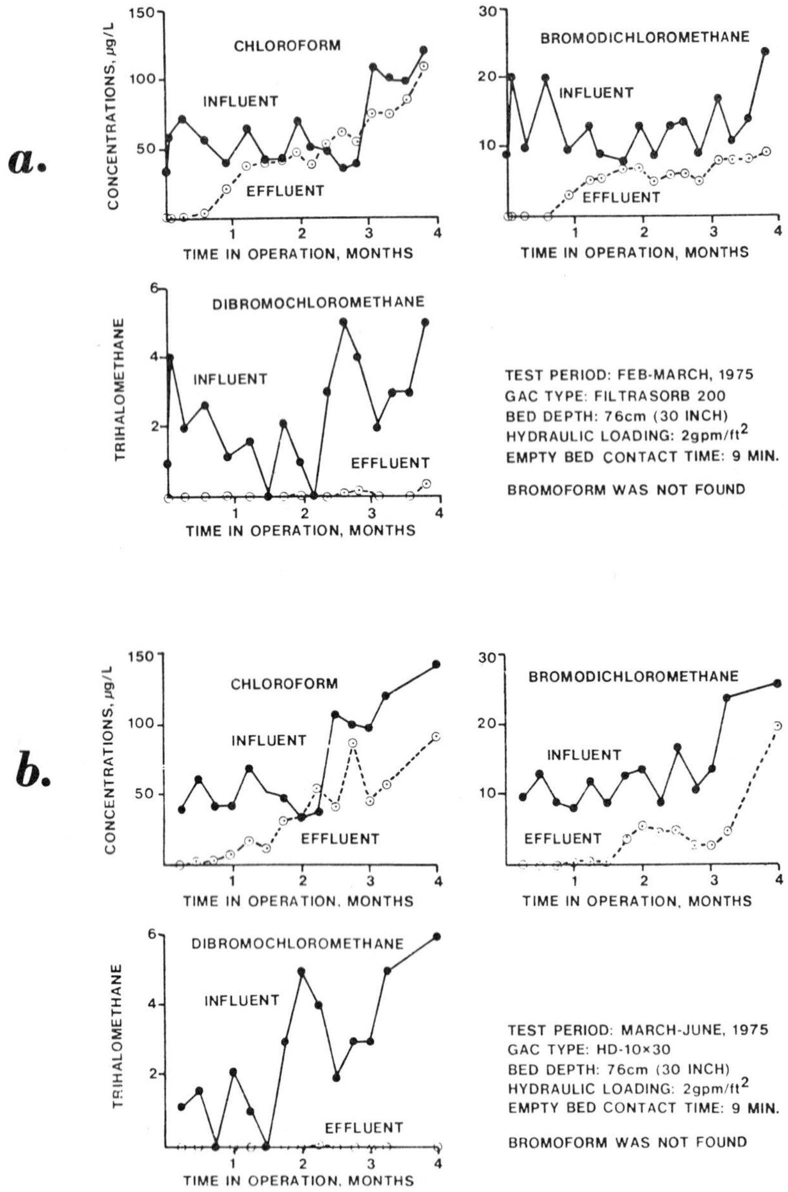

(a) Removal by coal-base granular activated carbon
(b) Removal by lignite-base granular activated carbon

LITERATURE REVIEW

Taste and Odor Control

Many water treatment plants in the United States are practicing taste and odor control by the addition of powdered activated carbon, which also removes some organic matter. Currently (Fall 1977) however, about 35 plants are using granular activated carbon, either alone or on top of some sand, as both a filter media for particulate control and an adsorption media for organic contaminant control. This type of system is hereafter called a sand replacement system and is used primarily for taste and odor control. A few water purveyors installed these granular activated carbon beds because of the organic pollution in their raw water as well as taste and odor problems.

In their 1964 paper, Woodward, Dostal, and Robeck (1) reported on five installations of granular activated carbon beds during the 1930s. A more recent survey (2) of the known granular activated carbon installations in this country shows that approximately 80% of the water treatment plants routinely using granular activated carbon beds have been installed since 1970.

Early Installations: A discussion of some of the early installations is of interest. In the fall of 1963 the management of the Nitro, West Virginia, water utility installed two beds of granular activated carbon to investigate treatment for removal of taste and odor. During the three-month testing period the Threshold Odor Number (TON) of the raw water varied between 500 and 1000. During this same period the TON of the influent water to the experimental granular activated carbon beds varied between 100 and 200, with one excursion to 400. These beds, with empty bed contact times (EBCT) of about eight minutes, were able to produce an odor-free water for as long as 26 days (3).

The success of these tests encouraged the water purveyor to convert the entire plant to granular activated carbon beds in 1965. Another interesting feature of this installation was the construction of an on-site 10,000 lb/day multiple-hearth reactivation furnace. From that time, until the plant was closed several years later (because the municipality obtained a different source of water), the granular activated carbon was reactivated approximately every six months.

The success of granular activated carbon beds for controlling taste and odor demonstrated at Nitro, West Virginia, has been duplicated at many other water utilities. For example, at Piqua, Ohio; Mt. Clemens, Michigan; Lawrence, Massachusetts (4); and Davenport, Iowa, the water purveyors were having difficulty providing their consumers with an acceptable drinking water in spite of the use of large doses of powdered activated carbon.

After conversion to granular activated carbon beds, these purveyors reported successful control of their taste and odor problems. Figure 5.7 shows that at Lawrence, Massachusetts, during a time period when the TON in the settled water was in the 6 to 10 range, the granular activated carbon beds controlled the effluent TON to acceptable levels for over a year.

When conversion was first contemplated, the Massachusetts State Health Department insisted that some sand be left in the bottom of the filters to avoid any problems with particulate (turbidity) breakthrough.

FIGURE 5.7: THRESHOLD ODOR NUMBER BREAKTHROUGH PATTERN FOR NEW ACTIVATED CARBON FILTER, LAWRENCE, MASS.

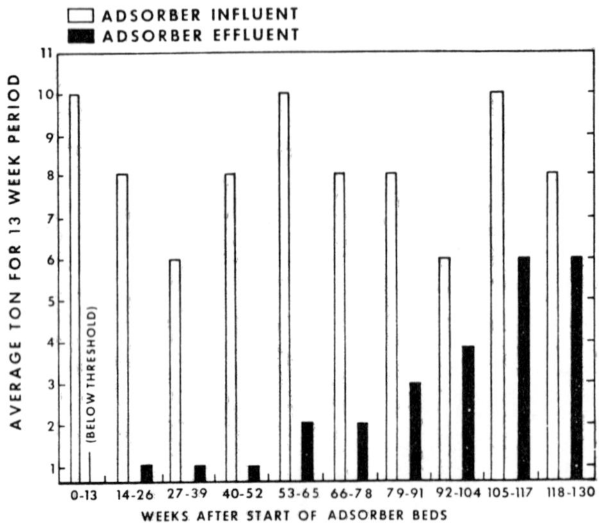

FIGURE 5.8: COMPARISON OF SAND vs GRANULAR ACTIVATED CARBON FOR TURBIDITY REMOVAL

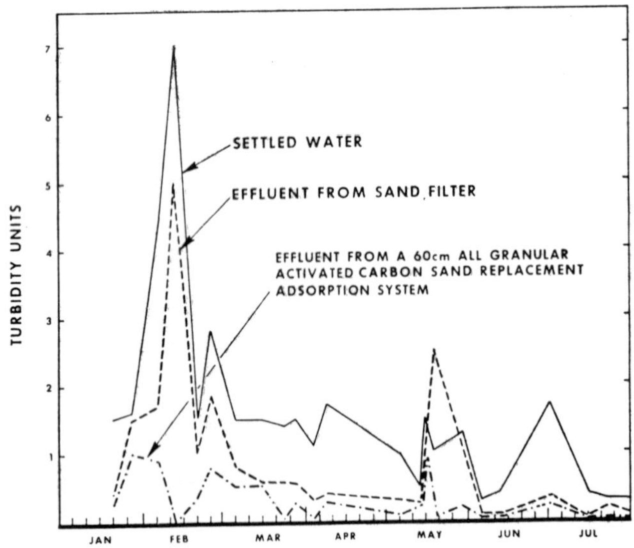

A six-month comparison of the turbidity in the effluent of a conventional sand filter and an all granular activated carbon system (see Figure 5.8) showed that their performance was equal. Pilot plant scale comparisons by the Water Supply Research Division Laboratory have confirmed this finding (5). When the entire plant was converted to granular activated carbon, no sand was left in the filter boxes.

In the survey cited above (2) the water purveyors were asked how long their current charge of granular activated carbon had been in service and whether or not the beds were still effective for taste and odor removal. The shortest effective life reported was 23 months and some beds had been in service for four years and were still effective. Reports from the United Kingdom (6) substantiate these findings from United States practice.

Removal of Organic Compounds as Measured by General Organic Parameters

The previous section detailed the performance of granular activated carbon beds for the removal of taste and odor causing compounds, but investigators have also been interested in the performance of the granular activated carbon adsorption process with respect to the removal of analytic parameters that would be reflective of the total organic content of water. As pointed out by Stevens and Symons (7), no analytic test currently available measures the total concentration of organic compounds in water.

Several tests have been proposed to approximate or be proportional to this parameter. A few of these are: (1) the organics-carbon adsorbable test producing a carbon chloroform extract (CCE), (2) chemical oxygen demand (COD), and (3) total organic carbon (TOC). These tests have been used to monitor various operating granular activated carbon adsorption beds.

In 1965, Dostal, Pierson, Hager and Robeck (3) reported that the percent removal of chemical oxygen demand declined from 72% at the start of the test to 56% after 31 days at the Nitro, West Virginia, water treatment plant, mentioned previously, while 97.5% of the odor was still being removed after 31 days. More recently, monitoring of operating granular activated carbon beds has been carried out at five field locations: Nitro, West Virginia; Piqua, Ohio; Mt. Clemens, Michigan; Lawrence, Massachusetts; and Davenport, Iowa.

This monitoring began in September 1968 and spanned several years. Because analytic capabilities have changed and improved gradually, the same monitoring techniques were not used at all of these installations. In general, however, the purpose of the monitoring was to compare the time of breakthrough (first detectable increase) of organic compounds as measured by a general organic parameter with the time of odor breakthrough.

In every case, the breakthrough of odor occurred much later than the breakthrough of the parameter measuring the removal of general organic compounds. Except for the Nitro, West Virginia situation, the control of odor was successful for years, whereas control of organic compounds as measured by a general organic parameter was successful only for weeks. Figure 5.9 from Lawrence, Massachusetts, is a typical data plot of the breakthrough curve for a general organic parameter.

FIGURE 5.9: CCE-m REMOVAL AT LAWRENCE, MASS. BY SAND REPLACEMENT TYPE GRANULATED ACTIVATED CARBON ADSORPTION UNIT

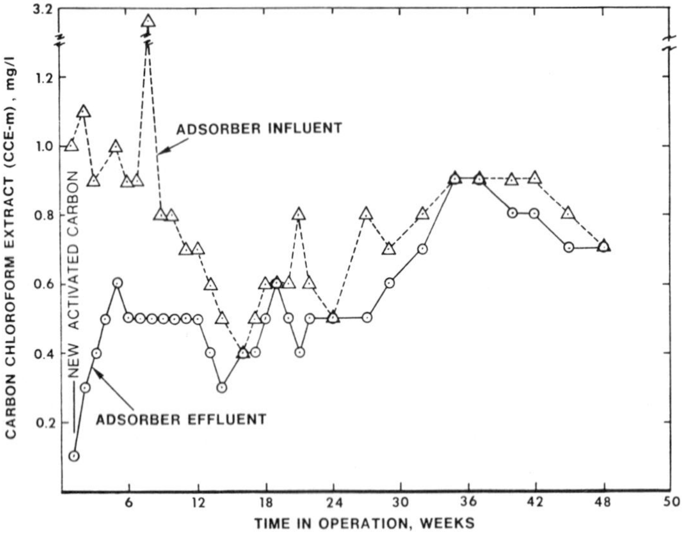

Figure 5.10 from Ford (6) indicates that experiences in the United Kingdom are similar to those in the United States. The experience in Germany is similar to that in the United Kingdom as shown by Heymann (9) of Duisburg who investigated the breakthrough pattern of the parameter dissolved organic carbon (DOC) through a granular activated carbon bed with sampling points at various depths.

FIGURE 5.10: REMOVAL OF DISSOLVED ORGANIC CARBON BY GRANULAR ACTIVATED CARBON

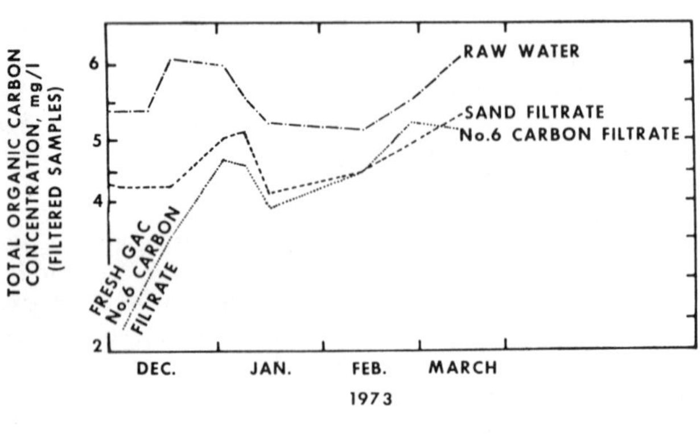

The maximum depth, corresponding to an empty bed contact time of four minutes, showed some increase in DOC after about four to five days. These reports lead to the generalized conclusion that if controlling the general organic content of drinking water is desirable, the useful life of a given charge of granular activated carbon in an adsorption bed will be much shorter than might be anticipated from performance based on taste and odor compound control.

Removal of Specific Organic Compounds

Early demonstrations of granular activated carbon's ability to remove specific organic compounds from water came from its use in an analytic procedure in which organics were adsorbed on granular activated carbon and desorbed (extracted) with a solvent, chloroform or ethyl alcohol. Analysis of these extracts revealed what organic compounds that were in the original sample were able to be adsorbed onto granular activated carbon under prescribed conditions, and then be desorbed with a solvent.

For example, in 1956, Middleton and Rosen (10) found the following organic compounds or classes of compounds in a carbon chloroform extract (CCE) from a surface water: substituted benzene compounds, polycyclic hydrocarbons, acrylonitrile, aldehydes, organic acids, DDT, kerosene, phenyl ether, alcohols, ketones, and esters.

In the same year, Rosen, Middleton and Taylor (11) isolated alkyl benzene sulfonate from a carbon-alcohol (ethyl) extract (CAE). In 1963 Rosen, Skeel and Ettinger (12) isolated the following organic compounds from a river water CCE: naphthalene, styrene, ethylbenzene, 2-ethylhexanol, diisobutyl carbinol, 2-methyl-5-ethylpyridine, tetralin, acetophenone, bis(2-chloroisopropyl) ether, bis(2-chloroethyl) ether, and phenyl methyl carbinol.

In 1970, Rosen, Mashni and Safferman (13) reported on finding both geosmin and 2-methyl isoborneol in a CCE from an Ohio Lake. In 1972 a report was published (14) concerning studies on organic contamination of drinking water conducted in the lower Mississippi River area. Organic compounds were collected on granular activated carbon and desorbed by both heat and solvents. One sample contained the following compounds: acetylene dichloride, benzene, carbon tetrachloride, chloroform, 1,2-dichloroethane, dimethyl sulfoxide, ethylbenzene, methyl chloride, propylbenzene, toluene, and vinylbenzene.

A second series of samples contained the following organic materials:

acetone
acetophenone
benzene
bromobenzene
bromochlorobenzene
bromoform
bromophenyl phenyl ether
  (positional isomer?)
butylbenzene
α-camphanone
chlorobenzene
chloroethyl ether
chloromethyl ether
chloroform
chloronitrobenzene
chloropyridine
dibromobenzene
dichlorobenzene
  (positional isomer?)
1,2-dichloroethane
dichloroethyl ether
dimethoxybenzene
2,6-dinitrotoluene
endo-2-camphanol
ethylbenzene
exo-2-camphanol

hexachlorobenzene
isocyanic acid
methylbiphenyl
methyl chloride
nitrobenzene
o-methoxyphenol
p-menth-1-en-8-ol
tetrachloroethylene
toluene
1,1,2-trichloroethane
vinylbenzene

Also in 1972 Kloepfer and Fairless (15) found the following compounds in CCEs from drinking water taken from the Ohio River at Evansville, Indiana, using gas chromatographic-mass spectrometric techniques.

bromodichloromethane
dibromochloromethane
bromoform
ethylbenzene
styrene
hexachloroethane
chlorohydroxybenzophenone
toluene
tetrachloroethylene
xylene
bis(2-chloroisopropyl) ether
bis(2-chloroethyl) ether
hexachlorobenzene

Finally, Symons and Stevens (7) reported on organic compound identification from an ethyl alcohol extract of granular activated carbon (CAE) that included the following:

trichlorobiphenyl
tetrachlorobiphenyl
pentachlorobiphenyl
hexadecane
octadecane
eicosane
docosane
diethylphthalate (plasticizers)
di(2-ethylhexyl) phthalate (plasticizers)
methyl ester of benzoic acid (probably esterified during extraction)
methyl ester of benzenesulfonic acid
methyl ester of palmitic acid (probably esterified during extraction)
methyl ester of stearic acid (probably esterified during extraction)
methyl ester of lignoceric acid (probably esterified during extraction)

These reports show some of the types of organic compounds that can be adsorbed from water by granular activated carbon as evaluated by this technique.

Below are typical references reporting on the removal of specific organic compounds in conjunction with other organics by granular activated carbon beds in use at operating water treatment plants. In 1965, Robeck, Dostal, Cohen and Kreissl (16) demonstrated that coal-base granular activated carbon, partially exhausted for CCE-hf removal, could reduce the concentration of endrin, dieldrin, lindane, 2,4,5-T ester, DDT, and parathion dosed into river water.

In the same year, Dostal, Pierson, Hager and Robeck (3) showed that the seven compounds listed below that were present in the Kanawha River water after aeration could be reduced to below detectable concentrations by fresh (2-day old) granular activated carbon beds. These compounds were bis(2-chloroethyl) ether, 2-ethylhexanol, bis(2-chloroisopropyl) ether, α-methylbenzyl alcohol, acetophenone, isophorone and tetralin.

Forty days later, however, all of these compounds with the exception of acetophenone were detected at a bed depth equal to an empty bed contact time of about eight minutes. Providing an additional two minutes of empty bed contact time did remove these seven compounds at this time (40 days), although another organic compound, ethylbenzene, was detected at a bed depth equal to 15 minutes of empty bed contact time.

Steiglitz, et al (17) reported on removals by two-month-old granular activated carbon beds of 61 organic compounds amenable to analyses by the Grob closed loop stripping and gas chromatographic procedures (18). The granular activated carbon appeared to have lost its effectiveness for adsorbing compounds eluting from the chromatograph early, such as chloroform and trichloroethylene, but was still quite effective for later eluting chlorinated aromatics. Some questions remain as to documentation of analytical recovery efficiencies and the operation of the adsorption beds, however.

The authors claim that some higher concentrations of a few aliphatic and aromatic hydrocarbons in the adsorber effluent can be explained by chromatographic or biological effects on the adsorber. Dissolved organic chlorine (DOCl) was still being reduced from 85 to 2 μg/l through the adsorber at this time. The sum, as organic chlorine, of the concentration of chloroform, carbon tetrachloride, trichloroethylene, and tetrachloroethylene in the treated water accounted for all the DOCl. To summarize, after two months of operation, all of the typical low molecular weight chlorinated compounds were breaking through the bed although DOCl was significantly removed from a chlorinated influent.

In 1965 Kolle, Sontheimer, and Stieglitz (19) reported on studies of pilot granular activated carbon adsorbers receiving ozonated prefiltered Rhine River water. Two or three adsorbers in series were used, each one meter in depth. Each meter of depth represented four minutes of empty bed contact time. After six months of operation, granular activated carbon samples were taken from the top of each section and from the bottom of the last section. These activated carbon samples were extracted with dioxane and the extracts analyzed for specific organics. The following organics were found in various concentrations:

chloroform	bis(2-chloroisopropyl) ether
1,2-dichloroethane	o-dichlorobenzene
1,2-dichloropropane	hexachlorobutadiene
tetrachloroethylene	hexachlorocyclohexane
trichloroethylene	tris(2-chloroethyl) phosphate

According to the authors, the substances identified in these extracts can be classified into three groups—the aromatic chlorohydrocarbons (o-dichlorobenzene), that were completely adsorbed by the uppermost layer of the adsorber, the aliphatic chlorohydrocarbons (hexachlorobutadiene and hexachlorocyclohexane) that were much less strongly adsorbed, but still are adsorbed well enough so that the lowest filter layers do not contain these substances, and the oxygen-containing organic chlorine compounds, bis(2-chloroisopropyl) ether and tris(2-chloroethyl) phosphate, that were breaking through the activated carbon beds and were identified in the bottom layer of the filter.

In addition to the above information, chloroform, 1,2-dichloroethane, and 1,2-dichloropropane were present at various levels throughout the adsorbers, although tri- and tetrachloroethylene were confined to the upper and middle layers of the three-layered adsorber. The presence of these first three compounds throughout the depth of the adsorber would tend to weaken the authors' gross classification of the adsorption of aliphatic hydrocarbons (see above).

In 1977 Suffet, et al (20) reported on the performance of granular activated carbon and some adsorbent resins for the removal of trace organics from Philadelphia drinking water. Suffet used computer-reconstructed gas chromatographic profiles plus mass spectrometric identifications to assess the ability of the adsorbents to remove 27 identified organic compounds.

For the activated carbon column (Calgon F-400, 9.7 minute empty bed contact time) in one experiment, the adsorbent was shown to be quite effective for removal of most of the compounds identified although exhaustion was noted for the organic compounds with lower boiling points by the 18th week of the run. Gas chromatographic profile analysis of the F-400 column effluent indicates chloroform and trichloroacetone first broke through after three weeks and dibromochloromethane and tetrachloroethane first broke through after four weeks.

In a second experiment, breakthrough patterns were presented for one, two, seven, nine and fifteen weeks of the run plus the respective organic profiles for the influent water. Twenty-nine compounds were identified from the gas chromatographic profiles for this experiment. Empty bed contact time in this experiment was 7.3 to 7.5 minutes. Again, detection in the effluent of organic species with lower boiling points occurred sooner than organic compounds with higher boiling points. Suffet (20) cautions the reader, however, that the data are largely qualitative and interpretation of results is complicated by the highly variable nature of the organic content of the influent to the adsorbent column.

In 1977 McCarty, et al (21) reported on the performance of "Water Factory 21" for removal of organic materials. Water Factory 21 is an advanced waste treatment facility designed to reclaim wastewater to provide injection water needed for a seawater barrier system to protect ground waters in Orange County, California. Part of the treatment train includes packed-bed, upflow pressure adsorbers filled with Calgon Filtrasorb 300. The empty bed contact time is 30 minutes.

During a period when the plant was operating on a continuous basis, single activated carbon adsorber influent and effluent samples were taken and subjected to a rather rigorous organic analysis by closed loop stripping (18). Relative influent and effluent concentrations were reported for sixteen compounds and absolute concentrations for twelve of these. The general trend was toward removal of these compounds to widely varying degrees.

An examination of the gas chromatograms verifies this trend for a large number of unidentified compounds (see section on "Unidentified Compounds," page 192) for results of a similar comparison). The meaning of the results presented by McCarty, et al, however, are difficult to interpret in the context of predicting activated carbon adsorber life for removal of the specific compounds identified, because few data are presented on the condition of the activated carbon itself (time in-place, reactivation frequency, and so forth). Measurement of influent-effluent organic concentration profiles vs time is part of the planned future Water Factory 21 work, however.

Most of the information on reducing various concentrations of trace organics has been gathered through laboratory studies and pilot-scale experiments. The National Interim Primary Drinking Water Regulations (22) established maximum contaminant levels for six organic chemicals: endrin, lindane, methoxychlor, toxaphene, 2,4-D and 2,4,5-TP (Silvex). These six specific organic contaminants can be grouped under the general term pesticides. The "Manual of Treatment Techniques for Meeting the Interim Primary Drinking Water Regulations" (23) reported that adsorption on granular activated carbon is the most effective treatment process for reducing the concentrations of these contaminants.

The U.S. EPA library in Cincinnati made a computer search of the literature on the subject of adsorption of organic contaminants on granular activated carbon. Listed below in Table 5.2 are 50 organic compounds in addition to those reported in the text that have been reported to be reduced in concentration through granular activated carbon treatment. Only those studies where the concentrations of the specific organics before treatment were below the 1 mg/l level were included, thus eliminating studies on industrial wastes where the concentrations are usually much higher. Even these concentrations are higher than usually found in source waters, making direct extrapolation or prediction of adsorption behavior speculative when low concentrations are present. These data are based on both isotherm and column type studies.

TABLE 5.2: ADDITIONAL ORGANIC COMPOUNDS THAT HAVE BEEN REPORTED IN THE LITERATURE TO BE ADSORBABLE ON GRANULAR ACTIVATED CARBON

acetophenone (29)	MS-222 (32)
aldrin (39)(43)(49)	malathion (43)(49)
Baygon (36)	methoxychlor (49)(95)
α-BHC (43)(49)(50)(54)	nitrobenzene (38)
benzocaine (32)	oil (fuel) (26)(35)(37)(45)
benzoic acid (29)	paraquat (30)(51)(52)
butyric acid (29)	parathion (50)
dibrom (32)	phenylacetic acid (29)
dieldrin (25)(39)(43)(49)	phenols (33)(34)(40)(42)(45)
di(n-butyl) phthalate (48)	p-nitrophenol (42)
di(2-ethylhexyl) phthalate (48)	propionic acid (29)
diuron (50)	pyridine (29)
diquat (30)(51)	PCB (39)(41)(49)
dimenthoate (49)(50)	rotenone (24)(32)
m-dinitrobenzene (38)	Sevin (36)
DDT (27)(39)(43)(46)(49)	simazin (28)(31)(50)
endosulfan (49)(52)	strychnine (53)
endrin (50)(53)	3-trifluoromethyl-4-nitrophenol (TFM) (32)
gasoline (26)	2,4-D (44)(50)
heptachlor (49)	2,4-dinitrophenol (44)
heptachlor epoxide (49)	2,4,5-T (ester) (43)(50)
hexachlorobenzene (49)	toxaphene (24)(50)
juglone (32)	tetrachlorobenzene (49)
lindane (27)(31)(49)(50)	Telodrin (49)
linuron (49)	triazine (28)

A report by Giusti, et al (55) includes a list of 12 alcohols, 8 aldehydes, 11 amines, 4 pyridines (and morpholines), 8 aromatics (benzene derivatives), 11 esters, 3 ethers, 14 glycols and glycol ethers, 2 halogenated hydrocarbon solvents, 10 ketones, 8 acids, and 2 oxides that were studied. Single dose studies (isotherms not determined) were carried out on each of the compounds with 100 ml of a 1000 mg/l solution being dosed with 0.5 gram of activated carbon (equivalent dose 5000 mg/l). Concentrations of solute dosed were less where solubilities so dictated. These data were used to draw conclusions and test hypotheses about effects of pH, polarity, functional groups, molecular weight, and other differences in physical and chemical characteristics, insofar as amenability to adsorption is concerned. These data might relate to relative adsorbability of the respective compounds at these very high concentrations, but do not evaluate competitive effects of mixtures or the relative effects of 5 to 6 order of magnitude lower concentrations of organic matter and adsorbent that are experienced in the drinking water treatment situation.

Isotherm tests were run for only five compounds at varying pH and although linear, the isotherms were determined only at high concentrations. Parallel column studies and multisolute isotherm studies were conducted on these compounds and the authors claim a fairly high level of predictability of the column capacities from the isotherm data. Whether this translates well to lower concentrations (where isotherms may become nonlinear) or to real systems where the solute species number in the hundreds, cannot be determined from these studies.

In summary, although the individual literature citations are often vague on critical details of the study, and data have been collected under a variety of circumstances, many atypical, in total they do demonstrate that granular activated carbon is correctly described as a broad-spectrum adsorbent.

CURRENT WATER SUPPLY RESEARCH DIVISION RESEARCH FINDINGS

In an attempt to provide additional information not currently in the literature on the performance of granular activated carbon as an adsorption medium, the Water Supply Research Division has an active in-house and extramural research program on this subject. The results of these experiments, many still on-going, are summarized below. Detailed papers on many of these projects will eventually be published in the technical press.

Class I Compounds (Taste and Odor Producing)

Snoeyink at the University of Illinois, working under EPA sponsorship, found that the odorous compounds 2-methylisoborneol (MIB) and geosmin are both strongly adsorbed by activated carbon (56)(57). When present, humic substances significantly reduce the capacity of activated carbon for adsorption of these compounds, more so before equilibrium is achieved than at equilibrium. Commercial humic acid (HA) and the humic substances from well water each had differing competitive effects on MIB. The capacity of activated carbon for geosmin adsorption was reduced by commercial HA to a greater extent than was observed for MIB. The performance of laboratory columns was consistent with the isotherm results. Application of distilled water to a partially saturated activated carbon bed resulted in almost no elution of MIB indicating that it was strongly adsorbed.

Using the data collected in this study and assuming, (1) complete saturation of the activated carbon, (2) no desorption and (3) no biological activity, Snoeyink predicted the bed life for the reduction of MIB or geosmin from 10 μg/l to its threshold odor level of 0.1 μg/l in a seven to eight minute EBCT bed to be much greater (several months to years) than the predicted life for the reduction of humic substances from 5 to 1 mg/l (1 to 2 months). When both MIB, or geosmin, and humic substances must be removed, humic substance removal will control the life of the bed.

Chlorophenols are adsorbed very strongly by activated carbon at the μg/l level, which is near the threshold odor limit for these compounds. The extent of adsorption of 2,4-dichlorophenol (DCP) and 2,4,6-trichlorophenol (TCP) is a function of pH. The neutral species of these compounds predominate at pHs below the pK_a (pH at which the concentrations of the free acid and the acid anion are equal) values (7.85 and 6.00, respectively, at 25°C) and are adsorbed more strongly than the anionic species. As the number of chlorine atoms substituted on the phenol increases, the solubility of the neutral species decreases and the adsorbability increases. As substitution increases, the pK_a of the species is lowered, however.

When water containing phenol is chlorinated with low levels of chlorine, a mixture of chlorophenols will form and thus the extent of adsorption of one chlorophenol in the presence of another chlorophenol is an important consideration. Significant reductions in adsorption capacity (up to 50%) of one chlorophenol was caused by the presence of a second chlorophenol. Evaluation of the competitive effects of commercial HA, soil fulvic acid (FA) and leaf FA showed that the presence of these materials decreased the capacity of activated carbon for chlorophenol adsorption and that each of the materials competed somewhat differently.

Even in the presence of humic substances and another chlorophenol species, however, the adsorption capacity is even greater for chlorophenol than it is for MIB and the bed life for chlorophenol adsorption will be greater than for MIB and much greater than for humic substances.

In another study, a joint effort between the AWWA Research Foundation, 14 water utilities, the EPA Water Supply Research Division, and the Universities of Illinois, Iowa State and Missouri-Columbia, is under way to determine the removal of trace organics (particularly taste and odor compounds and haloforms) on granular activated carbon and polymeric adsorbents.

In this study, pilot scale columns containing seven different adsorbents are located at the Kansas City Water Treatment Plant, Kansas City, Missouri. The adsorbents include granular activated carbon made from bituminous coal, lignite, petroleum, and peat, and carbonaceous anion exchange resins. The adsorbers are presently arranged so that both a coal-base and a lignite-base granular activated carbon system can provide up to 33 minutes EBCT. The applied water is unstable because of precipitative lime softening and a lack of recarbonation, resulting in calcium carbonate deposition on the adsorbents. The effect of this will be evaluated.

Class II Compounds (Synthetic Source Water Contaminants)

Naphthalene: An experiment comparing the adsorption of naphthalene with that of background organic content of Cincinnati tap water on granular activated carbon beds was terminated during June 1975. During the 330-day span of the experiment, 18,500 liters of Cincinnati tap water spiked with an average concentration of 30 μg/l of naphthalene were passed through the granular activated carbon (NPOC) 50% removal front had penetrated the entire 27-inch length of the column by May 1975, the maximum penetration of the naphthalene 50% removal front was only about 6.5 inches (this was quite variable throughout the experiment). Therefore granular activated carbon columns can be expected to remove compounds of low polarity and solubility such as naphthalene for a much longer period of time than they can remove the diverse organic group represented by NPOC.

The variability of the depth of the naphthalene penetration into the adsorbent (1 to 6.5 inches) is something of a mystery. Suspected causes are (1) variable influent concentrations, (2) variable constitution of NPOC (competitive adsorption), (3) biological activity, (4) variable flow, (5) variable temperature, or (6) influence of backwashing.

Other Polynuclear Aromatic Hydrocarbons: Snoeyink, at the University of Illinois, reported (57) that limited experimentation with the polynuclear aromatic hydrocarbon (PAH) anthracene led to the conclusion that no significant association between its adsorption and that of humic substances occurred. Thus the possibility of PAH passage through granular activated carbon beds because of its association with the less adsorbable humic substances is not a cause for concern.

Carbon Tetrachloride: The adsorption of this contaminant on granular activated carbon beds was observed in the U.S. EPA Cincinnati Laboratory during the fall and winter of 1967-77 when carbon tetrachloride was in the tap water. At that time, granular activated carbon columns that had been in service for two months were being monitored for trihalomethane reduction. Table 5.3 shows the monthly average carbon tetrachloride concentrations in the tap water and corresponding monthly average concentrations in the effluent. In spite of being in service for two months, the granular activated carbon was effective for three to five months, probably because this contaminant was new to the adsorbent when it occurred during the third month. Low level desorption of the carbon tetrachloride continued for several months after the contaminant disappeared from the influent.

TABLE 5.3: ADSORPTION OF CARBON TETRACHLORIDE ON GRANULAR ACTIVATED CARBON

	Time in Service, months										
	1	2	3	4	5	6	7	8	9	10	11
Monthly average influent, µg/l	NF	NF	5	8	13	15	44	NF	NF	NF	NF
Monthly average effluent, µg/l	NF	NF	NF	NF	NF	<6*	14	16	13	6	6

Note: Filtrasorb 400 granular activated carbon
10 minute empty bed contact time
NF = None found
*Single value during month

During the first week of January 1977 a six-inch diameter by thirty-inch depth bed (11 minutes EBCT) of virgin Calgon Filtrasorb 400 granular activated carbon was placed in upflow operation treating Cincinnati tap water. Coincidently with the start of operation, the concentration of carbon tetrachloride in the tap water began to increase.

Figure 5.11 shows the performance of the adsorbent bed during the first 43 weeks of 1977. During the first seven weeks, the granular activated carbon effectively removed carbon tetrachloride during the period of extremely high influent concentrations. As the influent concentration returned to the limit of detection (0.1 μg/l), the adsorbent, acting under the influence of the adsorption-desorption equilibrium phenomenon, began to desorb carbon tetrachloride. This desorption, which was monitored approximately every six weeks, has continued for about nine months.

FIGURE 5.11: PERFORMANCE OF GRANULAR ACTIVATED CARBON BED FOR CARBON TETRACHLORIDE REMOVAL

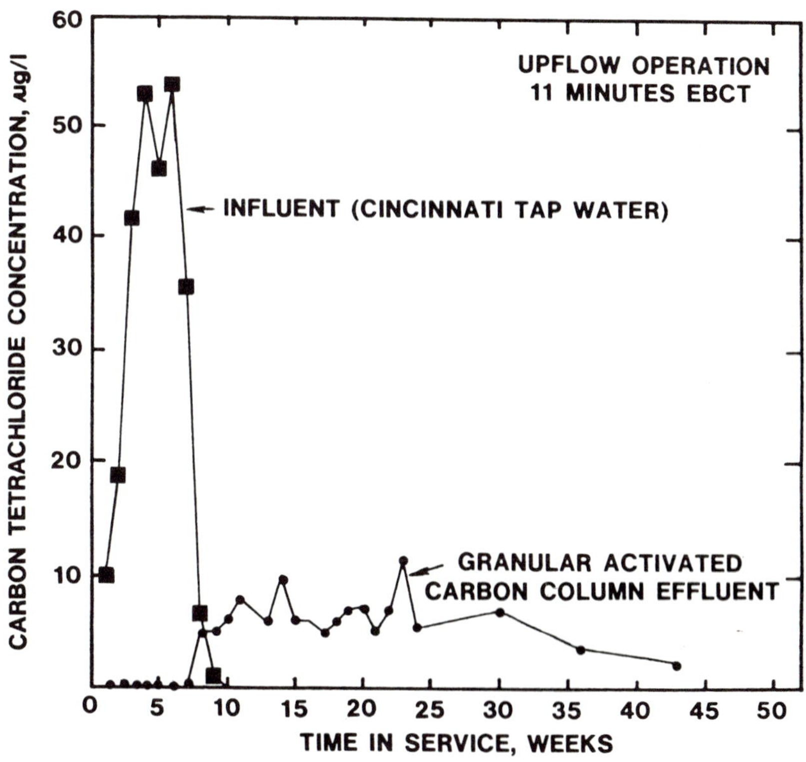

Figure 5.11 is important because it illustrates the protection a granular activated carbon adsorption barrier can afford the consumer during periods of sudden contamination of the raw water supply.

These data also demonstrate that granular activated carbon reactivation will probably be necessary fairly frequently to afford continuing consumer protection and guard against desorption of unwanted contaminants.

Trichloroethylene, Tetrachloroethylene and Similar Chlorinated Organics: Because several incidences of tri- and tetrachloroethylene contamination of drinking water supplies have been reported, the Water Supply Research Division is studying the effectiveness of granular activated carbon beds for removing these pollutants.

In most instances the water source is ground water and the contamination can be be linked to some industrial activity involving the present or past use of cleaning solvents in the aquifer recharge area. Columns containing granular activated carbon (10 minute EBCT) have been installed in three locations in New England and the first few weeks of data indicate that tri- and tetrachloroethylene are very well removed when the adsorbent is fresh (see Table 5.1).

As indicated in the footnote to Table 5.1, the contaminant, 1,1-dichloroethane, was not as well removed. At another location (not shown in the table) the tetrachloroethylene contamination has exceeded 2500 μg/l, yet the granular activated carbon bed produced an effluent with less than 0.5 μg/l tetrachloroethylene, for 15 weeks.

Studies with Humic Acid and Carbon Tetrachloride, Dieldrin, PCB, Benzene, and p-Dichlorobenzene: Weber, at the University of Michigan, is studying the adsorption of several specific organic compounds on granular activated carbon.

The main thrust of this work is to investigate the possibility of predicting activated carbon column adsorber performance on the basis of mathematical parameters obtained from simple batch kinetic and equilibrium (isotherm) studies.

Although the studies include only the use of model systems (humic acids as the competitive species), certain general conclusions about relative activated carbon adsorber operating life can be drawn on the basis of some of the data obtained to date from the modeling studies.

The results of the work with carbon tetrachloride support the in-house work reported above in that an adsorber life of several weeks to months can be expected. The work with dieldrin indicates that this compound is very strongly adsorbed, supporting evidence to that effect previously reported in the literature (23).

PCB mixtures resemble pesticides in properties and are also strongly adsorbed. Adsorber operating life for PCBs are likely to be on the order of months to one year. Adsorber operating periods for treatment of benzene and p-dichlorobenzene are likely to range somewhere between carbon tetrachloride and the PCB mixtures.

Mixtures from Actual Waters: In an attempt to further evaluate the capability of granular activated carbon beds to remove actual mixtures of raw water organic contaminants in competition with other organics, several studies are being conducted on actual waters at locations other than Cincinnati, Ohio.

One of these research projects, using bench scale pilot granular activated carbon columns is being performed with a ground water in Southern Florida. The granular activated carbon columns are one inch in diameter and contain 30 inches of activated carbon. The bed was loaded at 3 gpm/ft^2, which resulted in a 6.2 minute EBCT. The influent to the granular activated carbon column was untreated ground water for the first phase of the study and finished water for the second and third phases. The source has a color of approximately 50 color units, a TOC concentration of about 10 mg/l, and a pH of about 7.1.

The project includes routine analysis for the concentration of 19 organic compounds, total organic carbon, and trihalomethane formation potential, both before and after the granular activated carbon column. Data show that at these short contact times, some organic compounds did break through fairly early in the test, but that the chlorinated aromatic compounds were well removed. Longer empty bed contact time tests are now underway, as will be reported later in this section.

Another actual plant site research project is being performed using full-scale granular activated carbon beds located at a water treatment plant in the lower Mississippi Valley. In this project one bed is used as a post-filter adsorber and another as a sand replacement system in parallel operation. The post-filter adsorber is an existing conventional rapid sand filter that had the 30 inches of sand replaced by 30 inches of 12 x 40 mesh granular activated carbon.

The post-filter adsorber is in series after a rapid sand filter, so that the granular activated carbon received coagulated, settled, softened, and filtered water. The empty bed contact time was 27 minutes at the start of the test and 20 minutes at the end because of granular activated carbon loss, caused by inadvertent excessive backwashing.

The sand replacement system is an existing rapid sand filter that had the top 24 inches of the 30 inches of sand replaced by 24 inches of 12 x 40 granular activated carbon. This unit receives coagulated, settled and softened water directly from the precipitator of the full-scale plant. The empty bed contact time was 24 minutes at the start of the test and 18 minutes at the end because of granular activated carbon loss caused by inadvertent excessive backwashing.

The project includes routine analysis for the concentration of at least 35 organic substances and total organic carbon and trihalomethane formation potential both before and after the granular activated carbon beds. Table 5.4 shows the effluent concentration for each substance found, as contrasted to the influent concentrations listed at the bottom of the table for the duration of the study, 25+ weeks. For this report only the post-filter adsorber data will be shown as an example. Note, for this project, too, all data are preliminary in nature as the gas chromatograph-mass spectrometry confirmations had not yet been evaluated for consistency of identification at the time of writing (Fall 1977).

Within the precautions stated, Table 5.4 can be used to show the trends for removal by the granular activated carbon bed. The adsorber removed the ambient concentrations of toluene, benzene and trichloroethylene more efficiently than other substances consistently present during the first 10+ weeks of operation. The 1,2-dichloroethane was removed at an average of greater than 80% for about eight weeks, but the effluent concentration consistently exceeded 0.5 μg/l after six weeks. Desorption occurred from about the 16th week to the end of the test. Preliminary data show that trace concentrations (ng/l) of chlorinated hydrocarbon insecticides were generally reduced to below detectable concentrations throughout the 25+ week study. Finally, the sand replacement system performed similarly to the post-filter adsorber, but the removals were not as long-lasting.

TABLE 5.4: REMOVALS OF TENTATIVELY IDENTIFIED ORGANICS BY POST-FILTER GRANULAR ACTIVATED CARBON ADSORBER*

Time Weeks	Benzene Effluent Concentration μg/l	1,2-Dichloroethane Effluent Concentration μg/l	Trichloroethylene Effluent Concentration μg/l	Toluene Effluent Concentration μg/l
0	ND	0.4	ND	ND
1	0.1	0.2	ND	0.1
2+	ND	ND	ND	ND
3	ND	0.1	ND	ND
4	ND	ND	ND	ND
5	0.1	0.3	ND	0.1
6	ND	1.1	ND	ND
7	ND	0.7	ND	ND
8	ND	0.8	ND	ND
9	ND	1.3	0.4	ND
10+	ND	1.4	ND	ND
11+	0.1	1.0	ND	0.2
12+	ND	2.0	ND	ND
13+	ND	3.5	ND	ND
15+	ND	4.5	ND	ND
20+	ND	10.1	ND	ND
25+	ND	9.4	ND	ND
Avg. Influent Conc. μg/l	0.3	8.0	0.3	0.2
Range Influent Conc. μg/l	ND–2.1	1.2–23.7	ND–0.9	ND–0.9

ND = Not detected

*Empty bed contact time, 27 min. Start, 20 min. End

A third study on the performance of granular activated carbon beds is being conducted in full scale at a water treatment plant in the upper Ohio Valley. As this project is just starting, the data are preliminary, mainly because the identification of specific organics thus far is based only on gas chromatographic retention times without mass spectrometry confirmation.

The following generalizations are made after 52 days of virgin granular activated carbon being on-line. Compounds for which granular activated carbon adsorption may serve as a control mechanism and their influent concentration ranges are 1,1,1-trichloroethane or carbon tetrachloride or both (0.17 to 1.14 μg/l) (not resolved); 1,2-dichloropropane or trans-1,3-dichloropropylene or both (0.10 to 0.64 μg/l) (not resolved); trichloroethylene (0.16 to 0.96 μg/l); cis-1,3-dichloropropylene or 1,1,2-trichloroethane or both (0.15 to 24.0 μg/l) (not resolved); dichloroiodomethane (0.10 to 1.22 μg/l) and chlorobenzene (0.24 to 3.36 μg/l). Although present occasionally, these compounds were not consistently present in the influent to the adsorber, therefore removals could not be calculated.

Unidentified Compounds: After concentration, gas chromatography can be used to separate many organic compounds, producing a gas chromatogram in which the separated organic compounds are represented by peaks on a chart. Although unidentified, the absence of certain peaks after a given type of treatment gives an indication of the success of the treatment. In 1972, influent and effluent carbon chloroform extracts (8) from an operating granular activated carbon sand replacement system in Lawrence, Massachusetts were compared by gas chromatography. Figure 5.12 shows the reduction of many of the organic peaks when the granular activated carbon was fresh, but after 16 weeks of operation the influent and effluent gas chromatograms were similar.

FIGURE 5.12: COMPARISON OF GAS CHROMATOGRAMS OF CCE-m's COLLECTED DURING FIRST WEEK OF OPERATION AT LAWRENCE, MASS.

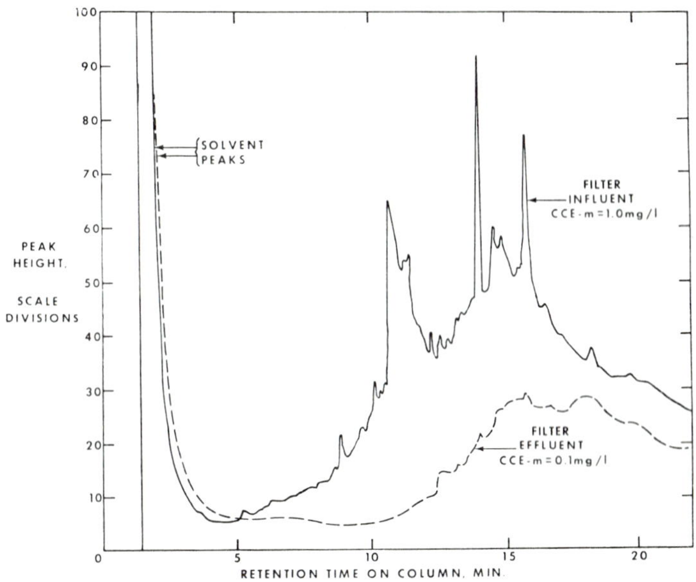

Recently, in the WSRD laboratories this approach was repeated using improved analytic procedures. Weekly one gallon samples of influent and effluent from a virgin granular activated carbon bed (about nine minute EBCT), receiving coagulated, settled, and dual media filtered Ohio River water, were collected and extracted with one 250 ml and two 100 ml portions of redistilled methylene chloride. After concentration of the extract, 2 μg of anthracene in methylene chloride was added as an internal standard.

In-house analyses are being conducted on a gas chromatograph employing a 30 meter SP-2100 wall coated glass capillary column. Injections are made at 20°C, and after five minutes the oven temperature is programmed at 2°/min to 240°C. Detection limits for anthracene (internal standard) using a flame ionization detector is approximately 0.4 ng or approximately 20 ng/l for similarly responding compounds from the original 4 liter sample. Preliminary results indicate that after one week almost all of the compounds present in the influent were not detectable in the effluent.

Further, no major new peaks occurred, indicating that organics detectable by this procedure were not leaching off the granular activated carbon. Analysis of later samples showed: (1) that the influent quality was very variable and (2) that although some peaks began occurring in the effluent it always contained fewer peaks than the influent, even after three months. These studies are continuing. Note that, because the influent to the pilot plant (Ohio River water) is stored in open tanks for several days prior to treatment, relatively volatile chlorinated organics were not present in the influent to the granular activated carbon bed.

Summary: Taken in total, these results, which were collected under conditions representative of those found in drinking water, support the concept of granular activated carbon adsorption being a broad-spectrum organic removal unit process. These data do, however, indicate the variation adsorbability of different organic compounds and show the importance of site specific evaluations of granular activated carbon performance.

Class III Compounds (Disinfection By-Product Precursors)

Because no direct measurement exists for Class III organics (for example, trihalomethane precursors), the degree of precursor removal can be judged by comparing trihalomethane concentrations upon chlorination of an untreated control water (called the terminal trihalomethane concentration) (58) to similar data collected on a treated water after similar chlorination and storage.

For example, if the effluent from a sand filter that was chlorinated and stored for two days yielded 50 μg/l chloroform and the same effluent passed through an adsorbent, then chlorinated and stored under similar conditions produced 25 μg/l chloroform, the adsorbent would be 50% effective in removing chloroform formation potential. This example assumes that no trihalomethanes were present in the filter effluents, that is the terminal trihalomethane concentration and trihalomethane formation potential are equal.

Using Ohio River water as the source, Love et al (59) observed that the relative effectiveness of granular activated carbon adsorbers to prevent the formation of trihalomethanes was highest for chloroform and lowest for dibromochloromethane. A granular activated carbon bed receiving coagulated, settled, but undisinfected water was initially effective for preventing trihalomethane formation upon subsequent chlorination.

Similar pilot plant studies have been carried out by Sylvia at the Lawrence Experiment station in Lawrence, Massachusetts under a WSRD research contract. The Merrimac River receives considerable industrial contamination upstream from the Lawrence study site; however, relative to the Ohio River, the Merrimac has a very low turbidity (2 to 4 NTU vs 10 to 75+ NTU, for the Ohio River at Cincinnati), and experimentation has shown that trihalomethane precursors in the Merrimac are only slightly reduced in concentration through coagulation and settling, yet are removed for long periods by granular activated carbon adsorption.

For example, three different types of granular activated carbons exposed to treated yet undisinfected Merrimac River water were found to reduce the trihalomethane formation potential (2 day) of 40 to 60 μg/l to less than 1 μg/l after six months of operation. Details on this will be contained in the final project report, which was due in early 1978.

Work by Snoeyink (57) at the University of Illinois on the adsorption of humic and fulvic acids, major trihalomethane precursors, relates to this phase of the overall problem of organic contamination. He found that activated carbon adsorbed humic substances in all cases that were studied, but the adsorption properties of the substances from different sources varied widely as did the extent to which they competed with selected trace organics for adsorption sites on activated carbon.

Humic substances from leaf and soil extract, a well water, and a commercial source were examined in detail. Extent of adsorption depended upon solubility, with the less soluble humic acid (HA) fraction being more adsorbable than the fulvic acid (FA) fraction from the same source. The lower molecular weight species from a given FA or HA fraction are more adsorbable than the high molecular weight species, presumably because more surface area is accessible to them.

The adsorption characteristics of the humic substances are also dependent on the method of analysis used to quantify them. The species that fluoresce the most were found to be the lower molecular weight species and these adsorb best. Ultraviolet absorbing species did not adsorb as well as those that fluoresce. Solution pH and phosphate concentration also had a marked effect on adsorbability of the humic materials, with adsorption generally improving with decreasing pH and increasing phosphate concentration.

The trihalomethane formation potential of the humic substances varied widely from source to source, with only one exception, but no dependence on molecular weight was found for fractions of FA or HA. This work reemphasizes the need for on-site pilot studies to determine adsorbability for that particular location.

The removal of disinfection (chlorination) by-product precursors was also studied at the three projects operating at the actual water treatment plants cited above. The same variability in the treatability of trihalomethane precursors as noted in the previous studies was shown.

Class IV Compounds (Disinfection By-Products)

To determine the effectiveness of granular activated carbon in removing trihalomethanes, Cincinnati tap water, which contains these compounds, was exposed to granular activated carbon columns at various hydraulic loadings (different contact times). Chloroform was reduced 90% or more for about three weeks (10 min EBCT), then breakthrough was steady until the adsorber was exhausted at about the ninth or tenth week.

The trihalomethanes containing bromine were effectively reduced by the granular activated carbon for 30 or more weeks. Figure 5.13 shows the difference in effectiveness of six different types of granular activated carbon for removing all of the trihalomethanes summed together (59). In earlier field studies where the applied water and the effluent from granular activated carbon beds were sampled, the findings regarding trihalomethane reductions were very similar to the laboratory results.

FIGURE 5.13: REMOVAL OF TRIHALOMETHANES FROM CINCINNATI, OHIO TAP WATER USING SIX TYPES OF GRANULAR ACTIVATED CARBON (EMPTY BED CONTACT TIME = 10 MIN)

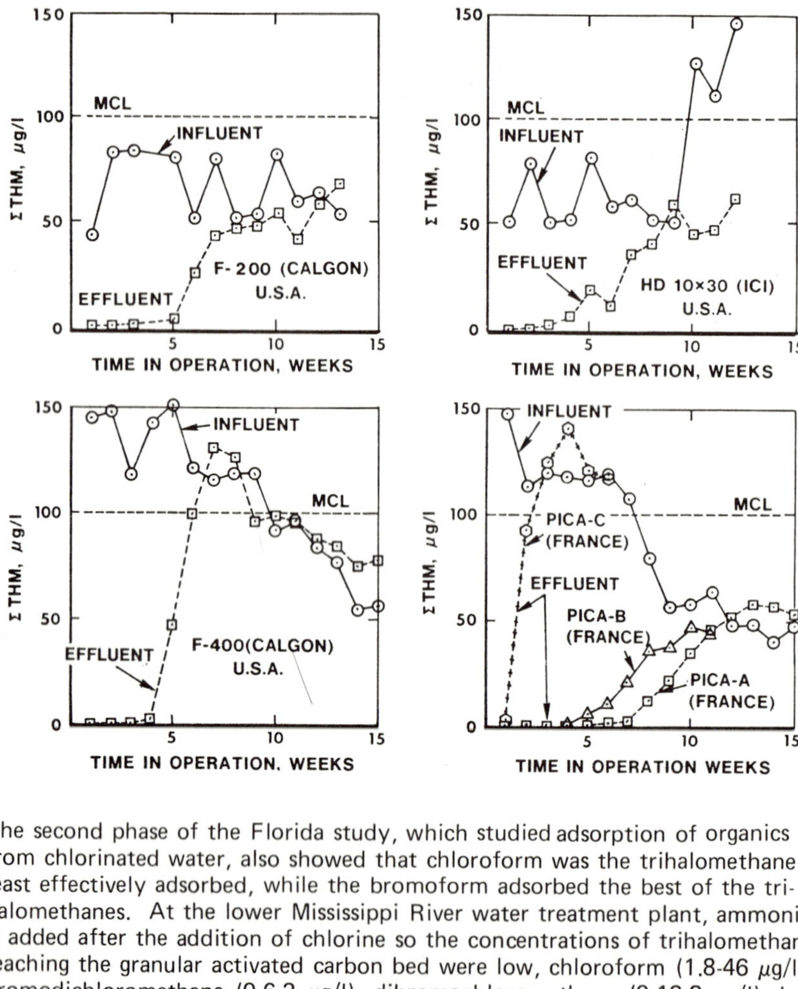

The second phase of the Florida study, which studied adsorption of organics from chlorinated water, also showed that chloroform was the trihalomethane least effectively adsorbed, while the bromoform adsorbed the best of the trihalomethanes. At the lower Mississippi River water treatment plant, ammonia is added after the addition of chlorine so the concentrations of trihalomethanes reaching the granular activated carbon bed were low, chloroform (1.8-46 µg/l), bromodichloromethane (0-6.2 µg/l), dibromochloromethane (0-12.8 µg/l), bromoform (0). Under these conditions the bromine-containing trihalomethanes were well removed, although chloroform began to appear consistently in the adsorber effluent at low concentrations after the fifth week. Adsorption-desorption cycles began after 13 weeks.

Figure 5.14, data from the water treatment plant in the upper Ohio Valley, shows a breakthrough pattern for the trihalomethanes very similar to the pilot plant data described above and shown in Figure 5.13. Here too, the bromine-containing trihalomethanes were removed better than chloroform.

FIGURE 5.14: REMOVAL OF TRIHALOMETHANES BY GRANULAR ACTIVATED CARBON BEDS–6.5 MIN EMPTY BED CONTACT TIME–VIRGIN WVW 14 x 40 GRANULAR ACTIVATED CARBON

Upper Ohio Valley Water Treatment Plant

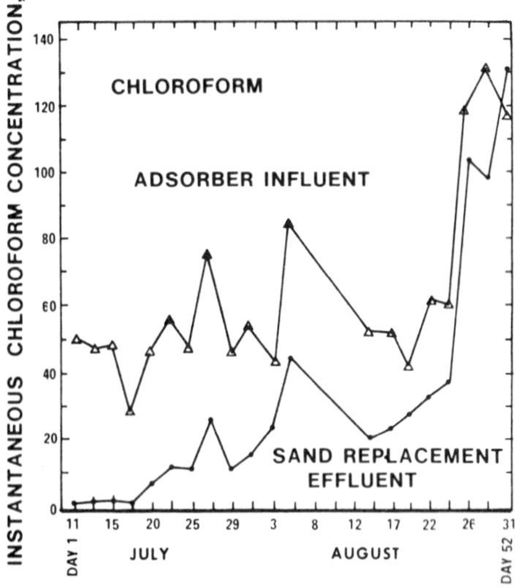

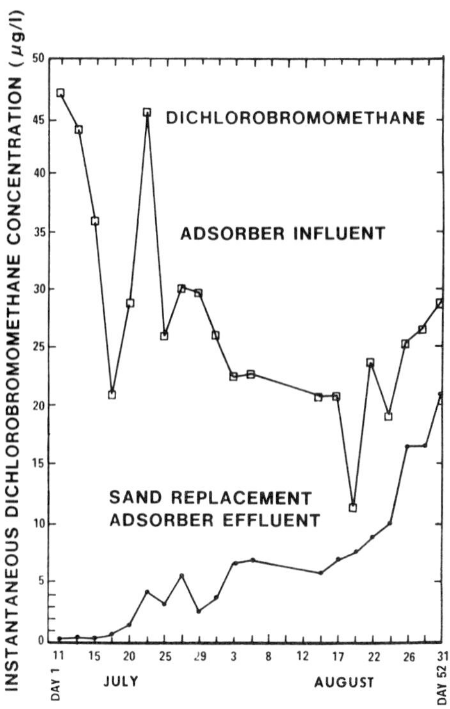

(continued)

FIGURE 5.14: (continued)

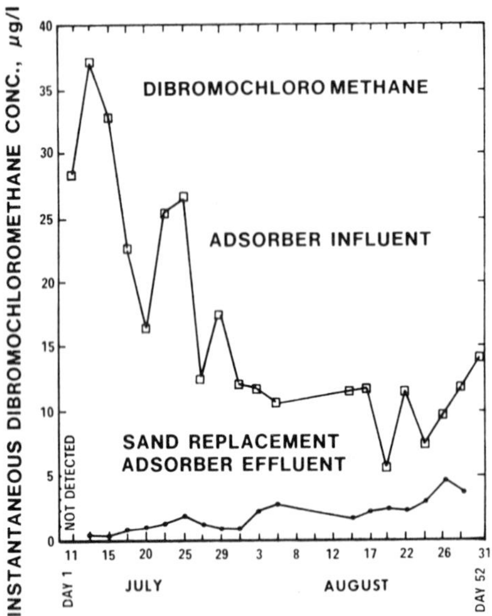

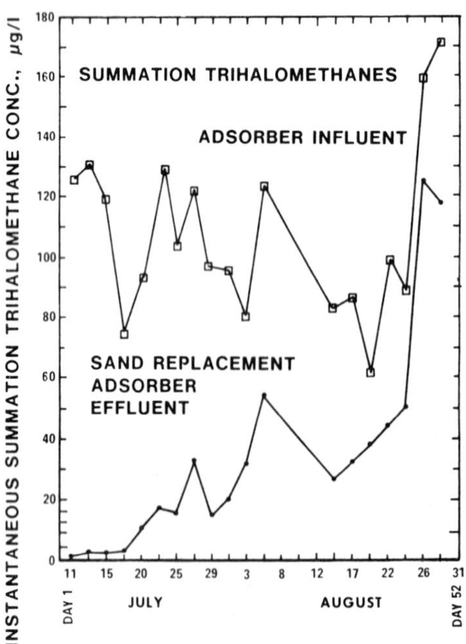

In summary, if the aromatics, taste and odor compounds, and certain pesticides can be categorized as strongly adsorbed onto granular activated carbon, then chloroform is located near the other end of the adsorption spectrum.

General Organic Parameters

Although not directly related to any single class of organic compounds discussed above, the use of an easily measured general parameter to monitor the performance of an adsorption treatment system is appealing from the standpoint of convenience and cost. The Water Supply Research Division pilot plant studies routinely include the following general organic measurements: nonpurgeable organic carbon (NPOC) (60), ultraviolet absorbance (UV) at 254 nm (60), and fluorescence–both the emission scan (EMS) (60) and the fixed wavelength procedure described by Silvia as the rapid fluorometric method (RFM) (61).

Figure 5.15 shows the pattern of each general organic parameter for the first 15 weeks after a fresh granular activated carbon adsorber was put into service. In an attempt to develop a simple test to predict the organic carbon content of effluents, UV, RFM and EMS data were correlated to NPOC. Table 5.5 summarizes the regression analysis. The most promising relationship thus far is between RFM and NPOC. A clearer understanding of these relationships and others, such as between some general parameter and trihalomethane (Class IV) or by-product precursor (Class III) concentrations is being sought through additional statistical analyses and should be available in the near future.

TABLE 5.5: RELATIONSHIP BETWEEN NPOC AND OTHER GENERAL ORGANIC PARAMETERS FOR MONITORING A GRANULAR ACTIVATED CARBON ADSORBER

	UV	RFM	EMS
(R^2) Coefficient of determination	0.22	0.67	0.52
(R) Correlation coefficient	0.47	0.82	0.72
Significant at 95% by F and t test	Yes	Yes	Yes
95% Confidence interval, mg/l	±0.73	±0.48	±0.58

Thus far, the best predictor of NPOC, RFM, has such a high 95% confidence interval (+0.48 mg/l) that its potential use is discouraging. Sylvia (61), however, obtained better results correlating RFM measurements with granular activated carbon bed performance as measured by CCE-m, so some possibilities for this approach may exist in certain waters.

As noted above, in addition to correlating a more easily measured general organic parameter against NPOC, the use of a general parameter to predict the performance of a granular activated carbon bed with respect to breakthrough of a specific organic compound or group of specific organic compounds is also an attractive concept. In the lower Mississippi River water treatment plant study, four general parameters have been correlated to total trihalomethanes, trihalomethane formation potential, and the sum of all of the nontrihalomethane chlorinated organic compounds in the adsorber effluent. Except for UV absorbance, the correlation coefficients have good statistical significance. The confidence limits on these data are, however, not yet available.

FIGURE 5.15: COMPARISON OF VARIOUS GENERAL ORGANIC PARAMETERS AS POSSIBILITIES FOR OPERATIONAL CONTROL OF GRANULAR ACTIVATED CARBON BEDS

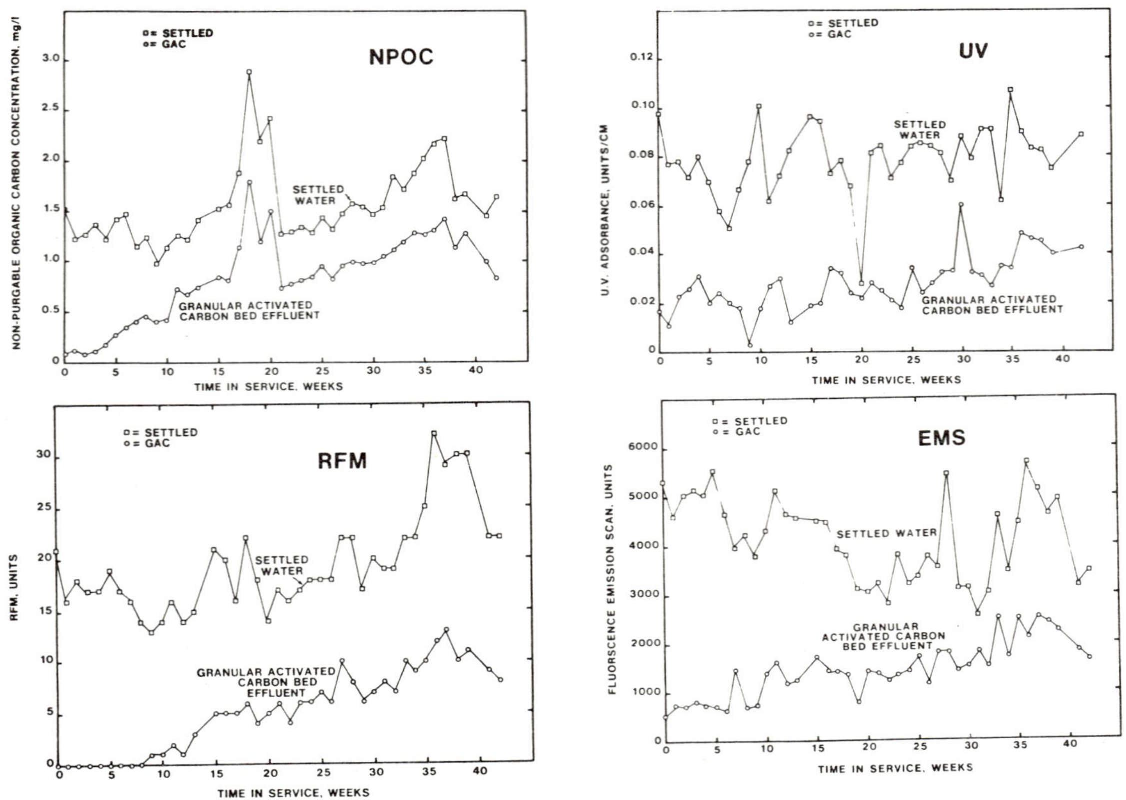

Influence of Empty Bed Contact Time: One phase of the project studying the use of adsorbents to remove organics from a southern Florida groundwater includes the effect of empty bed contact time on organic removal. Finished water from the treatment plant was diverted to four pilot granular activated carbon columns connected in series. Each column contained 30 inches of 12 x 40 mesh activated carbon and was operated at a hydraulic loading of 3 gpm/ft^2. The nominal empty bed contact time for each column was 6.2 minutes and thus the contact times were 6.2 minutes, 12.4 minutes, 18.6 minutes, and 24.8 minutes, respectively for the four columns. Twenty purgeable halogenated organic substances were monitored at the inlet and outlet of each column. Also TOC and Terminal THM concentration data were collected.

The ability to achieve a given quality of effluent for longer time periods (larger water volumes) was shown to be related to longer empty bed contact times (larger activated carbon volumes). In order to assess the period of time between activated carbon reactivations for a given empty bed contact time, three performance criteria (criterion No. 1 = low MW halogenated organic compounds excluding THMs; criteria No. 2 and No. 3 = TOC; discussed thoroughly in Chapter 7) were applied. Whenever a performance criterion was consistently exceeded, the first time at which the criterion was exceeded was used and the corresponding operating time and water volume treated calculated. The concentrations between data points were taken as linear.

Table 5.6 presenting the maximum duration of time in days of operation and liters treated when complying with each criterion shows that the TOC criterion is the most limiting of the three at all contact times for the conditions tested. TOC would have limited the operating time to 1.5 days, 4.5 days, 11.2 days and 18.6 days for the respective empty bed contact times.

TABLE 5.6: DURATION OF RUN ALLOWED BY EACH CRITERION

Empty Bed Contact Time	Criterion No. 1 (Vinyl Chloride)		Criterion No. 1 (cis-1,2-Dichloroethylene)		Criterion No. 2 (TOC)		. . . MCL	
minutes	days	liters	days	liters	days	liters	days	liters
6.2	3.5	313.8	17.5	1573.3	1.5	132.5	2.3	204.8
12.4	10.8	966.4	60.5	5437.4	4.5	402.3	9.3	832.7
18.6	35.7	3196.1	–	–	11.2	999.6	14.5	1297.1
24.8	35.3	3096.5	–	–	18.6	1653.7	19.5	1733.2

– = not yet exceeded criterion

Table 5.7 is a different display of the prior data in which the activated carbon volumes used and water volumes treated are recalculated using the 6.2 minute EBCT data as unity and presenting all data as ratios. Thus for the most stringent criterion, TOC, if two times the activated carbon volume is used, three times the amount of water can be treated before the criterion is exceeded. Further, again doubling the empty bed contact time increases the length of time of operation over six more times, a favorable improvement. Similar beneficial results occurred relative to criterion one and the MCL requirements with a doubling of the base empty bed contact time, but redoubling produced only a proportional increase in operating time according to the MCL requirement.

TABLE 5.7: RATIO OF VOLUME OF WATER TREATED IN COMPLIANCE WITH EACH CRITERION

Empty Bed Contact Time minutes	Ratio of Activated Carbon Required	Criterion No. 1 Specific Organics (VC)	(cis)	Criterion No. 2 TOC	MCL
6.2	1	1	1	1	1
12.4	2	3.1	3.5	3.0	4.1
18.6	3	10.2	–	7.5	6.3
24.8	4	9.9	–	12.5	8.5

VC = Vinyl chloride
cis = cis-1,2-Dichloroethylene
– = Not yet exceeded criterion

This evaluation of empty bed contact time shows that increased quantities of activated carbon, at fixed flow conditions, resulted in the ability to treat more water before exceeding a given criterion. In each case the first doubling of empty bed contact time resulted in adsorber operating times that were proportionately greater than the increase in empty bed contact times for all criteria. This has an important impact on the economics of treatment, see Chapter 8.

Two in-house studies on the influence of empty bed contact time on adsorber operating times, Figures 5.16 and 5.17, showed that for the two organic compounds, chloroform and carbon tetrachloride, the time to reach a given breakthrough point was approximately proportional to the empty bed contact time. Because of these different results in two different locations, this type of information should be collected in the location under study.

FIGURE 5.16: EFFECT OF EMPTY BED CONTACT TIME ON CHLOROFORM REMOVAL FROM TAP WATER

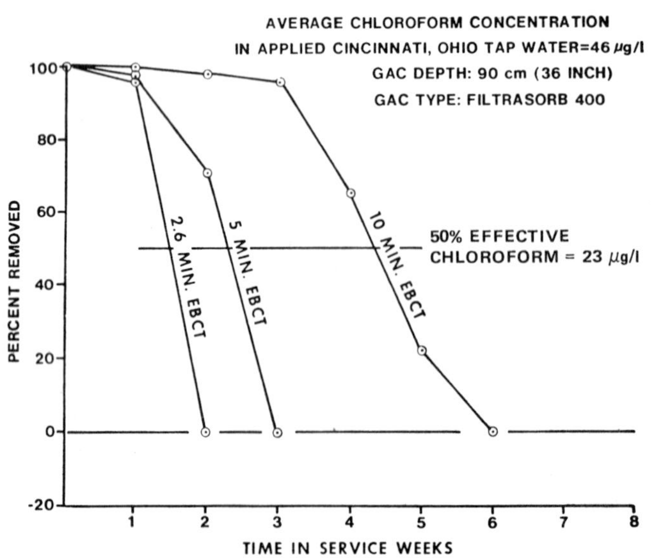

FIGURE 5.17: INFLUENCE OF EMPTY BED CONTACT TIME ON CARBON TETRACHLORIDE REMOVAL

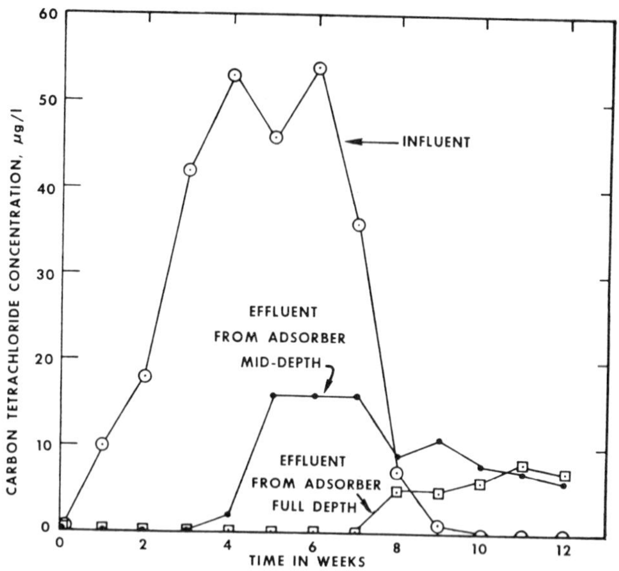

Biological Growth and Endotoxin Production

Standard Plate Count: Controlling bacterial populations (and particularly killing or inactivating pathogenic microorganisms) is a primary goal of water treatment. Some concern, therefore, has been expressed about the possibility of bacteria proliferating within granular activated carbon beds. To investigate this, the bacterial quality of the untreated and treated water from the WSRD pilot plant was routinely monitored over a four-month period by the staff of the WSRD Microbiological Treatment Branch using the standard plate count (SPC) test. Samples were also collected intermittently and analyzed for total and fecal coliforms; however, these indicator organisms seldom survived the coagulation and settling processes and were never detected in the filter or adsorber effluents.

The monthly average SPC (expressed as the geometric mean) for the pilot plant studies (see Table 5.8), in general, shows a two log (99%) reduction in the bacterial count through the treatment plant. Note that no disinfectant was added anywhere in the treatment process. Although an increase in bacterial populations was expected a priori, the SPC of the effluent from the granular activated carbon adsorber was slightly lower than the SPC in the companion sample taken from the dual media filter effluent. No attempt was made to measure attached growths. An attempt was made, however, to isolate and identify the predominant populations in the pilot plant. From the adsorber effluent, five or six different types of colonies could be recognized and two genera, Flavobacterium and Xanthomonas were identified from smear plates.

TABLE 5.8: MONTHLY MEAN (G_m) STANDARD PLATE COUNT*

Time in Operation, months	Raw	Settled	Effluent from Dual Media Filter	Effluent from Granular Activated Carbon Adsorber
1	19,600	1650	137	63
2	12,000	1000	270	72
3	7,170	790	80	29
4	6,680	700	100	37

Note: All bacterial counts are No./1 ml
*Pilot Plant Studies–Ohio River Water

Endotoxins: Endotoxins are lipopolysaccharide-protein complexes produced in the cell walls of Gram-negative bacteria. The lipopolysaccharide portion of the complex is pyrogenic and as little as 1 μg can produce fever in a 700 kg horse (62). Concern, therefore, has been expressed regarding the possible formation of endotoxins in granular activated carbon adsorbers because of bacteriological activity. For a six-month period in 1977, the U.S. EPA Health Effects Research Laboratory (HERL) monitored bacterial endotoxin concentrations in untreated and treated water from the WSRD organics removal pilot plant. These were companion samples with those collected for SPC analyses.

Using the Limulus lysate bioassay (63), HERL scientists observed a marked reduction in pyrogenic activity as a result of chemical coagulation (and settling) and a slight additional decrease through filtration by either dual media or granular activated carbon. The encouraging finding was that no increase in pyrogenic activity occurred in the effluent from the granular activated carbon bed. These data are shown in Table 5.9.

TABLE 5.9: MONTHLY MEAN ENDOTOXIN CONCENTRATIONS IN OHIO RIVER WATER*

Time in Operation, months	Endotoxin Concentration, μg/l			
	Raw	Coagulated/Settled	Dual Media Effluent	Granular Activated Carbon Adsorbent Effluent
2	158	16	16	9
3	236	63	7	6
4	205	36	41	11
5	500	66	16	15
6	45	20	5	4
7	35	11	11	11

*Pilot Plant Studies

OZONE ENHANCED GAC ADSORPTION

One organic removal unit process being used in some locations has not yet been discussed in this report. It has been receiving much attention since the fall of 1977; therefore, although much still needs to be known about the process, a discussion of a variation of granular activated carbon adsorption for organic contaminant control in which ozonation precedes the granular activated carbon treatment processes is included here. These two processes in combination are frequently called biological activated carbon.

As part of a research grant with the WSRD, Dr. Rip Rice, in collaboration with Public Technology, Inc., Washington, D.C. prepared a chapter of the final report on the "Status of Ozonation and Chlorine Dioxide Technologies for Treatment of Municipal Water Supplies" project summarizing what is known about biological activated carbon. The first few pages of that chapter, with some editing, are presented below.

Introduction

In a recent article that discusses the use of granular activated carbon in water treatment, McCreary and Snoeyink (64) state that "beds of granular activated carbon (GAC) are a convenient place for microorganisms to grow because bacteria attach themselves to the irregular external surfaces of the activated carbon particles and are very difficult to dislodge via backwashing procedures." In the presence of soluble carbonaceous matter, which serves as food for these organisms, and in the absence of oxygen, anaerobic bacteria can develop.

There are numerous instances in which sulfidic odors have been reported emanating from granular activated carbon columns used for the removal of dissolved organic materials contained in sewage treatment plant effluents (65)(66) and drinking water supplies (67). On the other hand, with sufficient dissolved oxygen and carbonaceous matter, the bacteria that develop in activated carbon beds will be aerobic. These do not produce sulfidic odors.

Many of the advantages of biological granular activated carbon (BAC) were first recognized by German water treatment scientists in the 1960s in drinking water plants along the Rhine River in the Düsseldorf area. Subsequently, BAC processes also have been installed in Swiss and French drinking water treatment plants and are subjects of active pilot studies in Holland and Belgium. In the United States, the U.S. Environmental Protection Agency's Water Supply Research Division in Cincinnati, Ohio has been testing a pilot BAC column since late in 1976.

Fundamental Principles

Granular activated carbon is made biologically active by the deliberate introduction of sufficient dissolved oxygen (DO) to aqueous streams just before they are passed through GAC columns. As long as the water contains sufficient DO to maintain aerobicity of the bacteria and sufficient dissolved carbon to provide food, the aerobic bacteria will thrive in this environment. Eberhardt (68) has likened bacterial activity in such an ideal environment to a "herd of cows grazing in a luscious meadow." Preozonation can convert larger, less biodegradable organic molecules into smaller, more biodegradable organics, for example, into acetic and oxalic acids. Sontheimer (69) has summarized the German findings to date which have led to the current theories of operation of BAC.

Although aerobic bacteria are necessary to obtain the benefits from BAC, so also is the adsorptive capacity of the GAC for the dissolved organic materials that will serve as food for these bacteria. This means that the surface area and pore volume of the activated carbon should be high. Stated another way, the organic materials present in solution should be adsorbable onto the activated carbon column, because the contact times of solutions with the activated carbon particles in the columns or beds are normally short (15 to 30 minutes). This does not necessarily give the bacteria sufficient time to degrade larger organic carbon molecules, ideally to carbon dioxide and water. Therefore, retaining the dissolved organic molecules in the activated carbon columns so that the bacteria then will have sufficient time to degrade them is important, even though the actual contact times involved are rather short.

Many organic materials are readily adsorbed onto GAC, but many others are not. For example, high molecular weight natural humic acids, so prevalent in drinking water supplies, are not readily adsorbed by activated carbon (70). If solutions of these nonsorbable organic materials are ozonized before passage through the GAC columns, they are converted to more readily biodegradable organic materials (71)(72). At the same time, ozonation introduces a large quantity of oxygen into the water, which promotes aerobic bacterial growth.

Advantages of Biological Activated Carbon

European pilot studies in drinking water treatment plants by many workers (68) (70)(73)(74)(75)(76)(77)(78)(79) have shown that ozonation followed by granular activated carbon adsorption results in:

More effective removal of dissolved organics from solution by the BAC;

Increased operating life of the activated carbon columns before having to be reactivated especially if the GAC can be kept free of halogenated organics;

Biological conversion of ammonia in the GAC columns;

Use of less ozone for removing a given amount of organics than using ozonation alone (BAC is more cost-effective over ozonation in removing dissolved organic carbon);

Filtrates from BAC columns in drinking water plants can be treated with small quantities (0.1 to 0.5 mg/l) of chlorine or chlorine dioxide, which produces drinking water of acceptable bacterial quality (zero coliforms) and provides a residual disinfectant for distribution systems.

Independent studies on physical-chemical treated sewage at the Cleveland Regional Sewer District (72), and in Israel (80) have confirmed these advantages with respect to removing organic materials.

European Background

Introduction of granular activated carbon into European drinking water treatment practices occurred after World War II. Its initial application was for dechlorination, then for taste and odor removal (81). Many surface waters containing ammonia undergo break-point chlorination at the beginning of the treatment process. This technique effectively removes ammonia, but produces considerable amounts of residual chlorine and chlorinated products in the water (79).

German water treatment objectives are to process surface waters to the same quality as that of natural groundwater (which does not have to be treated in many cases). Therefore, waters treated by break-point chlorination have to be dechlorinated before they are treated further or distributed (82). Schalekamp (74) points out that in Switzerland a residual chlorine concentration of only 0.05 mg/l is permitted in finished drinking water. Therefore, the raw water chlorine dose is removed by granular activated carbon prior to the addition of 0.05 mg/l of chlorine dioxide for final disinfection.

Combinations of ozone and granular activated carbon first were installed in the Düsseldorf area in the early 1960s (81) but nearly ten years passed until the biological activity in the activated carbon columns was recognized as being beneficial. By the early 1960s, the lower Rhine River water quality had declined, and advantage was taken of filtration of organic materials from the river water through the sand banks of the Rhine. Wells were dug into the river banks and water is drawn from these wells as the treatment plant raw water.

In the intervening years since the introduction of ozone/activated carbon at Düsseldorf, the beneficial effects of biological activity in the activated carbon columns have been recognized, characterized and optimized. After ozonation, the water is allowed to stand for 20 to 30 minutes to allow the more refractory organic compounds sufficient time to react with residual ozone. This retention time also allows residual ozone to be utilized, rather than simply to be destroyed when passed through the activated carbon column.

On the other hand, the Rhine River also contains considerable amounts of chlorinated organic materials that are not removed during river bank filtration. These halogenated materials also are more resistant to oxidation by ozone than are nonhalogenated organics and thus are less likely to be converted into easily biodegradable materials. In addition, halogenated organics are variably adsorbed onto activated carbon (19)(83).

Combining the variable adsorptivity of halogenated organic compounds on granular activated carbon with their lesser reactivity upon ozonation and their lower biodegradability means that breakthrough of halogenated organic compounds occurs sooner than does breakthrough of nonhalogenated organic compounds, even if the granular activated carbon columns are biologically activated. Thus German water works along the Rhine in the Düsseldorf area monitor their activated carbon column capacities for TOCl (84)(85)(86) as well as by DOC (87) and/or UV absorption. Activated carbon columns at three Düsseldorf plants along the Rhine (Flehe, Am Staad, Holthausen) are backwashed every four to six weeks and reactivated every six months.

When Düsseldorf activated carbons are thermally reactivated, however, only some 80% of the activated carbon charge is taken out of the columns. This leaves a portion of biologically active activated carbon in the column so that the level of bioactivity will not drop significantly when fresh or reactivated activated carbon is added. With fresh activated carbon columns, about 15 days of operation usually are required for biological activity to build up to an effective "steady state," particularly for ammonia removal (88).

WSRD Pilot Plant Results

In late 1976, WSRD began a pilot plant study on the use of ozone to extend the life of a granular activated carbon adsorber used as a sand replacement system as measured by Total Organic Carbon, and Trihalomethane Formation Potential (two day, 25°C) concentrations. This study was prompted by the successful results from pilot- and full-scale BAC tests conducted in Europe (principally West Germany) cited above.

In this test a granular activated carbon adsorber receiving settled ozonated water and a granular activated carbon adsorber receiving settled unozonated water were operated for a 10 month study period. Both units had about a nine minute empty bed contact time. Ozone dosages applied to the settled water (20 minute plug flow contact time) and dissolved ozone concentrations prior to filtration or adsorption are shown in Figure 5.18.

FIGURE 5.18: OZONE CONCENTRATIONS

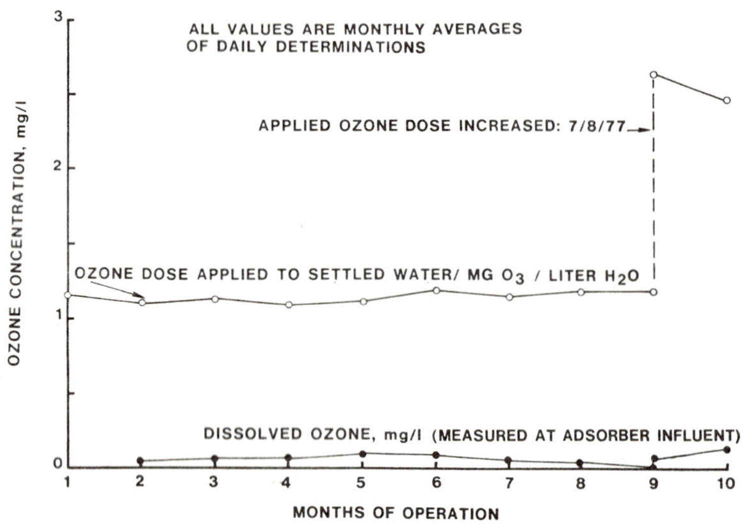

Because of an increase in ozone demand within the system and the resultant loss of ozone residual, the applied ozone dose was increased following ozone contactor cleaning early in the ninth month of operation.

Figure 5.19 compares the TOC concentrations in both effluents and illustrates the extension in bed life afforded by the combination of ozone followed by granular activated carbon. Operating time was extended from about three and three quarters months to about seven and one half months according to Criterion 2, increase of 0.5 mg/l of TOC concentration.

Figure 5.20 shows a similar trend for the removal of trihalomethane formation potential, although the MCL requirement was never exceeded in either effluent during this study.

The results of this study are encouraging enough to prompt additional in-house and extramural research on biological activated carbon, particularly to determine whether or not the use of ozonation prior to adsorption improves granular activated carbon performance with respect to Criterion 1, the control of low molecular weight halogenated organic compounds.

FIGURE 5.19: INFLUENCE OF OZONATION PRIOR TO ADSORPTION ON ORGANIC CARBON REMOVAL

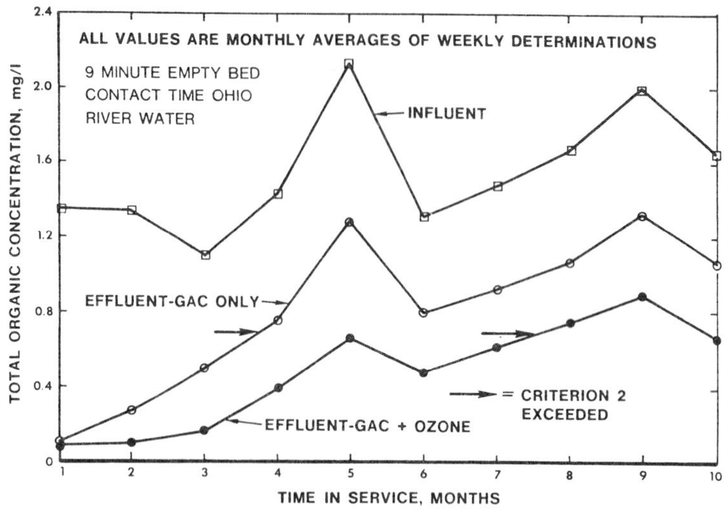

FIGURE 5.20: INFLUENCE OF OZONATION PRIOR TO ADSORPTION ON TRIHALOMETHANE FORMATION POTENTIAL REMOVAL

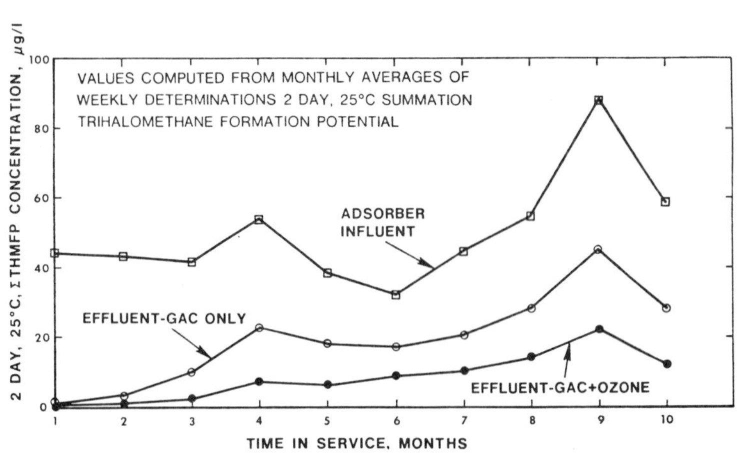

REFERENCES

(1) Woodward, R.L., Dostal, K.A. and Robeck, G.G., "Granular-Activated Carbon Beds for Odor Removal," *JAWWA,* 56, No. 3, 287-295 (May 1964).

(2) Symons, J.M., "Summary of Granular Activated Carbon Practice Data," Water Supply Research Division, U.S. Environmental Protection Agency, Cincinnati, Ohio (Feb. 17, 1976).

(3) Dostal, K.A., Pierson, R.C., Hager, D.G. and Robeck, G.G., "Carbon Bed Design Criteria Study at Nitro, West Virginia," *JAWWA,* 57, No. 5, 663-674 (May 1965).

(4) Love, O.T., Jr., Robeck, G.G., Symons, J.M. and Buelow, R.W., "Experience with Activated Carbon in the U.S.A.," in *Papers and Proceedings of a Water Research Association Conference at the University of Reading, United Kingdom,* Paper 13, 279-312 (February 1974).

(5) Love, O.T., Jr. and Symons, J.M., "Evaluating the Need for a Sand Layer in a Granular Activated Carbon Filter," Water Supply Research Division, U.S. Environmental Protection Agency, Cincinnati, Ohio (April 22, 1977).

(6) Ford, D.B., "The Use of Granular Activated Carbon Filtration for Taste and Odour Control," in *Papers and Proceedings of a Water Research Association Conference at the University of Reading, United Kingdom,* Paper 12, 263-278 (February 1974).

(7) Stevens, A.A. and Symons, J.M., "Measurement of Organics in Drinking Water," in *Proceedings of the AWWA Water Quality Technology Conference* (Dec. 1973), American Water Works Association, Denver, CO (1974).

(8) Buelow, R.W., Carswell, J.K. and Symons, J.M., "An Improved Method for Determining Organics by Activated Carbon Adsorption and Solvent Extraction," *JAWWA*, 65, No. 1, 57-72 and No. 3, 195-199 (Jan., March, 1973).

(9) Heymann, E., "Practical Experience in the Use of Flocculation Filtration Connected in Series to Granular Activated Carbon Filters," in *Translation of Reports on Special Problems of Water Technology,* Volume 9, *Adsorption,* Conference held in Karlsruhe, Federal Republic of Germany, 1975, 112-127, USEPA-600, 9-76-030, Cincinnati, Ohio (December 1976).

(10) Middleton, F.M. and Rosen, A.A., "Organic Contaminants Affecting the Quality of Water," *Public Health Reports,* 71, 1125-1133 (November 1956).

(11) Rosen, A.A., Middleton, F.M. and Taylor, N.W., "Identification of Anionic Synthetic Detergents in Foams and Surface Waters," *JAWWA,* 48, 1321-1330 (October 1956).

(12) Rosen, A.A., Skeel, R.T. and Ettinger, M.B., "Relationship of River Water Odor to Specific Organic Contaminants," *JWPCF* 35, No. 6, 777-782 (January 1963).

(13) Rosen, A.A., Mashni, C.I. and Safferman, R.S., "Recent Developments in the Chemistry of Odour in Water: The Cause of Earthy/Musty Odour," *Water Treatment and Examination,* 19, No. 1, 106-119 (1970).

(14) "Industrial Pollution of the Lower Mississippi River in Louisiana," Surveillance and Analysis Division, U.S. Environmental Protection Agency, Region VI, Dallas, Texas (April 1972).

(15) Kloepfer, R.D. and Fairless, B.J., "Characterization of Organic Components in a Municipal Water Supply," *Environmental Science and Technology,* 6, 1036-1037 (Nov. 1972).

(16) Robeck, G.G., Dostal, K.A., Cohen, J. and Kreissl, J.F., "Effectiveness of Water Treatment Processes in Pesticide Removal," *JAWWA,* 57, No. 2, 181-199 (February 1965).

(17) Stieglitz, L., Roth, W., Kuhn, W. and Leger, W., "The Behavior of Organohalides in the Treatment of Drinking Water," *Vom Wasser,* 47, 347-377 (1976).

(18) Grob, K. and Grob, C., "Organic Substances in Potable Water and in Its Precursor. Part II. Applications in the Area of Zurich," *Chromatography,* 90, 303-313 (1974); "Part III. The Closed-Loop Stripping Procedure Compared with Rapid Liquid Extraction," *Chromatography,* 106, 299-315 (1975).

(19) Kolle, W., Sontheimer, H. and Stieglitz, L., "Quantification Test of Activated Carbon for Waterworks Based on Its Adsorption Properties for Organic Chlorine, *Vom Wasser,* 44, 203-217 (1975).

(20) Suffet, I.H., Radziul, J.V., Cairo, P.R., and Coyle, J.T., "Evaluation of the Capability of Granular Activated Carbon and Resin to Remove Trace Organics from Treated Drinking Water," in *Proceedings of Second Conference on Water Chlorination—Environmental Impact and Health Effects,* Gatlinburg, Tennessee (Oct 31-Nov. 4, 1977).

(21) McCarty, P.L., Reinhard, M. and Argo, P.G., "Organic Removal by Advanced Wastewater Treatment," Presented at 97th Annual AWWA Conference, Anaheim, California (May 8-13, 1977).

(22) Anon., "National Interim Primary Drinking Water Regulations," Office of Water Supply, U.S. Environmental Protection Agency, EPA-570/9-76-003, Washington, D.C.

(23) Sorg, T.J., Logsdon, G.S. and Love, O.T., Jr., "Manual of Treatment Techniques for Meeting the Interim Primary Drinking Water Regulations," Water Supply Research

Division, U.S. Environmental Protection Agency, EPA-600/8-77-005, Cincinnati, Ohio (May 1977).
(24) Cohen, J.M., Kamphake, L.J., Lemke, A.E., Henderson, C. and Woodward, R.L., "Effects of Fish Poison on Water Supplies, Part 1. Removal of Toxic Materials," *JAWWA* 52, 1551-1566 (1960).
(25) Goodrich, P.R. and Monke, E.J., "Insecticide Adsorption on Activated Carbon," *Trans. Amer. Soc. Agr. Engr. Gen. Ed.,* 13, No. 1, 56-60 (January-February 1970).
(26) Hyndshaw, A.Y., "Activated Carbon to Remove Organic Contaminants from Water," *JAWWA*, 64, No. 5, 309-311 (1972).
(27) Moergeli, B., "Removal of Pesticides from Drinking Water," *Sulzer Tech. Rev.,* 54, No. 2, 91-96 (1972). Also see *Gas, Wasser, Abwasser,* 52, No. 5, 142-146 (1972).
(28) Shevchenko, M.A., Marchenko, P.V., Taran, P.N. and Kravets, E.V., "Removal of Triazine-Series Herbicides from Natural Waters," *Vodopodgotovka Ochistka Prom. Stokov* (USSR), No. 9, 103-112 (1972).
(29) Spahn, H., Brauch, V., Schluender, E. and Sontheimer, H., "Active Carbon Fiters for Water Purification, I. Adsorption on a Single Particle," *Verfahrenstechnik* (Mainz), 8, No. 8, 224-231 (1974).
(30) Parkash, S., "Adsorption of Cationic Pesticides (Diquat and Paraquat) from Aqueous Solution by Activated Carbon," *Carbon,* 12, No. 5, 483-491 (1974).
(31) Shevehenko, M.A., Marchenko, P.V., Taran, P.N., and Kravets, E.V., "Removal of Some Pesticides from Drinking Water by Adsorption Methods," *Vodosnabzh. Sanit. Tekh.* (USSR), 10, 31-32 (1974).
(32) Dawson, V.K., Marking, L.L. and Bills, T.D., "Removal of Toxic Chemicals from Water with Activated Carbon," *Trans. Am. Fish Soc.,* 105, No. 1, 119-123 (1976).
(33) Thielemann, H., "Identification and Removal of Odor and Taste Imparting Substances in Filtered Groundwater," *Mikrochim Acta* (E. Germ.) No. 1, 79-80 (1972).
(34) Willey, B.F., "Surveillance and Control of Phenolic Tastes and Odors in Water to Prevent Their Effects on Taste and Flavor of Foods," *Amer. Chem. Soc. Symp. Series 26: Phenolic, Sulfur, Nitrogen Compd. Food Flavors Symp.,* 71-84 (1975).
(35) Engels, C., "The Removal of Extra Light Fuel Oil from Polluted Water," *Vom Wasser,* 45, 45-51 (1975).
(36) El-Dib, M.A., Ramadan, F.M. and Ismail, M., "Adsorption of Sevin and Baygon on Granular Activated Carbon," *Water Res.,* 9, No. 9, 795-798 (1975).
(37) Allen, D.H., *Evaluation of Activated Carbon for Fuel Oil Adsorption from a Potable Water Supply,* NTIS Report, AD-A037478 (1976).
(38) Koganovskii, A.M. and Leuchenko, T.M., "Calculations of Isotherms for Adsorption of Organic Nonelectrolytes from Aqueous Solutions," *Colloid J. (USSR),* 31, No. 3, 305-307 (May-June 1969).
(39) Bernardin, F.E., Jr., "Selecting and Specifying Activated Carbon Adsorption Systems," *Chem. Engr. (NY),* 83, No. 22, 77-82 (December 18, 1976).
(40) Zogorski, J.S., Faust, S.D. and Haas, J.H., Jr., "The Kinetics of Adsorption of Phenols by Granular Activated Carbon," *J. Colloid and Interface Science,* 55, No. 2, 329-341 (1976).
(41) Kondo, M., Kishi, T., Shimamoto, T., Nohihara, T. and Watabe, K., "Studies on the Adsorption Effect of Activated Charcoal on Polychlorinated Biphenyls in Water," *Memoirs of the Faculty of Pharmaceutical Sciences (Japan),* 18, 135-138 (1974).
(42) Snoeyink, V.L., Lai, H.T., Johnson, J.H. and Young, J.F., "Activated Carbon: Dechlorination and the Adsorption of Organic Compounds," *Amer. Chem. Soc. Symp. on the Chemistry of Water Supply, Treatment and Distribution,* Dallas, Texas (April 1973).
(43) Whitehouse, J.D., "A Study of the Removal of Pesticides from Water," *Univ. of Kentucky Water Resources Institute, Research Report No. 8,* Lexington, Ky. (1967). (Also available from NTIS, PB-242-971.)
(44) Zogorski, J.S. and Faust, S.D., "Removal of Phenols from Polluted Water," NTIS, PB-237-859 (June 1974).
(45) Gauntlett, R.B. and Packham, R.F., "The Removal of Organic Compounds in the Production of Potable Water," *Chemistry and Industry,* No. 17, 812-817 (1973).

(46) Rambow, C.A., "Effect of Flow Rate on Carbon Filter Performance," *JAWWA* 55, No. 8, 1037-1043 (August 1963).
(47) Aly, O.M. and Faust, S.D., "Removal of 2,4-D Derivatives from Natural Waters," *JAWWA,* 57, 221-230 (1965).
(48) Morita, M., Nakamura, H. and Mimura, S., "Phthalic Acid Esters in Water," *Water Res.,* 8, No. 10, 781-788 (1974).
(49) Schmidt, K., "Effectiveness of Standard Drinking Water Preparation for Eliminating Pesticides and Other Pollutants," *Gas. Wasserfach Wasser-Abwasser,* 115, No. 2, 72-76 (1974).
(50) Vrochinskiy, K.K., "Removal of Pesticides from Water by Various Water Preparation Methods," *Gig. Sanit. (USSR),* 38, No. 1, 76-78 (1973).
(51) Calderbank, A., "Environmental Considerations in the Development of Diquat and Paraquat on Aquatic Herbicides," *Outlook Agr.,* 7, No. 2, 51-54 (1972).
(52) Greve, P.A. and Wit., S.L., "Endosulfan in the Rhine River," *J. Water Poll. Contr. Federation,* 43, No. 12, 2338-2348 (1971).
(53) Stoltenberg, D.H., "How to Reclaim a Poisoned Lake," *Pub. Works,* 103, No. 3, 59-61 (1972).
(54) Hansen, R.E., "Organics Removal of Mount Clements, Michigan," Presented to Michigan Section AWWA, Ann Arbor, Michigan (January 1977).
(55) Giusti, D.M., Conway, R.A. and Lawson, C.T., "Activated Carbon Adsorption of Petrochemicals, *JWPCF,* 45, No. 5, 947-965 (May 1974).
(56) Herzing, D.R., Snoeyink, V.L. and Wood, N.F., "Activated Carbon Adsorption of the Odorous Compounds 2-Methylisoborneol and Geosmin," *JAWWA,* 69, No. 4, 223-228 (April 1977).
(57) Snoeyink, V.L., McCreary, J. and Murin, C.J., "Activated Carbon Adsorption of Trace Organic Compounds," Final Report on U.S. EPA Research Grant No. R803473, EPA-600/2-77-223, Cincinnati, Ohio (March 1978).
(58) Stevens, A.A. and Symons, J.M., "Measurement of Trihalomethane and Precursor Concentration Changes," *JAWWA,* 69, No. 10, 546-554 (October 1977).
(59) Love, O.T., Jr., Carswell, J.K., Miltner, R.J. and Symons, J.M., "Treatment for the Prevention or Removal of Trihalomethanes in Drinking Water, *JAWWA,* In Press.
(60) Symons, J.M., et al, "National Organics Reconnaissance Survey for Halogenated Organics in Drinking Water," *JAWWA,* 67, No. 11, Part 1, 634-647 (November 1975), Update, 67, No. 12, 708-709 (December 1975).
(61) Sylvia, A.E., Bancroft, D.A. and Miller, J.D., "Detection and Measurement of Microorganics in Drinking Water," in *Proceedings of the AWWA Water Quality Technology Conference,* Denver, CO (1975).
(62) Stanier, R.W., Adelberg, E.A. and Ingraham, J., *The Microbial World,* 4th Edition, Prentice-Hall, Chapter 29, "Microbial Pathogenicity," 791 (1976).
(63) Jorgensen, J.H., Lee, J.C. and Pahren, H.R., "Rapid Detection of Bacterial Endotoxins in Drinking Water and Renovated Wastewater," *Applied and Environmental Microbiology,* 32, No. 3 (September 1976).
(64) McCreary, J.J. and Snoeyink, V.L., "Granular Activated Carbon in Water Treatment," *JAWWA,* 69, No. 8, 437-444 (1977).
(65) Guirguis, W.A., Melnyk, P.B. and Harris, J.P., "The Negative Impact of Industrial Waste on Physical-Chemical Treatment," Presented at 31st Purdue Industrial Waste Conf., Lafayette, Indiana (May 1976).
(66) Directo, L.S., Chen, C.-L. and Kugelman, I.J., "Pilot Plant Study of Physical-Chemical Treatment," *J. Water Poll. Control Fed.,* 49, No. 10, 2081-2098 (1977).
(67) Monscvitz, J.T. and Ainsworth, L.D., "Detection and Control of Hydrogen Polysulfide in Water," *Public Works,* 101, 113 (1970).
(68) Eberhardt, M., "Experience with the Use of Biologically Effective Activated Carbon," in *Translation of Reports on Special Problems of Water Technology,* Vol. 9, *Adsorption,* H. Sontheimer, Ed., EPA Report EPA 600/9-76-030 (December 1976).
(69) Sontheimer, H., "Biological Treatment of Surface Waters in Activated Carbon Filters," presented at Seminar on Current Status of Wastewater Treatment and Disinfection with Ozone, Cincinnati, Ohio, Intl. Ozone Inst., Cleveland, Ohio (September 15, 1977).

(70) Kuhn, W., Sontheimer, H. and Kurz, R., "Use of Ozone and Chlorine in Water Works in the Federal Republic of Germany," Presented at Workshop on Ozone Chlorine Dioxide Oxidation Products of Organic Materials, Cincinnati, Ohio, Int'l. Ozone Inst., Cleveland, Ohio (November 1976).

(71) Guirguis, W.A., Jain, J.S., Hanna, Y.A. and Srivastava, P.K., "Ozone Application for Disinfection in the Westerly Advanced Wastewater Treatment Facility," in *Forum on Ozone Disinfection,* E.G. Fochtman, R.G. Rice and M.E. Browning, editors. Intl. Ozone Inst., Cleveland, Ohio, 363-381 (1976).

(72) Guirguis, W.A., Cooper, T., Harris, J. and Ungar, A., "Improved Performance of Activated Carbon by Preozonation," presented at 49th Annual Conf. Water Poll. Control Fed., Minneapolis, Minn. (October 1976).

(73) Scheidtmann, W., "Investigations of the Optimization of Pretreatment When Using Ozone," in *Translation of Reports on Special Problems of Water Technology,* Vol. 9, *Adsorption,* 98-111.

(74) Schalekamp, M., "Use of Activated Carbon in the Treatment of Lake Water," in *Translation of Reports of Special Problems of Water Technology,* Vol. 9, *Adsorption,* 128-159.

(75) Van Lier, W.C., Graveland, A., Rook, J.J. and Schultink, L.J., "Experiences with Pilot Plant Activated Carbon Filters in Dutch Waterworks," in *Translation of Reports on Special Problems of Water Technology,* Vol. 9, *Adsorption,* 160-181.

(76) Sontheimer, H., "Considerations on the Optimization of Activated Carbon Use in Waterworks," in *Translation of Reports on Special Problems of Water Technology,* Vol. 9, *Adsorption,* 208-214.

(77) Van der Kooij, D., "Some Investigations into the Presence and Behaviour of Bacteria in Activated Carbon Filters," in *Translation of Reports on Special Problems of Water Technology,* Vol. 9, *Adsorption,* 348-354.

(78) Gomella, C. and Versanne, D., "Le Role de l'Ozone dans la Nitrification Bacterienne de l'Azote Ammoniacal—Cas de l'Usine de la Chapelle Banlieue Sud de Rouene (Seine Maritime), France," Presented at 3rd Intl. Symp. on Ozone Technol., Paris, France, Intl. Ozone Int., Cleveland, Ohio (May 1977).

(79) Sontheimer, H., Heilker, E., Jekel, M., Nolte, H. and Vollmer, F.-H., "The 'Mulheim Process'—Experience with a New Process Scheme for Treating Polluted Surface Waters," submitted for publication in *JAWWA*.

(80) Wachs, A., Narkis, N. and Schneider, M., "Organic Matter Removal from Effluents by Lime Treatment, Ozonation and Biologically Extended Activated Carbon Treatment," Presented at 3rd Intl. Symp. on Ozone Technology, Paris, France, Intl. Ozone Inst., Cleveland, Ohio (May 1977).

(81) Hopf, W., "Versuche mit Aktivkohlen zur Aufbereitung des Düsseldorfer Trinkwassers," *Wasser/Abwasser,* 101, No. 14, 330-336 (1960).

(82) Sontheimer, H., Univ. Karlsruhe, Germany, Private Communication (1977).

(83) Kuhn, W. and Fuchs, F., "Untersuchungen zur Bedeutung der Organischen Chlorverbindungen und Ihrer Adsorbierbarkeit," *Vom Wasser,* 45, 217-232 (1975).

(84) Kuhn, W. and Sontheimer, H., "Einige Untersuchungen zur Bestimmung von Organischen Chlorverbindungen auf Aktivkohle," *Vom Wasser,* 41, 65-79 (1973).

(85) Kuhn, W. and Sontheimer, H., "Einfluss Chemischer Umsetzungen auf die Lage der Adsorptionsgleichgewichte an Aktivkohlen," *Vom Wasser,* 40, 115-123 (1973).

(86) Kuhn, W., "Untersuchungen auf Bestimmung von Organischen Chlorverbindungen auf Aktivkohle," Dissertation, Fak. f. Chemie-Ing. Wesen, Univ. Karlsruhe, Fed. Rep. Germany (1974).

(87) Wolfel, P. and Sontheimer, H., "Ein Neues Verfahren zur Bestimmung von Organisch Gebundenem Kohlenstoff in Wasser Durch Photochemische Oxidation," *Vom Wasser,* 43, 315-325 (1974).

(88) Poggenburg, W., Wasserwerk Düsseldorf, Private Communication (1977).

(89) Steiner IV, J. and Singley, J.E., "Methoxychlor Removal from Potable Water," Engineering Report, Department of Environmental Engineering Sciences, University of Florida, Gainesville, Florida (1977).

Design of Granulated Activated Carbon Systems

The material in this chapter is taken from the report entitled *Interim Treatment Guide for Controlling Organic Contaminants in Drinking Water Using Granular Activated Carbon,* compiled and edited by James M. Symons of the Water Supply Research Division of the Municipal Environmental Research Laboratory, Office of Research and Development of the U.S. EPA in Cincinnati, Ohio, and published in January 1978.

COLLECTION OF TREATABILITY DATA

Performance Criteria

Having chosen granular activated carbon as the best "broad-spectrum" organic removal unit process currently available (see Chapter 5), several questions still remain for a given utility required to install this form of treatment. These are: (1) the way in which the facility should be designed and operated to optimize performance; and (2) which of the currently commercially available granular activated carbons are best for a given location. To begin to answer these questions, the objective of treatment must be defined.

In general terms, the treatment process should remove, to the degree possible, the four classes of organic compounds previously defined: taste- and odor-producing components, source water synthetic contaminants, disinfection by-product precursors, and disinfection by-products. The fifth class, natural organic compounds that are not disinfection by-product precursors, are not easily controlled by granular activated carbon adsorption, but are of little public health concern. Further, the reactivation cycle should neither be uneconomically short, nor so long as to allow the desorption of previously adsorbed material.

Within these general constraints, how are the two basic questions to be answered? Some performance criteria for the adsorption process must be chosen so that it can be properly designed and operated. Experience has shown that wide differences exist in the adsorbability of various organic compounds. Some toxic chemicals such as pesticides, and polynuclear aromatic hydrocarbons are strongly adsorbed as are taste and odor compounds and should not be a limiting performance factor. Toward the other end of the adsorption spectrum are the low molecular weight halogenated compounds such as chloroform, carbon tetrachloride and so forth.

Because of the variability of adsorption among organic compounds and of adsorption competition among different organic compounds, the design and operation of an adsorption facility must be site specific. To allow the evaluation of granular activated carbon as a treatment process in any given location, performance criteria have been chosen, as detailed below, beyond the established maximum contaminant levels (MCLs) for trihalomethanes and pesticides. In a given location, any one of these performance factors may be the one that controls design and operation, because to assure maximum protection for consumers, all three must be met, as well as the MCLs.

Performance Criterion 1: Although EPA is not now setting MCLs for other specific organic compounds, reduction of exposure to known or suspected carcinogens in drinking water to the extent possible is encouraged. Because evidence is mounting that low molecular weight halogenated compounds such as carbon tetrachloride, tri- and tetrachloroethylene, vinyl chloride, and so forth are a health hazard, and are adsorbable by granular activated carbon, the treatment process should be designed and operated so that the granular activated carbon is reactivated when the concentration of any organic compound of this type (excluding the trihalomethanes) exceeds 0.5 μg/l in the effluent. Additionally, these organic compounds can be measured by a single analytic procedure (see Chapter 7).

Performance Criterion 2: This criterion leaves a gap in contaminant control. Neither known carcinogens, such as benzene (an unchlorinated compound), nor toxicants that are unable to be analyzed at this time would necessarily be controlled if only the above criterion were used. Therefore, Performance Criterion 2 must be related to some group parameter(s).

As a group parameter, organic carbon has merit, as any deterioration in granular activated carbon bed performance with respect to this parameter would indicate an increase in the concentration of some unknown organic compound(s) of unknown hazard in the treated water. Therefore, to maximize protection for the consumer, the treatment process should be designed and operated so that the granular activated carbon is reactivated when the increase in the concentration of total organic carbon (TOC) exceeds 0.5 mg/l over the initial effluent TOC concentration.

Performance Criterion 3: Finally, to assure that the granular activated carbon is performing adequately at the start of the test, Performance Criterion 3 is that the initial TOC removal [(Adsorber Influent TOC Concentration – Adsorber Effluent TOC Concentration)/Adsorber Influent TOC Concentration)] x 100, must exceed 50%.

In addition to these criteria, to meet the maximum contaminant level (MCL) portion of the Regulation, the treatment process should be designed and operated so that the granular activated carbon is reactivated when (1) the arithmetic sum (called "summation" trihalomethanes in the Guide and "total" trihalomethanes in the Regulations) of the terminal trihalomethane concentrations, a technique for estimating compliance with the Regulations, or (2) the pesticide concentrations (rarely involved) exceed the MCLs (0.10 mg/l proposed for the trihalomethanes) specified in the Interim Primary Drinking Water Regulations.

These performance criteria should be applied as follows. The simplest method of retrofitting a large water treatment plant with granular activated carbon would be to replace the present filter media with granular activated carbon (sand replacement) and add a reactivation facility. To investigate this mode of operation in a particular location, a small granular activated carbon column with an empty bed contact time, volume/flow (EBCT), similar to that available in the existing filter boxes during average flow should be fabricated (see Chapter 6) and used to treat settled water.

The effluent concentration of (1) any low molecular weight halogenated compounds present, (2) total organic carbon, and (3) terminal summation trihalomethane should be determined weekly to show how long good performance can be maintained with that particular adsorbent and EBCT. When Criterion 1 or 2 or the MCL are exceeded, using a 3-week running average, $[(W_i + W_{i+1} + W_{i+2})/3$, value assigned to $Week_{i+1}]$, the granular activated carbon would have to be reactivated. If Criterion 3 is exceeded initially, this EBCT is not adequate.

To investigate the use of a granular activated carbon system with a longer EBCT, a second system consisting of three pilot granular activated carbon columns in series each with a 10-minute EBCT should be operated simultaneously with the one described above on filtered water. Sampling after each of these three additional columns will show the influence of longer contact time on reactivation frequency as measured by all three criteria (see Chapter 7). Finally, because experience has shown that Terminal ΣTHM concentrations are likely to be greater in the summer and the concentration of the types of organic compounds related to Criterion 1 are often higher in source waters in the winter, collecting these pilot column data in both seasons would be very helpful.

Another major question beyond mode of operation, empty bed contact time, and reactivation frequency is which commercially available granular activated carbon is most cost effective for a given location. A utility wishing to evaluate the performance of competitive commercial products could do so by expanding the test procedure outlined above to include parallel units containing different kinds of adsorbents. Possibly the adsorbent suppliers would help in this evaluation.

Example Data

Data collected by the Water Supply Research Division in three locations, the lower Mississippi River, the central Ohio River, and a southern Florida groundwater will be used to show how these performance criteria would be applied. Tables 6.1 and 6.2 contain data from a full-scale installation in the lower Mississippi Valley, rather than the recommended pilot granular activated carbon columns, but do show that for this installation, Criteria 1 and 2 are exceeded after two plus weeks for the sand replacement system, while the reactivation frequency could be extended to six weeks through the use of postfilter adsorbers. The difference in EBCT is too small to account for this difference.

TABLE 6.1: APPLICATION OF PERFORMANCE CRITERIA FROM LOWER MISSISSIPPI PROJECT–SAND REPLACEMENT–AVERAGE EBCT–21 MIN.

	Criterion 1		Criterion 2	Criterion 3	MCL
Time, Weeks	1,2-Dichloroethane Conc.	Trichloroethylene Conc.	TOC Conc.	TOC Removal Percentage	Term. ΣTHM Conc.
Performance Criteria:	μg/ℓ < 0.5	μg/ℓ < 0.5	mg/ℓ < Δ0.5 over TOC_o	% Initial > 50%	mg/ℓ < 0.10
0	-	NF	0.8	77	-
1	0.25	NF	-		< 0.01
2+	0.7*	NF	2.0*		0.05
3	1.5	NF	2.0		0.07
4	3.7	NF	1.9		0.11*
5		NF			0.13
6		NF			
7		NF			
8		NF			
9		0.03			
10+		0.03			
11+		0.03			
12+		NF			
13+		NF			
14+		NF			
15+		NF			
16+		NF			

NF - None found

* - Performance criterion exceeded

Note: All data are three week running averages, except for TOC_o

TABLE 6.2: APPLICATION OF PERFORMANCE CRITERIA FROM LOWER MISSISSIPPI PROJECT–POSTFILTER ADSORBER–AVERAGE EBCT–24 MIN.

	Criterion 1		Criterion 2	Criterion 3	MCL
Time, Weeks,	1,2-Dichloroethane Conc.	Trichloroethylene Conc.	TOC Conc.	TOC Removal Percentage	Term. ΣTHM Conc.
Performance Criteria:	μg/ℓ < 0.5	μg/ℓ < 0.5	mg/ℓ < Δ0.5 over TOC_o	% Initial > 50%	mg/ℓ < 0.10
0	-	NF	0.8	73	-
1	0.2	NF	0.9		<0.01
2+	0.1	NF	0.8		0.01
3	0.0	NF	0.8		0.02
4	0.1	NF	0.8		0.02

(continued)

TABLE 6.2: (continued)

Time, Weeks,	Criterion 1 1,2-Dichloroethane Conc.	Criterion 1 Trichloroethylene Conc.	Criterion 2 TOC Conc.	Criterion 3 TOC Removal Percentage	MCL Term. ΣTHM Conc.
Performance Criteria:	μg/ℓ < 0.5	μg/ℓ < 0.5	mg/ℓ < Δ0.5 over TOC_o	% Initial > 50%	mg/ℓ < 0.10
5	0.5	NF	1.0		0.02
6	0.7*	NF	0.9		0.03
7	0.9	NF	1.0		0.04
8		NF	1.1		0.07
9		NF	1.3*		0.08
10+		0.1	1.3		0.09
11+		0.1			0.09
12+		0.1			0.09
13+		NF			0.09
14+		NF			0.09
15+		NF			0.10
16+		NF			-

NF - None Found

* - Performance criterion exceeded

Note: All data are three week running averages, except for TOC_o.

For the data in Table 6.3, collected with a pilot column operating on tap water from the central Ohio Valley, Criterion 1 controlled after 7 weeks, followed by Criterion 2 on the 8th week. The MCL was not exceeded until the 24th week.

TABLE 6.3: APPLICATION OF PERFORMANCE CRITERIA FROM OHIO RIVER TAP WATER—POSTFILTER ADSORBER—UPFLOW MODE—EBCT—11 MINUTES

Time, Weeks	Criterion 1 Carbon Tetrachloride Conc.	Criterion 2 TOC Conc.	Criterion 3 TOC Removal Percentage	MCL Term. ΣTHM Conc.
Performance Criteria	μg/ℓ < 0.5	mg/ℓ <Δ 0.5 over TOC_o	% Initial > 50%	mg/ℓ < 0.10
0	ND	ND	ND	ND
1	NF	0.34	77	NF
2	NF	0.28		< 0.01
3	NF	0.31		< 0.01
4	NF	0.40		< 0.01
5	NF	0.41		< 0.01
6	NF	0.57		< 0.01
7	1.7*	0.71		0.01
8	3.3	0.85*		0.02

(continued)

TABLE 6.3: (continued)

Time, Weeks	Criterion 1 Carbon Tetra- chloride Conc.	Criterion 2 TOC Conc.	Criterion 3 TOC Removal Percentage	MCL Term. ΣTHM Conc.
Performance Criteria	μg/ℓ < 0.5	mg/ℓ <Δ 0.5 over TOC_o	% Initial > 50%	mg/ℓ < 0.10
9	5.3	1.21		0.02
10				0.03
11				0.04
12				0.04
13				0.05
14				0.06
15				0.06
16				0.07
17				0.07
18				0.08
19				0.08
20				0.09
21				0.10
22				0.10
23				0.10
24				0.11*

ND - Not determined
NF - None found

* - Performance Criterion Exceeded

Note: All data are three week running averages, except for TOC_o.

The data in Tables 6.4, 6.5 and 6.6 were prepared from pilot granular activated carbon column studies conducted in southern Florida on a groundwater containing high concentrations of organic carbon and trihalomethane formation potential, as well as several low molecular weight halogenated organic compounds. These data show that the short EBCT does not provide good performance for very long and leads to the conclusion that longer EBCTs are necessary. Pilot columns with longer EBCTs are now being employed at this site. Tables 6.5 and 6.6 present data from repeat runs and indicate the general reproducibility of this approach.

TABLE 6.4: APPLICATION OF PERFORMANCE CRITERIA FROM SOUTHERN FLORIDA GROUNDWATER–RAW WATER APPLIED TO COLUMN–EBCT–6.2 MINUTES

Time, Weeks	Criterion 1 A Conc.	B Conc.	C Conc.	D Conc.	Criterion 2 TOC Conc.	Criterion 3 TOC Removal Percentage	MCL Term. ΣTHM Conc.
Performance Criteria	μg/ℓ <0.5	μg/ℓ <0.5	μg/ℓ <0.5	μg/ℓ <0.5	mg/ℓ < Δ0.5 over TOC_o	% Initial > 50%	mg/ℓ < 0.10
0	–	–	–	–	3.3	67	–
1	NF	NF	0.13	NF	4.9*		0.33*

(continued)

TABLE 6.4: (continued)

Time, Weeks	Criterion 1 A Conc.	B Conc.	C Conc.	D Conc.	Criterion 2 TOC Conc.	Criterion 3 TOC Removal Percentage	MCL Term. ΣTHM Conc.
Performance Criteria	μg/ℓ <0.5	μg/ℓ <0.5	μg/ℓ <0.5	μg/ℓ <0.5	mg/ℓ < Δ0.5 over TOC_o	% Initial > 50%	mg/ℓ < 0.10
2	NF	<0.01	1.0*	<0.01	6.1		0.51
3	<0.01	0.3	2.4	<0.01	7.6		0.59
4	<0.01	0.04	4.9	<0.01			
5	<0.01	0.03	5.8	NF			
6	<0.01	0.01		NF			
7	0.01	0.02		NF			
8	0.02	0.02		NF			
9	0.04	0.03		NF			
10	0.05	0.06		<0.01			
11	0.06	0.05		<0.01			
12	0.27	0.15		0.03			
13+	0.39	0.17		0.05			
14	0.67*	0.30		0.06			
15	0.70	0.30		0.06			
16	–	–		–			

NF = None found

* = Performance criterion exceeded

Note: All data are three week running average, except for TOC_o.

A = trans -1,2-Dichloroethylene

B = 1,1- Dichloroethane

C = cis-1,2-Dichloroethylene

D = Trichloroethylene

Note: These data are for illustrative purposes only and are not indicative of final design and operation based on longer EBCTs.

TABLE 6.5: APPLICATION OF PERFORMANCE CRITERIA FROM SOUTHERN FLORIDA GROUNDWATER–FILTERED WATER APPLIED TO COLUMN–EBCT–6.2 MINUTES

Time, Weeks	Criterion 1 A Conc.	B Conc.	C Conc.	D Conc.	E Conc.	Criterion 2 TOC Conc.	Criterion 3 TOC Removal Percentage	MCL Term. ΣTHM Conc.
Performance Criteria	μg/ℓ <0.5	μg/ℓ <0.5	μg/ℓ <0.5	μg/ℓ <0.5	μg/ℓ <0.5	mg/ℓ < Δ0.5 over TOC_o	% Initial >50%	mg/ℓ <0.10
0	NF	NF	NF	NF	ND	–	–	–
1	0.13	NF	3.57*	0.20	0.40	1.5	76	0.16*
2	0.13	0.18	4.30	0.21	0.38	4.3*		0.22

(continued)

TABLE 6.5: (continued)

Time, Weeks	Criterion 1 A Conc.	B Conc.	C Conc.	D Conc.	E Conc.	Criterion 2 TOC Conc.	Criterion 3 TOC Removal Percentage	MCL Term. ΣTHM Conc.
Performance Criteria	μg/ℓ <0.5	μg/ℓ <0.5	μg/ℓ <0.5	μg/ℓ <0.5	μg/ℓ <0.5	mg/ℓ < Δ0.5 over TOC_o	% Initial >50%	mg/ℓ <0.10
3	0.13	0.18	5.25	0.21	0.32	5.2		
4	0.04	0.39	4.65	0.01	1.68*	4.5		
5	0.04	0.21	6.85	<0.01	2.65	4.5		
6	0.04	0.21	9.19	<0.01	3.27	4.8		
7	–	–	–	–	–	–		

ND = Not Determined

NF = None Found

* = Performance criterion exceeded

Note: All data are three week running average, except for TOC_o.

A = trans-1,2-Dichloroethylene

B = 1,1-Dichloroethane

C = cis-1,2-Dichloroethylene

D = Trichloroethylene

E = Vinyl chloride

Note: These data are for illustrative purposes only and are not indicative of final design and operation based on longer EBCTs.

TABLE 6.6: APPLICATION OF PERFORMANCE CRITERIA FROM SOUTHERN FLORIDA GROUNDWATER–FILTERED WATER APPLIED TO COLUMN–EBCT–6.2 MINUTES

Time, Weeks,	Criterion 1 A Conc.	B Conc.	C Conc.	D Conc.	E Conc.	Criterion 2 TOC Conc.	Criterion 3 TOC Removal Percentage	MCL Term. ΣTHM Conc.
Performance Criteria	μg/ℓ <0.5	μg/ℓ <0.5	μg/ℓ <0.5	μg/ℓ <0.5	μg/ℓ <0.5	mg/ℓ <Δ0.5 over TOC_o	% Initial >50%	mg/ℓ <0.10
0	–	–	–	–	–	0.4	93	–
1	NF	NF	NF	NF	3.80*	2.3*		0.12*
2	NF	0.37	0.43	NF	3.80	3.5		0.23
3	NF	0.43	1.70*	NF	4.47	3.9		0.37
4	NF	0.43	5.40	NF	1.73	3.6		
5	NF	0.17	6.80	0.07	2.30	3.0		
6	NF	0.37	9.53	0.17	1.17	4.4		
7	0.2	0.60*	8.87	0.37	2.00	–		
8	–	–	–	–	–	–		

NF = None found

(continued)

TABLE 6.6: (continued)

* = Performance criterion exceeded

Note: All data are three week running average, except for TOC_o.

A = trans -1,2-Dichloroethylene

B = 1,1-Dichloroethane

C = cis-1,2-Dichloroethylene

D = Trichloroethylene

E = Vinyl chloride

Note: These data are for illustrative purposes only and are not indicative of final design and operation based on longer EBCTs.

Calculation of Adsorbent Use Rate

The data collected by the procedure summarized above can be used to calculate the use rate (dose) in weight per unit volume of water of granular activated carbon needed to meet the controlling criterion for any empty bed contact time. Adsorbent use rate is calculated by dividing the dry weight of adsorbent for a given empty bed contact time by the total volume of water passing through the column(s) until each criterion and the MCL is exceeded.

If the adsorbent use rate decreases with increasing empty bed contact time then a more proportional improvement in performance is gained by increasing the empty bed contact time. Chapter 9 on economic analysis will show the importance of knowing whether or not this proportion between empty bed contact time and performance occurs in a given location. Further, the adsorbent use rate can be used to estimate one of the major components of cost associated with a granular activated carbon adsorption treatment system, the cost of adsorbent reactivation. The data in Table 6.7, based on data per treatment of a southern Florida groundwater, can be used as an example of this type of analysis.

TABLE 6.7: ADSORBENT USE RATES FOR TREATING A SOUTHERN FLORIDA GROUNDWATER

Empty Bed Contact Time Minutes	Criterion 1 VC Use, mg/l	Criterion 1 Cis Use, mg/l	Criterion 2 TOC Use, mg/l	MCL Use, mg/l
6.2	560	90	1,340	860
12.4	370	70	880	430
18.6	170	NR	530	410
24.8	230	NR	430	410

VC = vinyl chloride
Cis = cis-1,2-dichloroethylene
NR = criterion not reached

These data show that the adsorbent use rate declined for all criteria as the empty bed contact time increased from 6.2 to 18.6 minutes (more than proportional performance improvement), but that the next increment of empty bed contact time to 24.8 minutes did not lower the adsorbent use rate for the MCL requirement. Therefore, for these data, an empty bed contact time of about 19 minutes was optimum.

Summary

In summary several points are important:

(1) Granular activated carbon treatment performance criteria have been selected that will permit collecting the data from pilot granular activated carbon columns that is necessary for:
   (a) the economical design and operation of a granular activated carbon adsorption unit process, and
   (b) choosing an adsorbent in any given location requiring organic contaminant control.

(2) This scheme is also flexible enough to serve other purposes such as evaluation of the combination of ozone and granular activated carbon (biological activated carbon), including an evaluation of the microbiological content of the effluent, and the performance of reactivated carbon compared to virgin material, if a sample of reactivated carbon is available for study.

(3) Because of changing seasons and possible changes in the quality of source waters, continued monitoring of the full-scale treatment plant will be necessary for good process control.

(4) Although the pilot granular activated carbon column tests should give a good indication of full-scale plant performance, some adjustments in reactivation frequency should be expected.

(5) Although the above outlined approach may not be the ultimate solution, operation of the full-scale granular activated carbon beds to meet the established performance criteria consistently will provide a large measure of protection for consumers against the health hazards and esthetic problems of organic compounds in drinking water.

DESIGN OF PILOT GRANULAR ACTIVATED CARBON COLUMNS

The material in this section was taken from Appendix C, *Interim Guide for Organic Contaminants* (see bibliography).

Introduction

In Chapter 8 of this book, the costs of two types of GAC adsorption schemes will be discussed. The first scheme is to retrofit an existing water treatment plant by replacing the filtration media with granular activated carbon, and the second, called postfilter adsorption, requires separate contactors following the filtration step. The ability of either adsorption scheme to reduce the concentration of trace organics should be investigated on-site using a small manageable system before expending large sums of money on a process design that may give marginal performance. Trial experimentation (i.e., pilot studies) permits selecting the appropriate concept through improved engineering judgement. This section describes the design and operation of the EPA Water Supply Research

Division's experimental adsorption systems used in organics removal studies. This material is presented for guidance and as an example of one approach to gathering adsorption performance data.

Selection of Materials

One precaution taken by EPA in trace organic studies is to construct pilot scale equipment with stainless steel, Teflon, and glass whenever possible to minimize contamination from structural materials during experimentation. Whether or not materials such as rubber, acrylic resin, polyvinylchloride (PVC), polyethylene tubing, or similar products would compromise the experimental results is not known. To avoid this possibility, however, these materials should not be used if possible.

Adsorption Columns

Because of a limitation of available water, the WSRD pilot scale adsorbers consist of 3.8 cm (1.5 inch) diameter glass columns 153 cm (60 inches) in length. A schematic diagram of the experimental system is shown in Figure 6.1.

The columns are arranged so that both modes – sand replacement and postfiltration adsorption – can be studied simultaneously if desired. Granular activated carbon is placed in column 1 (see Figure 6.1) to a depth allowable in the existing filter boxes at the water treatment plant. Settled water should be applied to this unit at an approach velocity similar to that in the existing plant. Approximately 15 cm (6 inches) of graded gravel should be placed in the bottom of the columns as an aid in distributing the backwash water.

Some type of surface scrubbing (air or water scour) should be incorporated in the system because with granular activated carbon, like sand, most of the filtration occurs in the upper few centimeters of the bed and vigorous scouring insures adequate cleansing. The surface scrubber shown in Figure 6.1 is intended to slide on a Teflon ferrule so that it can be located close to the granular activated carbon surface for effective agitation. If headloss monitoring is desired, a U-tube manometer or a sensitive pressure gage can be included in the design. If, however, headloss is not monitored, a backwashing schedule such as that used within the existing plant can be employed.

Establishing Test Conditions

Like disinfection, the process of adsorption is very dependent upon contact time. The term "empty bed contact time" (EBCT) is commonly used to characterize this adsorption variable and is calculated by dividing the volume of medium "V" by the hydraulic loading "Q." Assuming that a filter box contains from 76 to 122 cm (30 to 48 inches) of medium and the hydraulic loadings range from 4 to 8 m/hr (2 to 4 gal/min/ft^2) then the EBCT in a sand replacement mode could typically vary between 1.5 to 15 minutes, with an average of 7 to 8 minutes.

The additional columns shown on Figure 6.1 can be operated concurrently in a postfilter adsorption mode to examine the effects of longer empty bed contact times. For example, columns 2, 3, and 4 (Figure 6.1) can be charged with granular activated carbon, exposed to filtered rather than settled water, and sampled in series to monitor the breakthrough (wave front) of specific organics (Performance Criterion 1) and the total organic carbon (Performance Criteria 2 and 3).

FIGURE 6.1: PILOT GRANULAR ACTIVATED CARBON COLUMNS

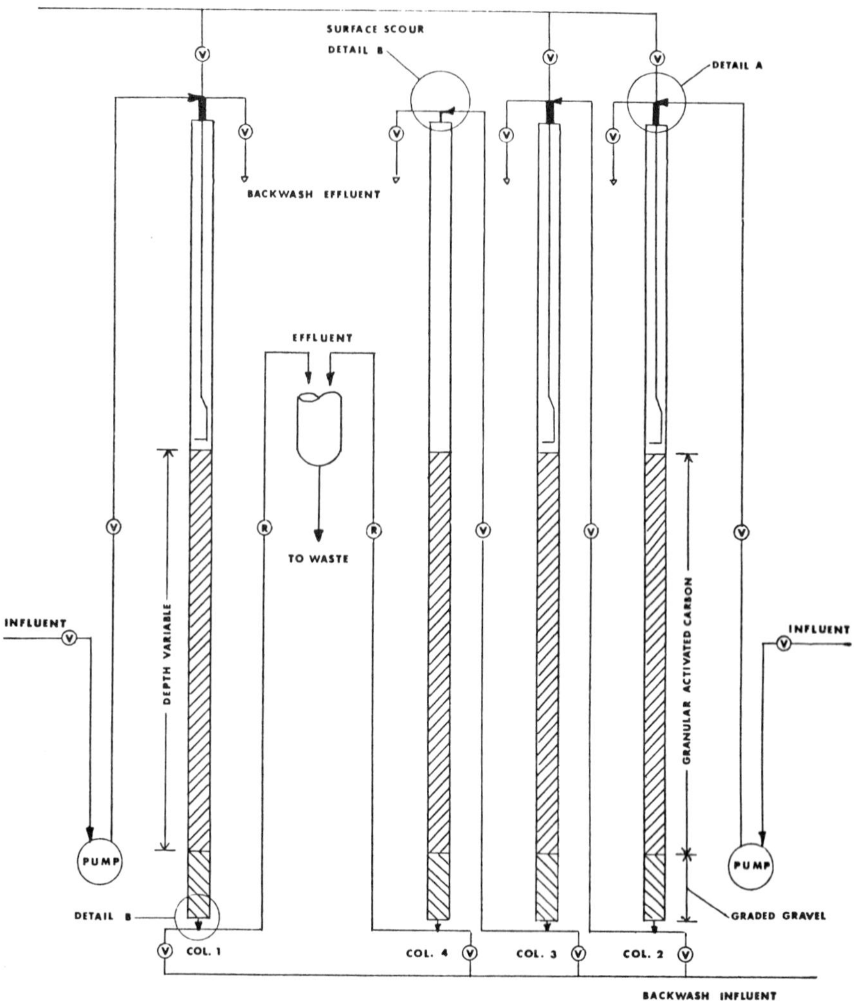

On the other hand, an investigator may desire to use column 2 as a sand filter receiving settled water and columns 3 and 4 as adsorbers. Another column in series would further increase the EBCT. Additional columns and the appropriate plumbing modifications would allow a utility to investigate in parallel, rather than in sequence, performance of different brands of commercially available granular activated carbon, should that be desirable. Finally, the choice as to whether or not to apply disinfected water to the adsorber will depend on how the final treatment scheme will be arranged. The engineer should have some experimental "breadth" to select the most satisfactory adsorption design.

Operation

The granular activated carbon must be wetted before it is put into service. This is accomplished by backwashing the material similar to the initial steps in using any granular medium. Gently tapping or bumping the columns might be necessary to insure that the medium is evenly wetted. Sufficient freeboard should exist to permit 50% bed expansion during backwash. The frequency of backwashing during the experimental study will vary, depending upon the same factors influencing full-scale filtration (e.g., headloss, turbidity, carryover floc from settling, etc.). Details on monitoring the system are given in Chapter 7, however, a routine operation schedule would include the following:

Parameter	Frequency
Flow adjustment, Q	Daily
Temperature, pH, turbidity	Investigator's discretion, based on variability of applied water
Low molecular weight halogenated organic compounds	Weekly
Organic carbon	Weekly
Terminal summation THM	Weekly

At most field installations no restriction in available water volume would exist as it does in the WSRD pilot plant. Therefore, larger, at least 10 cm (4 inch), diameter columns are recommended for use in the treatability studies proposed in the first part of this chapter.

Analytical Methods

ANALYTIC METHODOLOGY FOR MONITORING PILOT COLUMN TESTS

The material in this section is from Appendix B of the *Interim Guide for Organic Contaminants.* (See the bibliography for further details.)

Introduction

The methods needed to monitor the granular activated carbon treatment pilot studies can be divided into three specific categories: (1) low molecular weight halogenated organic compounds, excluding trihalomethanes; (2) organic carbon; and (3) terminal "summation trihalomethane" (the arithmetic sum of the concentration of the individual species of trihalomethane found in a given sample. This parameter is called "total trihalomethane" in the proposed regulation.)

Some inherent flexibility exists in each of these categories with regard to specific measurement methods, instruments selected, and so forth. This discussion will attempt to address some of the options available and relate them as closely as possible to present, proposed, or future U.S. EPA Standard Methods that are ultimately needed to meet compliance monitoring requirements.

Performance Criterion 1

This group of organic compounds–low MW halogenated organic compounds excluding THMs–has an operational definition. The intent is to include such low boiling (low molecular weight) halogenated organics as carbon tetrachloride, tri- and tetrachloroethylene, vinyl chloride, and others often found in contaminated drinking waters. This group is selected because of: (1) relative ease of analyses; (2) they are common contaminants; and (3) evidence suggests that they break through granular activated carbon adsorption systems earlier than higher molecular weight toxic materials and therefore as a monitor, place more stringent operating requirements on the process giving a higher level of protection to the consumer.

EPA researchers have determined that over 30 purgeable halogenated organic compounds can be detected and most resolved by use of the two gas chromatographic columns under the conditions listed for use in the approved purge and trap trihalomethane method (1) and when the "multipurpose" trap containing silica gel is used. Table 7.1 lists these compounds in order of increasing retention time when a 0.2% Carbowax 1500 on Carbopak C column is used.

TABLE 7.1: COMPOUNDS DETERMINABLE ON 0.2% CARBOWAX 1500 ON CARBOPAK C

Compound	
chloromethane	
bromomethane	
vinyl chloride	
chloroethane	
methylene chloride	
1,1-dichloroethylene	
bromochloromethane	
1,1-dichloroethane	
trans-1,2-dichloroethylene	
chloroform	
1,2-dichloroethane	
1,1,1- trichloroethane	
carbon tetrachloride	
bromodichloromethane	
1,2- dichloropropane	unresolved (with 2,3-dichloro-1-propene)
2,3-dichloro-1-propene	
trans-1,3-dichloropropene	
trichloroethylene	
1,1,2- trichloroethane	
dibromochloromethane	unresolved (1,1,2-trichloroethane, dibromochloromethane, *cis*-1,3-dichloro-1-propene)
cis-1,3-dichloro-1-propene	
1,2-dibromoethane	
2-bromo-1--chloropropane	
bromoform	unresolved (with 1,1,1,2-tetrachloroethane)
1,1,1,2-tetrachloroethane	
tetrachloroethylene	
1,1,2,2-tetrachloroethane	
1,4-dichlorobutane	
chlorobenzene	

A second column (n-octane on Porasil-c) changes the retention order (see Table 7.2), gives different separations, and allows detection of some organic compounds with higher boiling points. Compounds are listed by increasing retention time.

TABLE 7.2: COMPOUNDS DETERMINABLE ON n-OCTANE ON PORASIL-C

Compound	
vinyl chloride	unresolved (with chloromethane)
chloromethane	
bromomethane	
1,1-dichloroethylene	
chloroethane	
trans-1,2-dichloroethylene	
methylene chloride	
carbon tetrachloride	
chloroform	unresolved (with *cis*-1,2-dichloroethylene)
cis-1,2-dichloroethylene	
1,1-dichloroethane	unresolved (with bromochloromethane)
bromochloromethane	
1,1,1-trichloroethane	unresolved (with trichloroethylene)
trichloroethylene	
bromodichloromethane	

(continued)

TABLE 7.2: (continued)

Compound	
dibromomethane	unresolved (with tetrachloroethylene)
tetrachloroethylene	
1,2-dichloroethane	
dibromochloromethane	unresolved (with *trans*-1,3-dichloro-1-propene and 1,2-dichloropropane)
trans-1,3-dichloro-1-propene	
1,2-dichloropropane	
cis-1,3-dichloro-1-propene	unresolved (with 1,1,2-trichloroethane)
1,1,2-trichloroethane	
2-bromo-1-chloropropane	
chlorobenzene	
1,2-dibromoethane	
bromoform	
1-chloro-1-hexene	
chlorohexane	
1,1,2,2-tetrachloroethane	unresolved (with pentachloroethane)
pentachloroethane	
o-chlorotoluene	
m-dichlorobenzene	unresolved (with hexachloroethane and *p*-dichlorobenzene)
hexachloroethane	
p-dichlorobenzene	
1,4-dichlorobutane	unresolved (with *o*-dichlorobenzene)
o-dichlorobenzene	
hexachlorobutadiene	
1,2,4-trichlorobenzene	

Therefore, the operational definition of those low molecular weight halogenated compounds excluding trihalomethanes related to performance Criterion 1 is those compounds that can be detected by use of the U.S. EPA-approved trihalomethane method—purge and trap version (1). Determining these compounds as part of the granular activated carbon adsorption system process control program should not be a large analytic burden when the water utility required to use granular activated carbon treatment has selected the proper approach for the required measurement of trihalomethane concentrations.

The purge and trap method is selected for this analysis mainly because gas chromatographic conditions have not been well defined for resolution of as many compounds by the liquid-liquid extraction method (2), and because of the difficulty of solvent interference in the liquid-liquid extraction method with the detection of many compounds eluting sooner than chloroform. This is not to say, however, that these conditions could not be developed.

Additionally, care must be taken in selecting a particular purge-and-trap apparatus for introduction of the sample into the gas chromatograph, as a well designed apparatus is needed to eventually accomplish resolution of the same lower boiling compounds (vinyl chloride, bromomethane, and so forth). Details of this are to begin in a supplemental writeup extending the coverage of the EPA approved trihalomethane method to this larger group of compounds.

In spite of the resolving power of the two columns used for the purge and trap trihalomethane method (1), all possible purgeable organohalides and isomers cannot be identified unambiguously. Identifications by dual column GC-specific halogen detection are still presumptive. In order to avoid escalating analytical costs (e.g., mass spectrometric detection), these presumptive identifications are considered sufficiently accurate and granular activated carbon adsorption treatment should be adjusted to reduce any of these contaminants to less than 0.5 μg/l based on a comparison with an analytical standard of the compound of presumed identity, "Performance Criterion 1." Except in the case of some isomer differentiations, MS confirmation of compound identity might be useful enough to warrant the effort and could ultimately lead to a better designed treatment strategy.

Organic Carbon—Performance Criteria 2 and 3

Analytic Method: A method for the determination of organic carbon must include a step to remove or adjust for the presence of inorganic carbonate. Instruments on the market today, using manufacturers recommended procedures, do this adequately and efficiently. Many procedures, however, effect the removal of carbonate by acidification with mineral acid followed by exhaustive purging of carbon dioxide to waste. In this process, a purgeable fraction of organic carbon (POC) is lost.

Ideally, this purgeable fraction should be included in any total organic carbon (TOC) measurement. Research in WSRD laboratories, however, has shown that, in general, the POC concentration is only a small fraction of the TOC concentration and little organic carbon (<10 μg/l) would be lost by measuring only the nonpurgeable organic fraction (NPOC). Only occasionally have POC concentrations been found in the range of 0.1 to 0.3 mg/l. When the insignificance of POC can be demonstrated, only NPOC need be measured. Note: For these cases NPOC determinations can be used to satisfy the TOC requirements of Performance Criteria 2 and 3. This can simplify the sampling requirements and analysis and expand the options for methods and instruments that can be used.

Because of the requirements of Performance Criterion 2 (measure a change of 0.5 mg/l) and the commonly found low concentrations of organic carbon in a granular activated carbon column effluent (<0.1 mg/l) the method used to measure organic carbon must be more precise and accurate when compared to methods and instruments used in the wastewater treatment field, where concentrations of organic carbon are often in the hundreds of milligrams per liter.

Required precision and accuracy for measurement of organic carbon in granular carbon column effluents therefore must be in the range of 0.1 mg/l±0.05 mg/l or better. Although no EPA standard procedures now exist for this purpose, instruments are available that can be operated according to manufacturers instructions to achieve these goals. At least one of these is capable of measuring both POC and NPOC independently.

The instrument (and method) must also be capable of measuring TOC concentrations as high as those expected to occur in the granular activated carbon column influent (2 to 10 mg/l). This is well within the range of instruments capable of the required low level analysis.

Sampling: Because of the variability with time of concentrations of organic carbon in actual treatment situations, determining what constitutes a representative "baseline" (TOC_o) sample for all systems is difficult. Additionally, activated carbon fines emanating from new granular activated carbon adsorbers have been observed to influence organic carbon determinations. For these reasons, the first organic carbon measurements are recommended to be performed after one week of adsorber operation and be performed at least in triplicate twice during a 2-day period (six samples each on adsorber influent and effluent).

The average of effluent values not influenced by activated carbon fines (activated carbon is not visible, or samples do not have a NPOC value considerably higher than others of the set) is then the "baseline" (TOC_0) effluent concentration against which future effluent samples are compared, Performance Criterion 2, and determines the initial organic carbon removal percentage, Performance Criterion 3, when compared with a "baseline" influent concentration.

Sample containers should be the same as those used for sampling for trihalomethane analysis when the POC fraction is to be measured. Headspace requirements are not as important when the POC fraction has previously been shown to be insignificant.

TERMINAL SUMMATION TRIHALOMETHANES—MAXIMUM CONTAMINANT LEVELS

> The material in this section is from Appendix 4 of the *Interim Guide for Trihalomethanes.* (See the bibliography for further details.) It consists of a detailed method for the determination of terminal trihalomethane (TermTHM) concentrations in water. The principle of TermTHM measurement is the special case where the individual species TermTHM concentrations in the units μg/l are arithmetically summed.

General

Because their formation is not instantaneous, THM concentrations increase in the water as it flows through a water treatment plant (unless removed during treatment) to reach some value higher than that which would be observed if an analysis for THM species was performed immediately after sampling at the first point of chlorination. Further, the consumer is likely to receive water with THM concentrations higher than those leaving the plant because the reaction proceeds in the distribution system.

Additionally, not only are the concentrations of THM time dependent, but the rate of the reaction is dependent on pH, precursor concentration, nature of precursor(s), temperature, and to some degree free chlorine concentration early in the chlorination process (3)(4)(5). Finally, the ratio of chloroform to other trihalomethanes is highly dependent on the bromide content of the source water (6)(7). These factors are discussed in detail below.

Definitions

Basic to this discussion are these three important definitions:

Instantaneous THM (InstTHM) Concentration: The concentration of THM in the water at the moment of sampling – This may be expressed in terms of the individual species or their sum as ΣTHM. (The term "total" is used in the proposed regulation, rather than "summation.")

Terminal THM (TermTHM) Concentration: The concentration of THM that occurs at the termination of the measurement of this parameter – To measure TermTHM concentration, chlorine-precursor reaction conditions are selected according to the treatment practiced at the particular plant being evaluated. In general, a sample of water is chlorinated under these plant conditions and chloroform and other THM species are measured after a specified time period.

TermTHM concentration is equally important as a parameter for evaluating consumer risk as is the InstTHM concentration, but because this parameter is a measure of the sum of the amounts of THM species already present (instantaneous) and those formed during the reaction time, a third parameter must be defined that is useful for evaluating unit process performance for removal of unreacted precursor.

THM Formation Potential (THMFP): THMFP is measured as the increase in THM concentration that occurs during the storage period in the determination of the TermTHM concentration. The THMFP is obtained by subtraction of the InstTHM concentration from the TermTHM concentration either when TTHM or when the individual species data are used. THMFP is a measure of the portion of the total precursor material of most concern to the consumer remaining in the water at a given point in the treatment train. This parameter, when measured on unit process influent and effluent samples, can be used to determine the efficiency of that process for removal of that pertinent fraction of precursor material.

The distinction between THMFP and a total precursor parameter is important. Total precursor concentration is the concentration of all organic THM precursor materials present in the water that could react with halogen species under conditions that maximize the yield of trihalomethanes.

Because the identities of these organic compounds are not precisely known, total THM precursor concentration could also be expressed as concentrations of THM or concentration of TTHM obtained from that reaction. However, no standardized procedure for measuring this parameter exists, and considerable research will be required to establish the optimum conditions to assure the complete reaction of all precursor(s) to yield maximum trihalomethane concentrations.

Because the chlorination conditions for TermTHM concentration measurement are somewhat less than optimum for THM formation, the TermTHM concentration in that determination obtained will be somewhat less than the theoretical maximum THM concentration. Thus, the value obtained for TermTHM under these conditions will yield by subtraction of the InstTHM concentration, a THMFP smaller than the theoretical "total precursor" parameter.

Although the value obtained (THMFP) is not the total precursor concentration, it is an index of the concentration of materials of most concern relative to THM formation at a particular water treatment plant. Also, because controlling parameters (under treatment plant conditions) are easily measured at the operating plant, TermTHM concentration (and therefore THMFP) is a practical measurement.

Measurement of Instantaneous THM Concentrations

For this measurement the reaction of chlorine with precursor materials must be halted at the time of sampling with the goal being to measure only trihalomethanes present at the time of sampling. In the WSRD laboratory and in others, a small amount of reducing agent is added to the sample to react with the chlorine and thus render it unavailable for oxidation or substitution reactions. A small increase in trihalomethane concentrations upon storage after addition of reducing agent is usually observed, and this is probably caused by a slow hydrolysis of certain trihalo-intermediates. The hydrolysis step, of course, does not require the presence of chlorine.

The distinction should be made between this minor effect on the InstTHM concentration and the continued THM formation reaction when no reducing agent is added. The increase in THM concentration during storage after the addition of a reducing agent has, in our experience, amounted to only a few percent of the total value. The effect should be most noticeable at neutral pH because the hydrolysis step would be accelerated at high pH and be near completion soon after the intermediates are formed. Therefore, intermediates would be present in low concentration when the reducing agent is added (little change would be observed after dechlorination).

Procedure for InstTHM Determinations: Normally, the sample is taken, head space free in muffled vials, and sufficient sodium thiosulfate is added to the vial prior to filling to completely reduce any chlorine present in the sample. In the WSRD laboratory 1 ml of 0.1 N sodium thiosulfate for each 100 ml of sample has been used successfully. This is 0.5 ml or about 10 drops for a 50 ml serum vial.

The amount used is not critical because on a stoichiometric basis this is an excess of reducing agent, and the volume used is insignificant compared to the size of the sample. The filled bottle is then sealed with a Teflon faced septum held in place by a crimped top or screw cap and placed under refrigeration to retard microbiological activity while the samples await analysis. This procedure has been used routinely for about two years in this laboratory (3)(8)(9) without any apparent problems. Potassium ferrocyanide, sodium sulfite, and ascorbic acid, have also been used successfully as reducing agents for this application (4)(10)(11).

An alternate suggestion to obtain the InstTHM concentrations is to analyze the sample immediately after sampling, providing a chlorine residual will not interfere with the analysis, but this is often inconvenient or impossible. Another suggested option was recently described (12), in which the THM species are separated from the water sample upon sampling by sorption on a macroreticular resin column. Whether or not the hydrolysis of trichloro-intermediates will affect this result is not known, however.

The actual method of determination of the THM concentrations is not critical, and acceptable procedures vary widely. At the WSRD laboratory the method described in the NORS report (13) has been used continuously in research since that survey because of its convenience and reliability. For this analysis, the sealed sample prepared as above is brought to 25°C prior to opening in order to obtain reproducible purging efficiencies.

A 5 ml aliquot is then removed and transferred to a glass purging apparatus wherein the trihalomethanes are stripped from the aqueous phase by passage of a flow of helium upward through the sample. The trihalomethanes stripped in this manner are collected on a sorbent porous polymer material contained in a stainless steel trap that is placed in series with the purging device. The trihalomethanes are then thermally desorbed from the trapping material onto a gas chromatographic column. Finally, temperature programmed gas chromatography is carried out, and the concentrations of trihalomethanes are measured by use of a halogen-specific detector.

Measurement of Terminal THM Concentration and THM Formation Potential

These two parameters are discussed together because, as mentioned above, the measurement for TermTHM concentration together with the InstTHM concentration yields the THMFP by subtraction. The TermTHM concentration is measured by reacting chlorine with THM precursors in a given sample under certain controlled conditions that affect yield and rate of formation of the trihalomethanes and subsequently measuring the concentrations THM species produced.

Because for the reasons described earlier, this is not a total precursor concentration measurement, the selected conditions for this measurement must be the same as those experienced at the water treatment plant under study and be reproducible from sample to sample. Critical conditions to consider are: (1) time of reaction (time elapsed before halting the halogenation reaction with a reducing agent), (2) maintenance of a free chlorine residual, (3) temperature, (4) pH, (5) prevention of loss of the volatile products during the time of reaction, and (6) avoidance of contamination of reagents. These will be discussed below.

Effect of Time: Although a single measurement of THM concentrations after a storage period of several days in a bottle under appropriate conditions can give a useful determination of TermTHM concentration for that specified time, much more information can be gained from the reaction rate curve(s) obtained by plotting THM concentrations vs time (3)(8)(9). The rate curve(s) obtained by periodic measurement of THM concentrations of properly stored finished water can be used to estimate the future THM concentrations at any given time after water leaves the treatment plant.

This is particularly important when the goal of the analyst is to estimate ultimate consumer exposure to THM at different points along the distribution system. The THM concentration vs time curve is especially useful where the utility has a large variation in the time water is in various parts of the distribution system. The rate curve can also be used to estimate THM concentrations at any given time after water is taken from a sampling point within the plant when the purpose is to use the concentration obtained to calculate the THMFP at that point in treatment for evaluation of unit process effectiveness.

In any system, it is preferable to generate a rate curve, at least initially, so that the nature of the reaction that occurs at that location can be determined. For example, Figure 7.1 shows two hypothetical curves describing the rate of chloroform formation that might be expected for finished waters of distinctly different quality after leaving typical water treatment plants.

Curves A and B in Figure 7.1 represent two extreme situations that might occur. Although at time T, the chloroform concentrations are the same for both waters, the short term chloroform concentration is greater in Plant A and long term chloroform concentration is greater in Plant B.

TABLE 7.1: FORMATION OF CHLOROFORM UNDER WIDELY DIFFERENT TREATMENT PLANT CONDITIONS

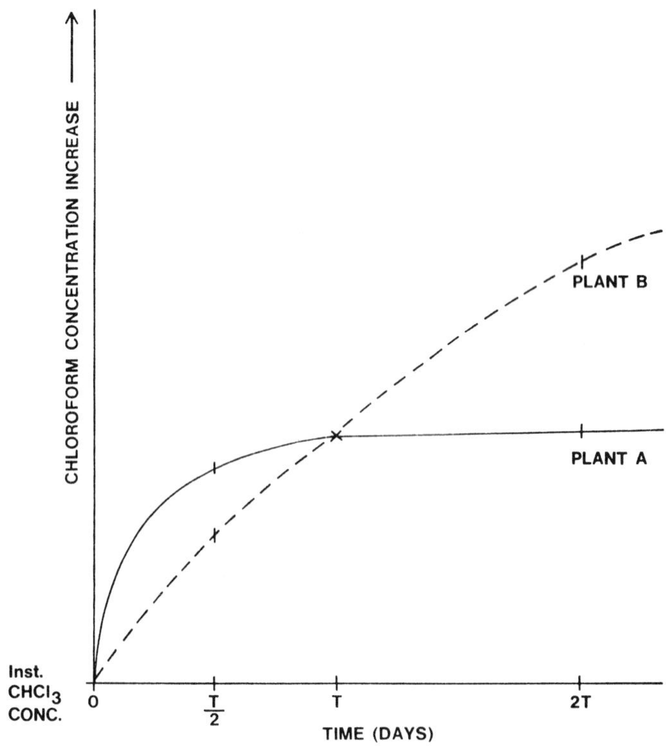

A Plant A curve would be expected where chloroform formation potential is relatively low but the precursor present is of the type that reacts quickly under the given conditions (3); that is, the final concentration of chloroform is reached early. A Plant B curve would be expected where chloroform formation potential is high, but the reaction with chlorine is slow, because of the nature of precursor or reaction conditions. Thus, these curves are more informative than a single chloroform determination performed at time T and the single measurement at each plant could easily be misinterpreted to mean that the plant situations were the same.

Good approximations of both curves are obtained by the selection of three or four points beyond time 0 (instantaneous value) such as 1/2T, T, 2T as shown, where 2T is equal to or slightly longer than the maximum distribution system residence time. These added analyses could cause the generation of rate curves to be time-consuming, especially if conditions are such that reactions are slow and the distribution system residence time is long. If the development of the rate curve is beyond the capability of a utility, the time for the determination of TermTHM concentration should be the longest residence time in the distribution system as this represents the most stringent condition for that utility.

Maintenance of Chlorine Residual: In conventional U.S. water treatment practice, maintenance of a free chlorine residual throughout the distribution system is often recommended or required. The continued reaction of precursor with chlorine to yield trihalomethanes depends on the maintenance of a free chlorine residual (1). Thus, one of the prime conditions necessary for THM formation is widely maintained. Again, with chloroform as an example, the "raw" water curve presented in Figure 7.2 shows the abrupt cessation of chloroform production as the chlorine became depleted. The chlorine dose was 8 mg/l at 25°C, pH 7.

FIGURE 7.2: EFFECT OF TREATMENTS ON CHLOROFORM FORMATION

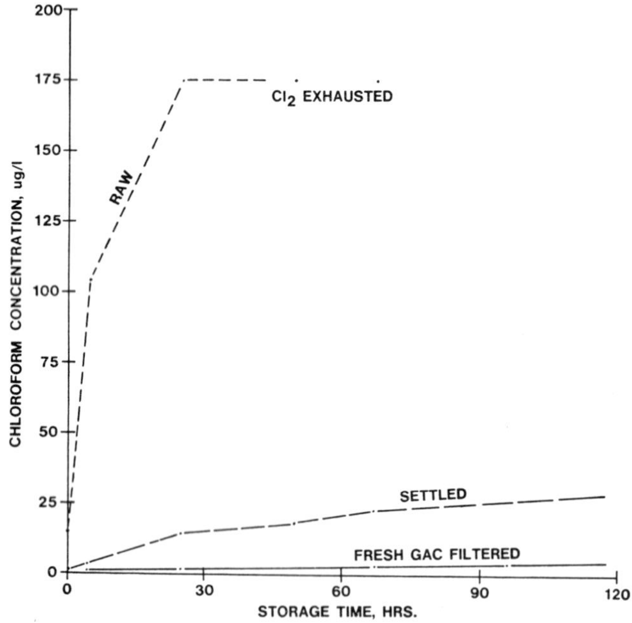

The 24-hour and all later samples gave the same chloroform concentrations, and chlorine residual determinations confirmed the lack of chlorine. Thus, the 24-hour and later chloroform concentrations could be misleading, assuming one of the conditions in the water utility under investigation was maintenance of a chlorine residual throughout the distribution system.

For example, a single 48-hour chloroform determination without an accompanying chlorine residual measurement would give a misleadingly low terminal chloroform concentration and chloroform formation potential. Thus, for evaluation of systems where free chlorination is practiced, to assure that these misleading results are not obtained, a chlorine residual measurement must always be performed at the time of THM analysis to assure that a free residual is present.

Work at the WSRD laboratory seems to indicate that TermTHM concentrations are not affected significantly by the amount of free chlorine present. This may be only because the concentrations usually are limited by the amounts of precursors present. Some work (5) has indicated that TTHM formation rates may be dependent on free chlorine concentration where the reaction is not precursor limited. In addition, Rook (4) has published data showing an increase in chloroform formation rate with an increase in chlorine dose. Since Rook did not show TTHM concentrations, whether this was caused by an increase in overall trihalomethane formation rate or simply a change in chlorine to bromide ratio cannot be determined (this effect of bromide concentration is discussed below).

Because of the uncertainty of the effect of chlorine concentration on reaction rate, the dose used in the TermTHM determination should be nearly the same as that used at the treatment plant, and because that dose is adequate for maintenance of a distribution system residual, it should be adequate to supply the required residual for the duration of the test.

Effect of Temperature: Upon chlorination of a natural water approximately twice as much chloroform can be formed in a given period of time at 25°C as is formed at 3°C (3)(4). This range of temperature is not uncommon, summer to winter, in U.S. surface waters. A need for close temperature control during the determination of TermTHM concentration is, therefore, indicated. Temperature is largely seasonally controlled, and, for a given system, an estimated average temperature of the distribution system is the logical choice for the controlled reaction temperature.

Effect of pH: The trihalomethane formation rate has been shown to increase with an increase in pH (3)(4). This increase is expected because the haloform reaction is base catalyzed (14). The selection of the pH for the controlled reaction during the TermTHM concentration determination is not as straightforward as that for reaction time and temperature discussed above, however. The variation of pH through an operating water treatment plant can be quite wide and the variation is operationally controlled.

If the determination of the TermTHM concentration and the THMFP for the finished water only is desired, pH selection is not a problem. The samples should be stored at the finished water pH. If, however, a comparison of the THMFP of the finished water with that of the source (raw) water, or with water at any stage of treatment to evaluate success of a unit process in reducing THMFP is desired, the selection of pH is more difficult.

The analyst must be sure that the same portion of the total precursor concentration (pH dependent) is reacting at each point, and that the reaction rate of chlorine with that material (also pH dependent) is the same at each point. Thus, all of the samples from each of the various sampling points must be chlorinated and stored buffered at a single selected pH value.

Therefore, because the THMFP test is designed to measure the portion of the total precursor that is significant in a given water as it leaves the treatment plant, the logical selection of the single reaction pH value is that of the finished water entering the distribution system, as with the choice of temperature.

This choice can lead to seemingly anomalous results, however. For example, where pH is high through a unit process the THMFP at that pH might be higher than THMFP for the same water measured at distribution system pH. Because the reactions in the distribution system occur at a lower pH, reactions which occur at the high pH at higher rates and involving a different portion of the total precursor are not important to the consumer. Therefore, that part of the total precursor should not be measured as THMFP and will not be if the samples are buffered at the distribution system pH.

Loss of Volatile Species: To prevent misleading losses of trihalomethanes produced during the reaction period, the reactions must be carried out in sealed, head-space-free containers. Container materials should be all glass or glass with Teflon lined caps. Standard glass stoppered reagent bottles filled to overflowing so as to set the stopper surface or the Teflon septum-sealed serum vials used for sampling for InstTHM determinations have been found suitable.

Effect of Bromide or Iodide Contamination: As mentioned earlier, bromide or iodide present in the water can, as a result of first reacting with chlorine, cause formation of THM species other than chloroform. In the case of bromide, the relative amounts of THM species formed have been shown (7) to be highly dependent on the bromide content of the water and the chlorine dose, presumably because these determine the ratio of bromine to chlorine available for competing reactions. Although the product ratios change, the effect of bromide may be small when TTHM is calculated, however.

Preliminary work indicates that equal amounts of bromine and chlorine substitution as trihalomethanes would be expected if the original bromide concentration is as little as 2% of the chlorine dose (7). Clearly, any bromide (or iodide) contamination of reagents used will cause a different ratio of THM species to be formed than would normally occur on chlorination of that water under plant conditions. Where individual THM species data are used, this effect could be the cause of misinterpretations of data.

For example, if only chloroform concentrations are reported and reagents are contaminated with bromide the TermTHM and THMFP tests would give lower chloroform concentrations than those observed in the plant or in distribution. However, measurement of all THM species would reveal that the bromine-containing species were present in higher concentrations and the TTHM concentrations approximately the same. As mentioned earlier, a change of chlorine dose where bromide content of the water is constant could cause the same effect.

Effect of Precursor Contamination: In the WSRD laboratory distilled, deionized activated-carbon-filtered water has been used for "blank" water for reagent preparation. At pH 7 the contribution of precursor in reagents has been small. At higher pH, however, blank values tend to be higher. Care should be taken to minimize volumes of reagents used in TermTHM measurements in order to avoid this contribution to the THM concentrations obtained.

Procedure for Terminal THM, and THM Formation Potential Determinations

A test for TermTHM concentrations and THMFP can be standardized in approach, but the conditions for sample treatment and storage, and therefore the portion of the total precursor concentration measured, will vary widely from system to system depending upon distribution system residence time, total chlorine demand of the sample, ambient temperature of the system, and pH of the finished water in the particular system under investigation. These variables must be chosen to match those in the system.

In work at the WSRD laboratory, a large (1 to 3 liters) sample of water is collected and the pH is adjusted to that selected with an appropriate inorganic (e.g., phosphate or borate) buffer. The final buffer strength is about 0.01 M. The sample is then chlorinated, if needed, by the addition of a previously standardized chlorine or hypochlorite solution. (Sufficient chlorine is added at this time to maintain a free residual for the duration of test period. Water leaving the treatment plant should already contain sufficient free chlorine and the chlorination step is not needed.) Several sample bottles are filled and capped head space free—two bottles for each point to be determined on the rate curve.

For example, four bottles are needed if only two points on the curve are to be determined, that is, initial (zero time) and final values. Two more bottles are required to determine each intermediate point. One of the "zero" time sample bottles contains sodium thiosulfate to immediately reduce the chlorine so only the InstTHM is measured. The other "zero" time sample has no reducing agent and is used for measurement of the chlorine residual. This entire sequence from sample collection to the capping of the bottles should be done as quickly as possible to avoid loss of InstTHM during the manipulations.

The samples, except "zero" times, are then stored at the selected temperature. After the preselected time, or times, if the reaction rate curve is to be determined, one sample bottle is opened and an aliquot is transferred by pouring into a smaller bottle containing sodium thiosulfate to prevent further reaction of precursor with chlorine. This smaller bottle is then quickly sealed head space free to await THM analysis. This measurement determines a THM concentration for the respective time on the rate curve, and the analytical procedure is exactly as described above for InstTHM determination. A second bottle is opened at the same preselected time(s) and the chlorine residual is measured. More details of this procedure are given in References (3)(13)(14).

After the above determinations are completed the THM concentration measured from the "zero" time sample is the InstTHM concentration; the THM concentration from the bottle stored for the longest time is TermTHM concentration. The difference between the InstTHM concentration and the TermTHM concentration yields the THMFP.

A much simpler approach to the determination of "free" (Inst) and "potential" (Term) haloforms has recently been suggested by Nicholson (16). Nicholson uses gas chromatography with direct aqueous injection of both purged and unpurged samples. The introduction to the hot injection port of the aqueous sample results in hydrolysis of halogenated intermediates to yield trihalomethanes. Thus, the direct aqueous injection method is said to measure "total potential haloforms," and subtraction of the prepurged sample value gives the "free haloform" (Inst.) concentration. The method does not, however, take into account the effects of a free chlorine residual over a period of days, nor has the relationship of the high temperature hydrolysis to treatment plant conditions been established.

Examples of the Use of Methods—Interpretation of Results

Some hypothetical examples will help to demonstrate the use of the two experimental determinations and the calculated THMFP to estimate both consumer exposure to trihalomethanes resulting from the chlorination process and the efficiencies of the various unit processes within the plant for removing precursor compounds during treatment. The efficiency of unit processes for removing chloroform or other trihalomethanes can also be estimated.

Simple Chlorination: The first example (Figure 7.3) represents the simplest case, a water treatment plant with chlorination only. Figure 7.3 depicts the relative values for the parameters that might be obtained if analyses were conducted for the InstTHM concentration and TermTHM concentration for source water **A**, plant clearwell **B**, and a theoretical point at the maximum residence time in the distribution system **C**.

FIGURE 7.3: TRIHALOMETHANES FORMED DURING WATER TREATMENT BY CHLORINATION ONLY

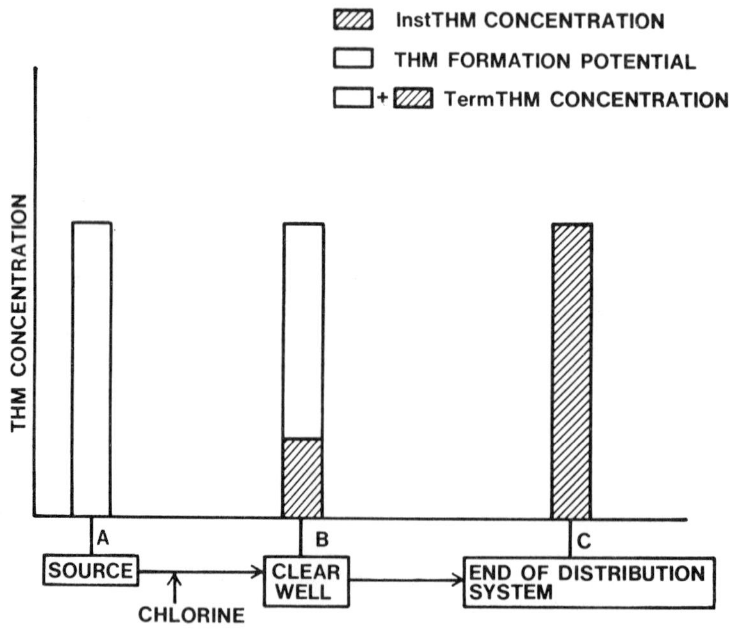

[Recall that for simplification the trihalomethanes are being discussed here as a group. Each bar could represent the single group index (TTHM), any one of the individual species, or be subdivided horizontally into four bars of different heights to represent all four commonly found trihalomethanes.] According to the bar graph, trihalomethanes were absent in the untreated source water (InstTHM was not found on analysis of source water), but the full THMFP was present and equal to the TermTHM concentration obtained experimentally. At the clearwell, some of the precursor measured as THMFP had reacted to form trihalomethanes (measured as InstTHM in finished water) leaving a smaller remaining THMFP. The remaining THMFP, plus InstTHM concentration equals the TermTHM concentration determined originally on the source water.

At point **C** the entire original THMFP had reacted to give an InstTHM concentration identical to the TermTHM concentration. No unit process exists at this plant that was effective for reduction of either TermTHM or InstTHM concentrations. The practice of chlorination itself converted THMFP to InstTHM, thereby causing a reduction in the THMFP. In assessing the THMFP removal by any unit process, care must be taken to separate removal of THMFP by conversion to InstTHM by chlorination and removal of THMFP by the unit process itself. Only at a point closer to the treatment plant than the maximum length in the distribution system is consumer exposure to THM lower than the TermTHM concentration shown in Figure 7.3.

Conventional Treatment: Shown in Figure 7.4, during conventional treatment with raw water chlorination, some THM is formed during rapid mixing and throughout the following treatment stages in the presence of chlorine. Thus, the InstTHM concentration increases as the water passes through rapid mixing, settling, and filtration–points **B**, **C**, and **D**.

FIGURE 7.4: TRIHALOMETHANES FORMED DURING CONVENTIONAL TREATMENT WITH RAW WATER CHLORINATION

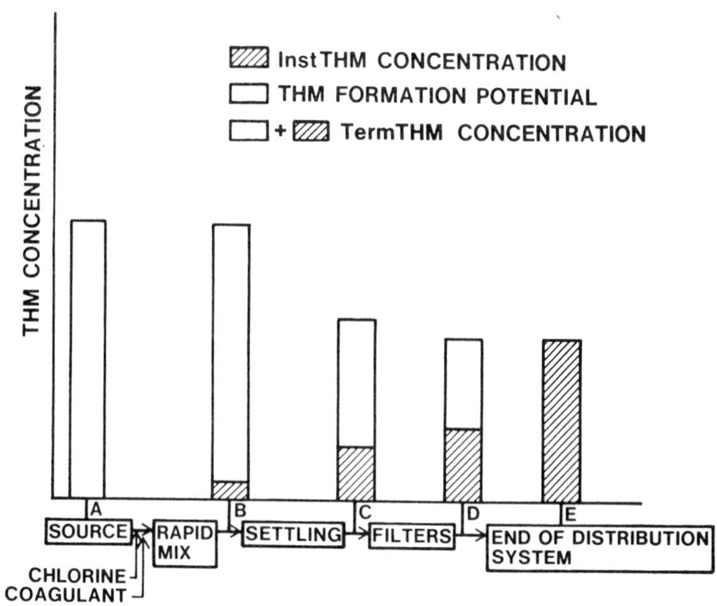

Coagulation and settling do reduce THMFP (i.e., precursor removal) so that parameter as well as TermTHM concentration declines from point **B** to **C**. Filtration removes a little more precursor material that is associated with the carryover floc so the THMFP declines slightly again from point **C** to **D**. The remaining THMFP is converted by chlorination to THM from point **D** to **E** and therefore the InstTHM concentration determined for a sample taken at that point in distribution equals the TermTHM concentration of the sample from point **D**.

Lime Softening: A case more complex than conventional treatment is illustrated in Figure 7.5. This example treatment plant employs lime softening, recarbonation after settling, and rapid sand filtration for treatment of the same water in the example shown in Figure 7.4.

FIGURE 7.5: TRIHALOMETHANES FORMED DURING TREATMENT BY LIME

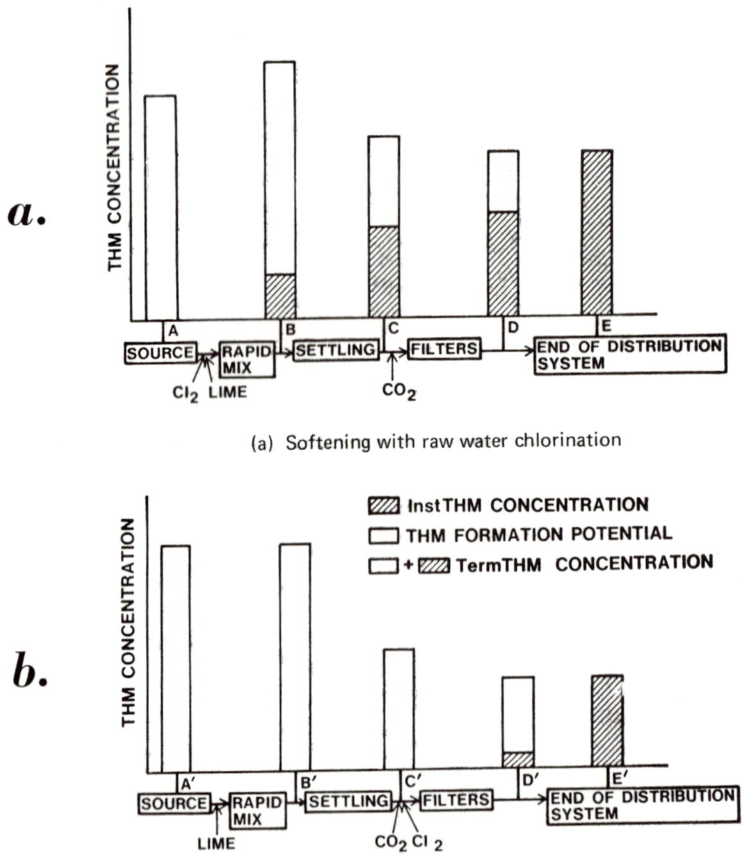

(a) Softening with raw water chlorination

(b) Softening with settled water chlorination

Two alternate points of chlorination are shown in Figure 7.5. The probable effect of raw water chlorination only is shown in the upper bar graph (Figure 7.5a); and chlorination after recarbonation only is shown in the lower bar graph (Figure 7.5b). The general explanation for the bar graphs of Figure 7.5 is the same as that for

Figure 7.4, except that some additional interesting effects of treatments are demonstrated.

Point **C** (after settling, before recarbonation) shows the effect of high pH on the reaction rate. In Figure 7.5a a large proportion of the original THMFP has been converted to THM as compared to Figure 7.4 where the pH in the settling basin was lower. Because the trihalomethanes produced through the lime softening process are carried through the treatment process, the TermTHM concentration measured after storage of samples **D** and **E** will include whatever InstTHM concentrations were formed at the accelerated rate during treatment. (This point will be elaborated on immediately below and when Figure 7.6 is discussed.)

The sum of THMFP and InstTHM concentrations (the TermTHM concentration) has declined between **B** and **C** because the settling during lime softening removes some precursor. From **C** to **D**, after recarbonation and sand filtration the rate of increase in InstTHM concentration is slowed as the pH is lowered so only a slight increase in InstTHM concentration is shown. Further, the filters remove a small additional amount of precursor(s) that is associated with floc particles, resulting in a slight decrease in the THMFP. The InstTHM concentration increases with time at the expense of THMFP between **D** and **E**.

In the lower bar graphs in Figure 7.5b, because of the location of the point of chlorination, no InstTHM concentration is shown until point **D** after recarbonation and chlorination. The InstTHM concentration increases during flow through the distribution system to point **E**. Note that the values for TermTHM concentration in the Figure 7.5b are lower at point **B′** through **E′** than those observed for their respective counterparts (**B** through **E**) in Figure 7.5a.

The first reason, as noted before, is that precursors that are otherwise removed during softening and settling are converted rapidly to trihalomethanes (shown as InstTHM) during the treatment process depicted in Figure 7.5a. The trihalomethanes are then carried through the treatment process.

The second reason is more directly related to the high pH during treatment and explains why the TermTHM concentration is higher at point **B** than at **B′** and is also higher at **B** than that measured at point **A**. Precursor materials that react at insignificant rates at the lower pH of the distribution system and during the TermTHM concentration test conducted at points **A**, **A′** and **B′** are converted rapidly to trihalomethanes during the treatment depicted by Figure 7.5a. This contribution of InstTHM concentration to TermTHM concentration at **B** does not appear in the TermTHM concentration measurement on water samples from **A**, **A′** or **B′**. These trihalomethanes are carried on to the subsequent sample points.

On the other hand, the THM Formation Potentials at points **C′** and **D′** are slightly higher than those at the corresponding points in Figure 7.5a. This is also caused by the rapid reaction of precursors at high pH. In this case, precursor materials represented by the THMFP in the lower scheme (test conducted at distribution system pH) reacted at a rapid rate during softening to form the InstTHM during the upper treatment process. That part of the THMFP material was therefore no longer available to appear as part of the calculated THMFP value on the upper bars **C** and **D**.

This reduction of THMFP might be expected to cause a reduction in TermTHM concentration through the upper treatment train, but the lessening of **C** and **D** THMFP is exactly compensated for by the corresponding increase in InstTHM concentration and therefore does not reduce the TermTHM concentration in that treatment mode. Of course, as noted above, settling and filtration does reduce the TermTHM concentration.

The above effects of the reaction of chlorine with precursor otherwise reduced in concentration during the softening process and the high reaction rate of chlorine at high pH with precursor unreactive at distribution system pH combine to cause the TermTHM concentrations to have higher values when source water chlorination is practiced. A more dramatic demonstration of these effects is shown in Figure 7.6.

FIGURE 7.6: TRIHALOMETHANES FORMED DURING TREATMENT BY LIME SOFTENING WITH LONG DETENTION TIME AT HIGH pH

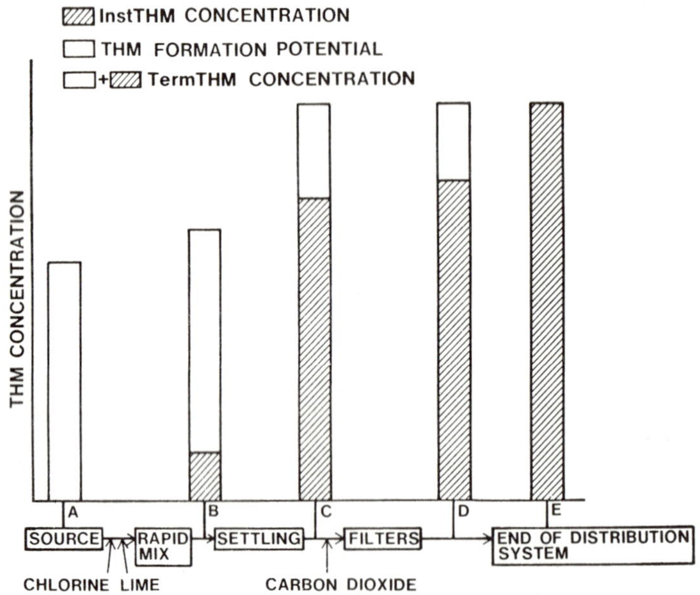

Because the TermTHM concentration determination is carried out at the distribution system pH, the InstTHM concentration after softening could theoretically exceed the TermTHM concentration established in the source water. This would occur if the detention time at high pH through softening and settling was long enough. Therefore, with sufficient contact time, the InstTHM concentration at the process effluent could exceed the TermTHM concentration measured under distribution system conditions at the influent to that process.

This type of result would not be inconsistent with the rationale for selecting testing conditions described in this paper, however. Reduction of THMFP through the unit process can still be directly calculated and that is the goal of the determination. The increase of InstTHM through the process is also easily calculated from the influent and effluent InstTHM concentrations.

Granular Activated Carbon Filtration/Adsorption: Figure 7.7 illustrates effects that might be observed at points through a water treatment plant that employs alum or iron coagulation, settling, and granular activated carbon (GAC) filtration/adsorption. Two alternate points of chlorination are shown.

FIGURE 7.7: TRIHALOMETHANES FORMED DURING TREATMENT BY GRANULAR ACTIVATED CARBON ADSORPTION

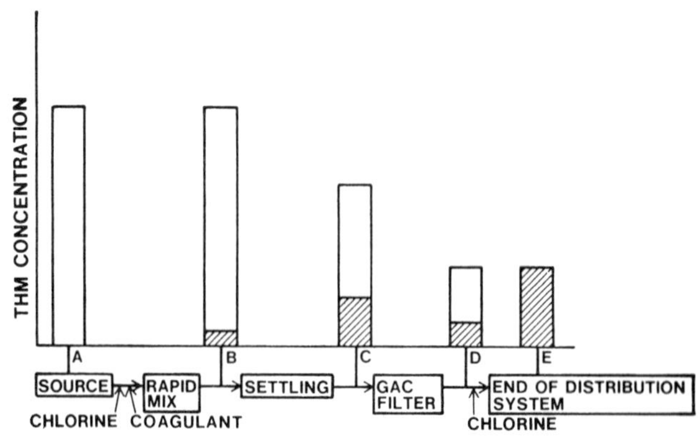

(a) With raw water chlorination

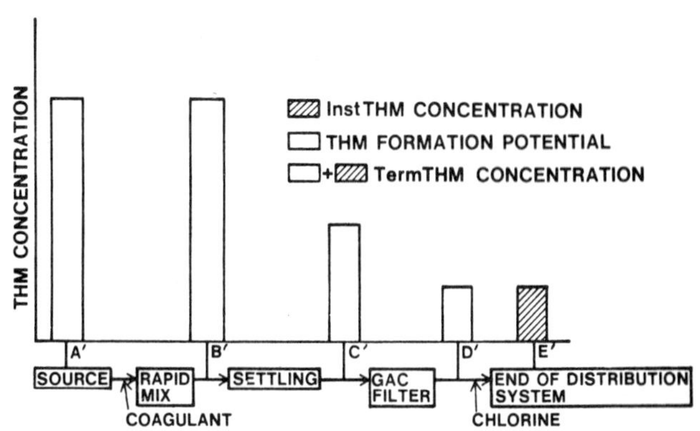

(b) With postchlorination only

Again, as in the lime-softening case, the TermTHM concentration at **C′** in Figure 7.7b is lower than the same parameter at point **C** in Figure 7.7a. Although the THMFP is similar at point **C′** to that at point **C**, the term THM value at point **C** is higher than at point **C′** because of the InstTHM concentrations resulting from the chlorination reaction through treatment (**A**-**C**). Here, however, the instantaneous value at **C** was caused only by the chlorination in the treatment plant of precursor that was present at points **A**, **B**, **A′**, and **B′** but was no longer present when the

TermTHM test was carried out on the chlorinated settled water at **C′** and was not influenced by a pH effect as noted previously. Because these precursors were removed during settling they were not available to contribute to the TermTHM concentration when sample **C′** was chlorinated.

The removal efficiencies of trihalomethanes and precursors vary widely with time in service of GAC filters (8)(9) and the efficiencies illustrated for this treatment step in Figure 7.7 are completely arbitrary. Note that although a 50% removal of precursor (as THMFP) and of THM (shown as InstTHM concentration) in the upper graph (Figure 7.7a) can seemingly be estimated (C minus D concentrations divided by C concentrations x 100), care must be taken in making such interpretations if this was a real observation.

THM is being produced during the time in a unit process at a rate dependent upon the physical and chemical factors mentioned previously and is removed at a competing rate according to the effectiveness of the process in use. The net rate is what is actually being measured through the process. For example, from Figure 7.7a, between points **C** and **D**, whether some amount of precursor THMFP between 0 and 50% reacted to produce THM that was efficiently adsorbed or a true 50% removal of precursor (THMFP) occurred cannot be determined. In other words, in the case of GAC adsorption systems, determining the separate adsorption efficiencies of THM and precursor in the mixed dynamic system is difficult.

At best, the effects can be described as 50% reductions as opposed to removal in InstTHM, TermTHM concentrations and THMFP with no connotation as to mechanism for this occurrence. This is not so complicated in the example shown in Figure 7.7b where no InstTHM is formed until after the filter adsorbs.

Summary of Examples: These four hypothetical examples are not designed as predictions of the success or failure of certain unit processes in a treatment train or to be indications of the relative effectiveness of those processes. The examples do serve to indicate the kinds of results that might be obtained when a plant is sampled for measurement of InstTHM and TermTHM concentrations and THMFP and when the results of these measurements are compiled for unit processes or whole plant evaluations. They also serve to illustrate the complexity of the design of in-plant studies of treatment processes for reduction of these parameters and some of the more important considerations pertinent to analysis of data obtained as a result of those studies.

Summary

Briefly, for the purge and trap (15) analysis, the sealed sample is brought to 25°C prior to opening in order to obtain reproducible purging efficiencies. A 5-ml aliquot then is removed and transferred to a glass purging apparatus, wherein the trihalomethanes are stripped from the aqueous phase by passage of a flow of helium upward through the sample. The trihalomethanes stripped in this manner are collected on a sorbent, porous polymer material contained in a stainless steel trap that is placed in series with the purging device.

The trihalomethanes are desorbed thermally from the trapping material onto a gas chromatographic column. Finally, temperature-programmed gas chromatography is carried out, and the concentrations of trihalomethanes are measured by use of a halogen-specific detector. The liquid-liquid extraction method (16) involves the extraction of a small volume of water with an even smaller volume of organic solvent, followed by gas chromatographic analysis of the extract using an electron-capture detector.

Weekly analysis of the effluent from a granular activated carbon adsorption system by this procedure will provide a good estimate of whether or not the delivered drinking water will be in compliance with the trihalomethane maximum contaminant limits as specified in the Interim Primary Drinking Water Regulations.

References

(1) "The Analysis of Trihalomethanes in Finished Waters by the Purge and Trap Method," Environmental Monitoring and Support Laboratory, Cincinnati, Ohio 45268 (Sept. 9, 1977).

(2) "The Analysis of Trihalomethanes in Drinking Water by Liquid/Liquid Extraction," Environmental Monitoring and Support Laboratory, Cincinnati, Ohio 45268.

(3) Stevens, A.A., Slocum, C.J., Seeger, D.R., and Robeck, G.G. (1975), "Chlorination of Organics in Drinking Water," Presented at the Conference on the Environmental Impact of Water Chlorination, Oak Ridge National Laboratory, Oak Ridge, Tennessee, Oct. 22-24.

(4) Rook, J.J. (1976), "Haloforms in Drinking Water," *Jour. AWWA,* 68, 168.

(5) Seeger, D.R. (1976), USEPA, Cincinnati, Personal Communication.

(6) Bunn, W.W., Haas, B.B., Deane, E.R., and Kleopfler, R.D. (1975), "Formation of Trihalomethanes by Chlorination of Surface Water," *Environmental Letters,* 10(3), 205-314 (1975).

(7) Moore, L. (1976), USEPA, Cincinnati, Personal Communication.

(8) Love, O.T., Jr., Carswell, J.K., Stevens, A.A., and Symons, J.M. (1975), "Treatment of Drinking Water for Prevention and Removal of Chlorinated Organic Compounds," An EPA Progress Report, Presented at 95th Annual Conference AWWA, Minneapolis, Minnesota, June 8-13.

(9) Love, O.T., Jr., Carswell, J.K., Stevens, A.A., and Symons, J.M. (1975), "Pilot Plant Studies and Measurement of Organics," Presented at 1975 Water Quality Technology Conference, American Water Works Association, Atlanta, Georgia, Dec. 8-10.

(10) Kopfler, F.C., Melton, R.G., Lingg, R.D., and Coleman, W.E., "GC/MS Determination of Volatiles for the National Organics Reconnaissance Survey (NORS) of Drinking Water," in *Identification and Analysis of Organic Pollutants in Water,* Keith, L.H., Ed., Ann Arbor Science Publishers, Inc., Ann Arbor, Michigan (1975), Chapter 6.

(11) Kissinger, L.D., Fritz, J.S. (1976), "Analytical Notes–Analysis of Drinking Water for Haloforms," *Jour. AWWA,* 68, 435.

(12) Fritz, J.S., as reported in *C&EN* (April 12, 1976), p. 35. See also: Stevens, A.A., and Kopfler, F.C., "Analyzing Drinking Water," *C&EN* (June 21, 1976), p. 5.

(13) Symons, J.M., Bellar, T.A., Carswell, J.K., DeMarco, J., Kropp, K.L., Robeck, G.G., Seeger, D.R., Slocum, C.J., Smith, B.L., and Stevens, A.A., "1975 National Organics Reconnaissance Survey for Halogenated Organics in Drinking Water," Water Supply Research Laboratory and Methods Development and Quality Assurance Laboratory, National Environmental Research Center, USEPA, Cincinnati, Ohio, *Jour. AWWA,* 67:634.

(14) Coleman, W.E., Lingg, R.D., Melton, R.G., and Kopfler, F.C., "The Occurrence of Volatile Organics in Five Drinking Water Supplies Using Gas Chromatography/Mass Spectrometer (GC/MS)," in *Identification and Analysis of Organic Pollutants in Water,* Keith, L.H., Ed., Ann Arbor Science Publishers, Inc., Ann Arbor, Michigan (1975), Chapter 21.

(15) Bellar, T.A., and Lichtenberg, J.J. (1974), "Determining Volatile Organics at the μg/l Level in Water by Gas Chromatography," *Jour. AWWA,* 66:739.

(16) Nicholson, A.A., and Meresz, O., "The Occurrence and Determination of Free and Total Potential Haloforms in Drinking Water," Presented at the 27th Pittsburgh Conference on Analytical Chemistry and Applied Spectroscopy, Cleveland, Ohio, March 1976.

Systems for Reactivation of Granular Activated Carbon

COSTS INVOLVED IN GAC REACTIVATION

The material below is excerpted from Appendix 1 of the *Interim Guide for Trihalomethanes* (see bibliography). This appendix is an overview of the costs of setting up a water purification system using GAC. The emphasis of the material taken for this section, however, will be the costs of the system which are affected by reactivation of the GAC. [The figures used in this section are of 1975.]

Treating water with activated carbon involves two major and separate process operations: filtration and reactivation. The water comes in contact with the carbon by passing through a structure filled either with carbon granules or with a carbon slurry. The structure can be either a water treatment filter shell, in which the filter media has been replaced with GAC, or an independent carbon filtration system. The separate carbon filtration system usually consists of a number of columns or basins used as contactors that are connected to a reactivation system. The primary focus of this economic analysis will be on the use of GAC as a replacement for existing media in the filter shell. The economics of a separate contactor system will also be examined.

After a period of use, the carbon's adsorptive capacity is exhausted and it must then be taken out of service and reactivated by combustion of the organic adsorbate. Fresh carbon is routinely added to the system to replace that lost during hydraulic transport and reactivation. These losses include both attrition due to physical deterioration and burning during the reactivation process.

The approach taken in this analysis is first to evaluate the use of GAC with an on-site reactivation system assuming that the GAC will replace the media in the filter shell. Various levels of key design parameters will be established at standard levels with the intent of evaluating their effect on sensitivity of the cost of GAC systems. The "standard" system will consist of fixing a given set of design variables at predetermined levels. Secondly, an analysis will be made for the replacement of GAC in the filter shell but with off-site or regional reactivation. Finally, an evaluation will be made of the cost of a separate GAC contactor system with on-site reactivation.

The Cost of GAC as Filter Media Replacement

For purposes of this analysis, a water treatment plant is assumed to consist of an integral number of 1 mgd filters. For example, a 10 mgd water treatment system is assumed to consist of ten 1 mgd filters each with the following dimensions: 18.5 x 18.5 x 2.5 feet, yielding a total volume of 865 cubic feet per filter.

The standard values and the design parameters chosen for examination are shown in Table 8.1. All analyses performed will be based on the effect of changing the design variables around these standard values.

TABLE 8.1: DESIGN PARAMETERS FOR GRANULAR ACTIVATED CARBON

	Level
Design parameters (variable)	
Carbon cost	0.38 ¢/lb
Carbon loss per reactivation cycle	10%
Fuel cost	1.26 \$/MMBtu
Electrical power cost	0.01 ¢/kWh
Construction cost index	256.7
Wholesale price index	178.1
Direct hourly wage rate	5.19 \$/hr
Amortization rate	7%
Amortization period	20 yr
Design parameters (fixed)	
Contact time	4.5 min
Hydraulic loading rate	2 gal/min/ft^2
Volume per filter (1 mgd)	865 ft^3
Capacity factor	70%
Reactivation frequency	1.4 mo
Loss in adsorptive capacity during reactivation	0

The impact of three basic factors must be considered–economy of scale, load factor and reactivation frequency. Plant size, whether the system has a capacity of 1, 5, 10, 100 or 150 mgd, for example, has considerable effect on cost, assuming that the design variables are held at the levels shown in Table 8.1. The unit cost for a 1 mgd plant is approximately 44 cents per 1,000 gallons, while the unit cost of a 150 mgd plant is close to 5.5 cents per 1,000 gallons (1975 figures). The cost curve rises sharply between 10 and 5 mgd, jumping from 12 cents to 15.5 cents per 1,000 gallons.

Figure 8.1 depicts the cost for a 100 mgd plant, operating at a 70% capacity factor, with the period between reactivation varying between 0.5 and 18 months. As might be expected, lengthening the time between reactivation reduces the unit cost from 7.5 cents per 1,000 gallons at 0.5 months to 1.6 cents per 1,000 gallons at 18 months.

FIGURE 8.1: TOTAL UNIT COST FOR A 100 MGD PLANT VERSUS TIME BETWEEN REACTIVATIONS IN MONTHS

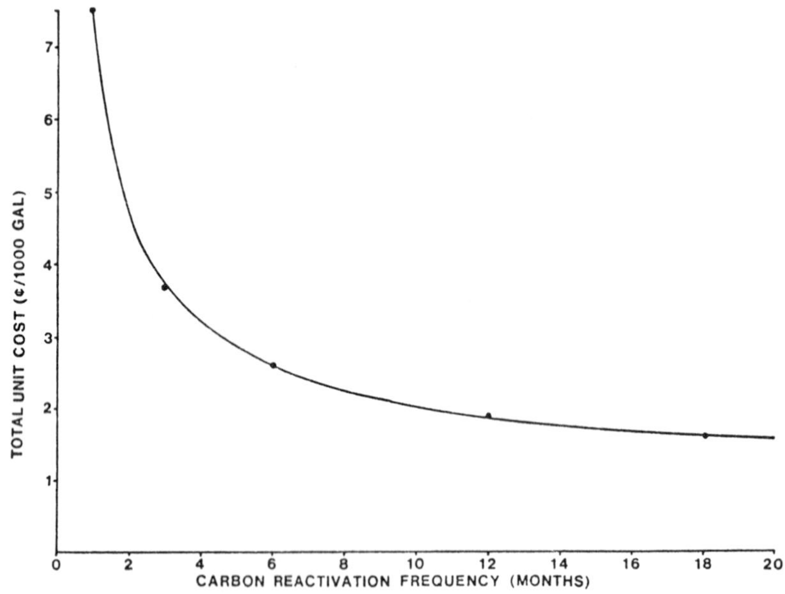

FIGURE 8.2: TOTAL UNIT COST FOR A 100 MGD PLANT VERSUS THE PRODUCT OF TIME BETWEEN REACTIVATIONS AND CAPACITY FACTOR

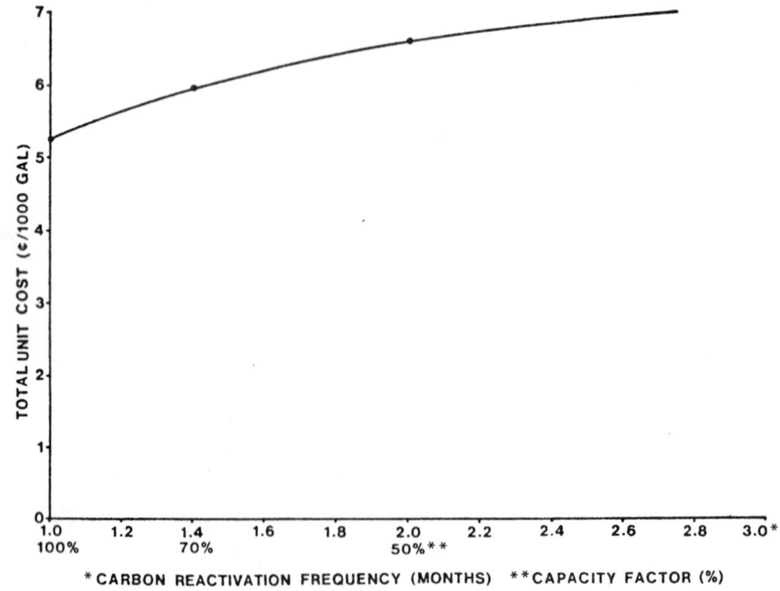

Figure 8.2 (on previous page) shows the interaction between capacity factor and reactivation frequency for a 100 mgd plant, in which it is assumed that the product of reactivation period and load factor is 1, and that as capacity factor decreases, the period between reactivations increases. For example, when capacity factor is 100%, the reactivation frequency is assumed to be 1 month, and when the capacity factor is 50%, the reactivation frequency is assumed to be 2 months. It can be seen that increasing the time between reactivation periods reduces unit costs; however, this reduced cost is offset by a reduced load factor which increases the unit cost. The net effect is an increased cost from 5.3 ¢/1,000 gal (100% load factor at 1 reactivation per month) to 6.2 ¢/1,000 gal (50% load factor at 1 reactivation every 2 months).

Cost Sensitivity to Design Changes

Having established the impact of these three variables (load factor, reactivation frequency, and economies of scale), it is possible to examine the sensitivity of cost to changes in the design parameters in Table 8.1. Some of these variables influence only operating and maintenance cost, some only amortized capital cost, and some of these variables affect both. The first group of variables to be examined that influence operating and maintenance cost are as follows: hourly wage rate ($/hr), carbon loss per reactivation cycle, fuel cost, wholesale price index, and electrical power cost.

Changes in hourly wage rate have a greater impact on the cost of small plants than on large plants. For example, it can be seen that as the hourly rate increased from 5.19 $/hr to 7 $/hr, the O&M cost for 1 mgd plant increases from slightly over 21 ¢/1,000 gal to slightly less than 28 ¢/1,000 gal. The same wage rate increase in a 150 mgd plant increases the O&M cost from approximately 4 ¢/1,000 gal to 4.5 ¢/1;000 gal.

O&M cost has been found to be very sensitive to changes in carbon loss but somewhat less sensitive to changes in fuel cost, wholesale price index, and electric power cost.

Several of the design parameters listed in Table 8.1 influence both amortized capital and O&M cost. These parameters are as follows: activated carbon cost, carbon reactivation frequency, and the interaction of carbon reactivation frequency and load factor.

Impact of Local Conditions: To illustrate how the sensitivity analysis can be applied to the standard values in order to study the impact of local conditions on costs, the following example has been constructed. If it were assumed that all of the standardized values were held at the levels shown in Table 8.1, with the exception of activated carbon loss, it would be possible to estimate the impact that changes in its value would have on the system cost. For example, compared to the standardized values when activated carbon loss is 15% for a 100 mgd plant, the percent change in O&M cost is +29.5%, but when it is at 5%, the change is −14.7%. The standardized values yield an amortized capital cost of 1.5 ¢/1,000 gal and an O&M cost of 4.5 ¢/1,000 gal. Therefore, as carbon loss affects only O&M cost, the change in total cost would be as follows:

(1) GAC system cost (15%) = 1.5 + [4.5 + 4.5 (.295)] = 7.3 ¢/1,000 gal

(2) GAC system cost (5%) = 1.5 + [4.5 − 4.5 (.147)] = 5.3 ¢/1,000 gal

Figure 8.3 illustrates the percent change in cost for 1, 5, 10, 100, and 150 mgd systems that results from various activated carbon losses.

FIGURE 8.3: PERCENT CHANGE IN PLANT COST VERSUS CARBON LOSS

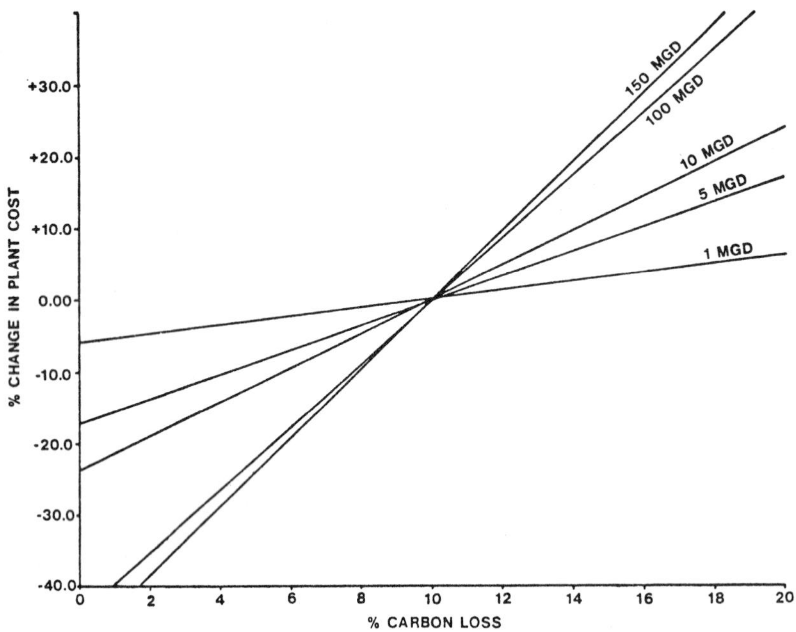

In equations (1) and (2) (shown on the previous page), amortized capital cost remains constant, as activated carbon loss affects only O&M cost. Some of the parameters, such as amortization period, have a multiplicative effect, as will be illustrated below. Assuming that amortization period, which affects only amortized capital cost, is in one case 10 years (150.9%) and in another case 30 years (85.4%), yields the following when compared with the standardized value of 1.5 ¢/1,000 gal:

(3) GAC system cost (10 years) = (1.5)(1.509) + 4.5 = 7.5 ¢/1,000 gal

(4) GAC system cost (30 years) = (1.5)(0.854) + 4.5 = 5.8 ¢/1,000 gal

Equations (3) and (4) illustrate how the multiplicative factor affects amortized capital cost.

Table 8.2 contains the percentage change in O&M cost that results from setting the level for each design parameter at 50%, and at 150% (for a 100 mgd plant), of standard values. Table 8.3 contains the same information for amortized capital cost (for a 100 mgd plant). It should be noted that some of the parameters, such as carbon cost, affect both amortized capital and O&M cost.

TABLE 8.2: DESIGN PARAMETERS AFFECTING O&M COSTS (100 MGD)

Additive Parameters	Values	% Change
Carbon loss per reactivation cycle (%)	15	29.6
	10	0
	5	14.7
Carbon cost (¢/lb)	54	16.9
	38	0
	19	-22.0
Fuel cost ($/MMBtu)	1.89	4.0
	1.26	0
	0.63	-3.8
Power cost (¢/kWh)	1.5	0.6
	1.0	0
	1.5	-0.6
Direct hourly wage rate ($/hr)	7.78	15.9
	5.19	0
	2.60	-15.9
Wholesale price index	267	6.0
	178	0
	89	-5.0
Multiplicative Parameters		
Reactivation frequency (wk between)	3	145.9
	6	100.0
	9	72.1
Capacity factor (%)	50	105.2
	70	100.0
	100	92.4

TABLE 8.3: DESIGN PARAMETERS AFFECTING AMORTIZED CAPITAL COSTS (100 MGD)

Additive Parameters	Values	% Change
Carbon cost (¢/lb)	54	16.9
	38	0
	19	-22.0
Construction cost index	385	49.5
	257	0
	125	-51.0
Multiplicative Parameters		
Amortization period (year)	10	150.9
	20	100.0
	30	85.4
Interest rate (%)	10.5	127.3
	7	100.0
	3.5	73.0
Reactivation frequency (wk between)	3	121.5
	6	100.0
	9	91.5
Capacity factor (%)	50	128.9
	70	100.0
	100	76.4

The net effect of setting the design value for each parameter at the high and low values is shown in Table 8.4.

TABLE 8.4: NET EFFECT FOR SETTING DESIGN PARAMETERS AT HIGH AND LOW LEVELS (100 MGD)

	High, %	Low, %
Additive factors for O&M		
Loss per reactivation	29.5	-14.7
Carbon cost	16.9	-22.0
Fuel cost	4.0	-3.8
Power cost	0.6	-0.6
Hourly wage rate	15.9	-15.9
Wholesale price index	0.6	-0.6
Sum	+67.5	-57.6
Additive factors for amortized capital cost		
Carbon cost	16.9	-22.0
Construction cost index	49.5	-51.0
Sum	+66.4	-73.0
Multiplicative factors for O&M		
Reactivation frequency	1.459	0.721
Hydraulic load	1.052	0.924
Product	1.535	0.666
Multiplicative factors for amortized capital cost		
Amortization period	1.509	0.854
Interest rate	1.273	0.730
Reactivation frequency	1.215	0.915
Hydraulic load factor	1.289	0.764
Product	3.008	0.436

Using the values shown in Table 8.4, the O&M and amortized capital costs are calculated as shown below for a 100 mgd plant.

	High	Low
Additive modifications		
Amortized capital cost (¢/1,000 gal)	1.5 + 1.5 (0.664)	1.5 - 1.5 (0.730)
Sum	2.5	0.4
O&M (¢/1,000 gal)	4.5 + 4.5 (0.675)	4.5 - 4.5 (.576)
Sum	7.5	1.91

Using the above values, the total cost can be calculated as follows:

Multiplicative modifications		
Amortized capital (¢/1,000 gal)	(2.5)(3.008)	(0.4)(0.436)
Product	7.5	0.17
O&M (¢/1,000 gal)	(7.5)(1.535)	(1.91)(0.666)
Product	11.5	1.27
Sum	19 ¢/1,000 gal	1.44 ¢/1,000 gal

Adding the final results for amortized capital and O&M costs yields a high value of 19 ¢/1,000 gal and a low value of 1.44 ¢/1,000 gal. These results illustrate the extremes which might result from localized conditions.

Alternatives to On-Site Reactivation

Disposal of Old GAC: As can be seen, unit costs associated with small treatment systems are extremely high. The bulk of the amortized capital costs are for on-site reactivation facilities, which suggests the possibility that for small plants some alternative to on-site reactivation should be explored. One possibility would be to dispose of exhausted activated carbon and to purchase new carbon. The cost of disposal for a plant operating at 70% capacity factor, with a period between reactivation of 1.4 months, is shown below:

$$\text{Disposal cost} = \frac{(865\ \text{ft}^3)\ (\text{no. filters})\ (30\ \text{lb/ft}^3)\ (8.57\ \text{reactivations/yr})\ (38\ ¢/\text{lb})}{(\text{flow in mgd})\ (365\ \text{days/yr})\ (0.70)}$$

$$\text{Disposal cost} = 33.1\ ¢/1{,}000\ \text{gal}$$

The above value can be compared to on-site reactivation costs for a 1, 5, 10, 100 or 150 mgd plant operating at 0.7 load factor with once-per-1.4-month reactivation. As shown in Figure 8.4, disposing of exhausted carbon is actually less expensive than on-site reactivation for small plants (2 mgd or less), although it is obviously more expensive for larger plants.

FIGURE 8.4: TOTAL UNIT COSTS VERSUS PLANT SIZE

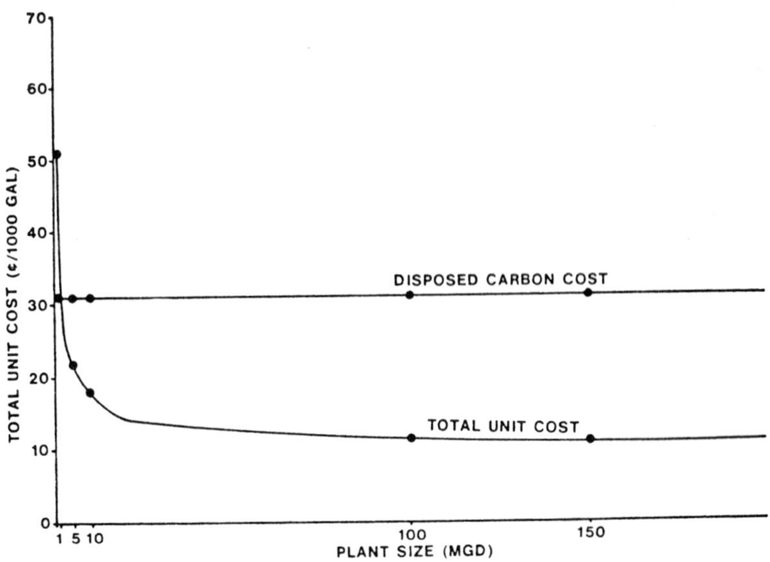

Another option that needs to be explored is that of regional reactivation. Regional reactivation consists of transporting the exhausted carbon to a central site where a reactivation furnace is located. This approach, which is particularly appropriate for small plants, will be explored in the following section.

Regional Reactivation: For the purposes of this analysis, the following sets of regional reactivation conditions will be examined:

(1) Regional reactivation systems consisting of off-site reactivation plants capable of processing carbon from the equivalent of 3,100 lb/day, 6,200 lb/day, 15,500 lb/day, 31,000 lb/day, and 62,000 lb/day reactivation facilities. These systems will be designated RP_1, RP_2, RP_3, RP_4 and RP_5, respectively.

(2) Costs for individual plants shipping to these systems will be examined, based on the assumption that a number of plants are sharing these reactivation systems simultaneously. For example, the cost in ¢/1,000 gal for a 1 mgd plant shipping carbon to a RP_1 and RP_2, and a 10 mgd plant transporting carbon to RP_4 and RP_5 systems will be calculated assuming that the various regional off-site reactivation systems are being used to capacity.

By transporting carbon to a regional reactivation center a small plant is able to take advantage of the economies of scale inherent in a larger reactivation furnace, although there is a debit associated with the transportation cost required to get the carbon to the site. The assumptions regarding the operation of the plants are the same as those in Table 8.1 (page 250); for example, 70% load factor and 1.4 months between reactivation.

The costs associated with the water treatment plant will be as follows: the initial activated carbon purchase (twice the capacity of the treatment plant, as one batch of carbon is being reactivated while the other is in use) and the make-up carbon (loss assumed at 10% per replacement cycle due to handling); transportation costs, which will be assumed as $0.10/ton-mile; and a proportionate share of the off-site reactivation costs which will consist of furnace capital and operating costs, assuming a 10% loss of activated carbon during the reactivation process.

Table 8.5 contains the costs associated with the initial carbon cost, and the carbon loss as well as the equivalent carbon reactivation requirements per day in lb/day for each plant size.

TABLE 8.5: CARBON COSTS AND REACTIVATION REQUIREMENTS FOR REGIONAL REACTIVATION SYSTEMS

Plant Size (mgd)	Initial Carbon Requirements (lb)	Annual Cost ($)	Make-up Carbon ($)	Reactivation Requirements (lb/day)
1	51,900	1,861.4	7,254	617.86
5	259,500	9,307.1	35,319	3,089.29
10	519,000	18,614.1	69,835	6,178.57
100	5,190,000	186,141.3	672,321	61,785.71
150	7,785,000	279,212.0	1,001,758	92,678.56

It is assumed in this analysis that the amortized capital and O&M costs for reactivation systems are divided equally among the number of plants shipping carbon to them. For example, if five 1 mgd plants are shipping to an RP_1 system, the cost will be higher than if ten 1 mgd plants are shipping carbon to an RP_2 system.

Transportation costs are calculated as follows for a 1 mgd plant shipping carbon to a reactivation plant for a 30-mile round trip:

$$\text{Transport cost} = \frac{(10\ ¢/\text{ton-mile})(865\ \text{ft}^3)(30\ \text{miles})(30\ \text{lb/ft}^3)(\frac{\text{ton}}{2{,}000})(8.57\ \text{reactivations/yr})}{(365\ \text{days})(1\ \text{mgd})(0.70)}$$

$$= 0.131\ ¢/1{,}000\ \text{gal}\ (30\text{-mile round-trip})$$

On a gallon per-mile basis, the cost is:

$$= (0.131\ ¢/1{,}000\ \text{gal})/(30\ \text{mile}) = 0.0044\ ¢/1{,}000\ \text{gal-mile}$$

The amortized capital cost and annual O&M cost for the off-site reactivation systems are shown in Table 8.6.

TABLE 8.6: AMORTIZED CAPITAL AND OPERATING COSTS FOR OFF-SITE REACTIVATION SYSTEMS

Reactivation System	Reactivation Requirement (lb/day)	Construction Cost ($)	Amortized Capital Cost ($)*	Annual O & M Cost ($)
RP_1	3,089.27	700,000	66,080	180,000
RP_2	6,178.57	820,000	77,408	240,000
RP_3	15,446.35	1,350,000	127,440	460,000
RP_4	30,892.85	1,630,000	153.872	750,000
RP_5	61,785.71	2,200,000	207,600	1,130,000

* 7 percent interest, 20-yr amortization period.

These costs can be assigned equally to the water treatment plants served. For example, RP_1 can serve five 1 mgd treatment plants and the total cost is divided by five; but for RP_2, which can serve ten 1 mgd plants, the total annual cost is divided by 10.

Table 8.7 contains the flow in million gallons per year and the total allocated cost plus the initial activated carbon and makeup costs for each plant reactivation configuration.

As can be seen from Table 8.7, the unit cost for a 1 mgd plant sending carbon to a regional reactivation furnace serving five plants is higher than for a 1 mgd plant sending carbon to a system serving ten plants due to economies of scale in the reactivation system. Transportation costs must also be considered, as in the discussion on the following page.

TABLE 8.7: REACTIVATION SYSTEMS COST FOR AN INDIVIDUAL PLANT

Regional Reactivation		Flow per Plant	Total Annual Cost	Unit Cost
Regional Reactivation Configuration	Total No. of Plants	(mil gal/yr)	($)	(¢/1000 gal)
1 - RP_1	5	225	56,470.00	25.1
1 - RP_2	10	225	38,995.00	17.3
5 - RP_3	5	1277.5	162,114.10	12.7
5 - RP_4	10	1277.5	135,013.10	10.6
10 - RP_4	5	2555.0	269,321.0	10.5
10 - RP_5	10	2555.0	222,217.1	8.7

Figure 8.5 shows the distance-related costs associated with the regional reactivation system for 1 mgd plants sending carbon to RP_1 and RP_2 reactivation systems, as compared to an on-site reactivation system, and the slope of the curve shows that carbon can be transported for many miles before an on-site system becomes cost effective.

FIGURE 8.5: COST OF TRANSPORTING CARBON FROM A 1 MGD PLANT TO REGIONAL REACTIVATION SITE VERSUS DISTANCE IN MILES

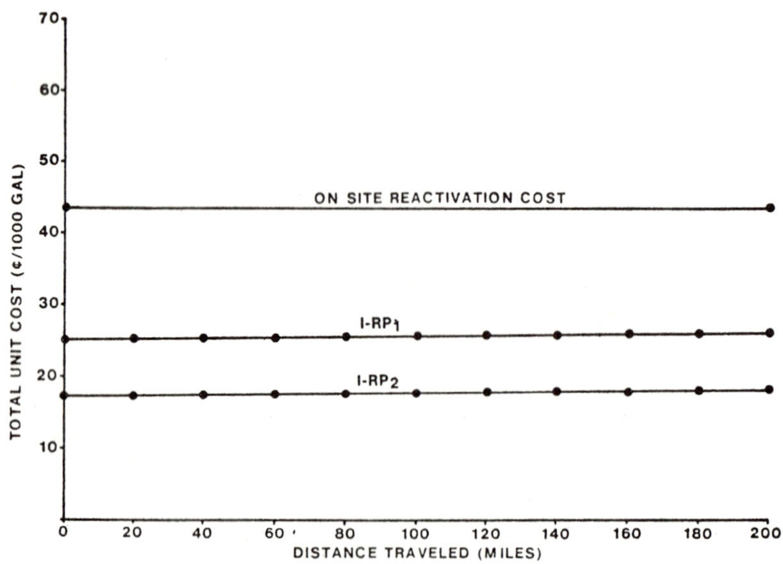

Figure 8.6 and figure 8.7 show similar conditions for 5 mgd plants and 10 mgd plants transporting to RP_3 and RP_4, and RP_4 and RP_5 systems, respectively.

It can be seen that for a 10 mgd plant it also is cost effective to transport spent carbon over relatively long distances; however, the gap between on-site and transporting off-site narrows at this level. Figure 8.8 shows the impact of variations in transportation cost on the total cost of a 10 mgd plant transporting waste to an RP_4 system.

FIGURE 8.6: COST OF TRANSPORTING CARBON FROM A 5 MGD PLANT TO REGIONAL REACTIVATION SITE VERSUS DISTANCE IN MILES

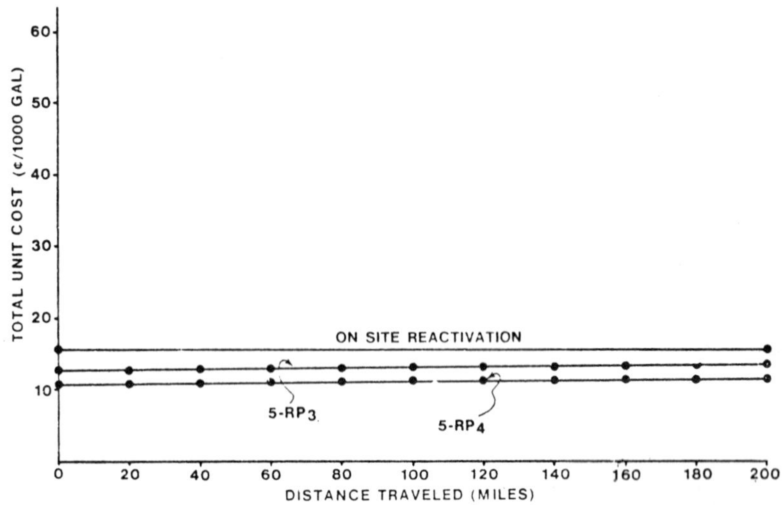

FIGURE 8.7: COST OF TRANSPORTING CARBON FROM A 10 MGD PLANT TO REGIONAL REACTIVATION SITE VERSUS DISTANCE IN MILES

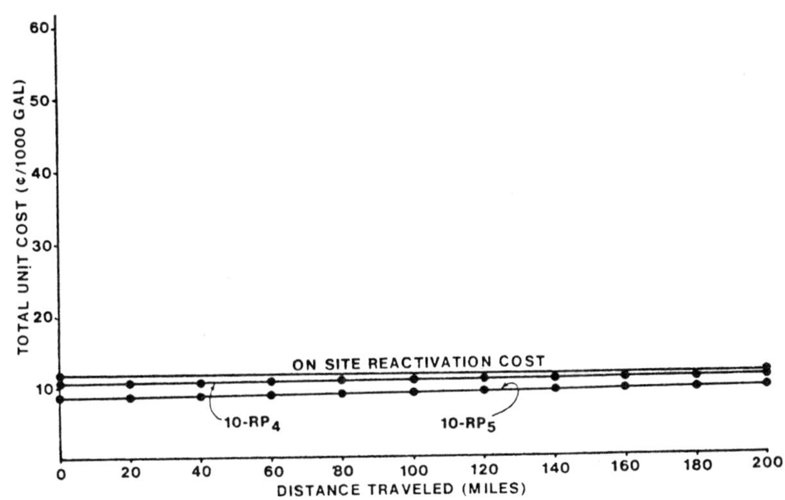

FIGURE 8.8: THE SENSITIVITY OF REACTIVATION COSTS TO TRANSPORTATION COST VARIATIONS

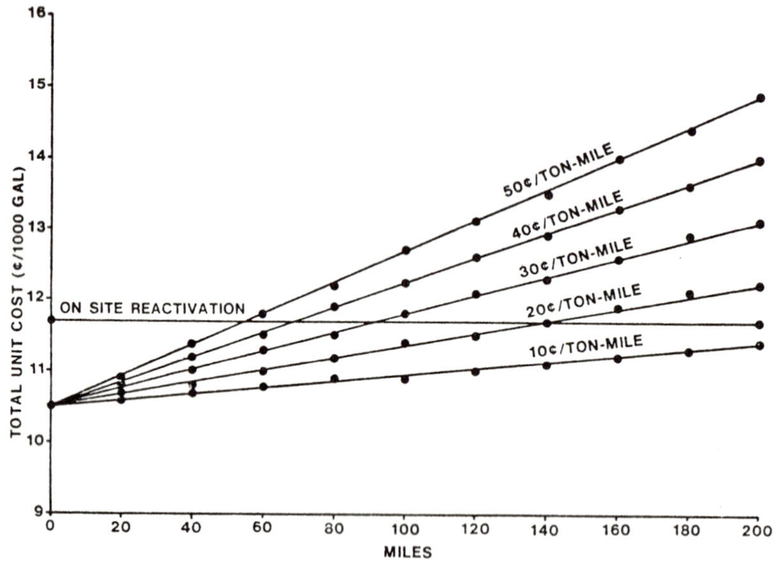

Separate Contactor System

In the discussion to this point the costs presented have been based on the assumption that carbon would replace sand in a filter plant. Therefore, no amortized capital and O&M costs associated with separate carbon contactors have been included in the analysis. It is very likely that operating in such a manner is inconvenient and inefficient, causing higher carbon losses due to excessive handling.

A contactor system may be tailored specifically for a given treatment plant operation. The assumptions used for the contactor system analysis are as follows (Table 8.8): 2 contactors connected in series, a contact time of 9 minutes and a corresponding recycle frequency of 1 per 2.8 months, bed depths of 20 feet, and a carbon loss of 5% per reactivation cycle.

TABLE 8.8: ASSUMPTIONS FOR SEPARATE CONTACTOR SYSTEMS

Item	Plant Capacity (mgd)				
	1	5	10	100	150
Number of Contactors	2	4	8	28	42
Hydraulic Loading (gal/min/sq ft)	4.87	5.42	5.42	5.57	5.57
Diameter of Contactors (ft)	8	12	12	20	20
Depth of Contactors (ft)	13	13	13	14	14

(continued)

TABLE 8.8: (continued)

Item	Plant Capacity (mgd) 1	5	10	100	150
Vol. of Granular Activated Carbon per Contactor (cu ft)	653.1	1469.5	1469.5	4396.0	4396.0
Apparent Contact Time (min)	9	9	9	9	9
Reactivation Frequency (months) at 70 percent Capacity	2.8	2.8	2.8	2.8	2.8
Activated Carbon Loss/ Reactivation (percent)	5	5	5	5	5

Figure 8.9 compares the costs for a separate contactor system versus replacement of carbon as filter media.

FIGURE 8.9: COMPARISON OF COSTS BETWEEN CONTACTOR SYSTEM AND MEDIA REPLACEMENT VERSUS PLANT CAPACITY

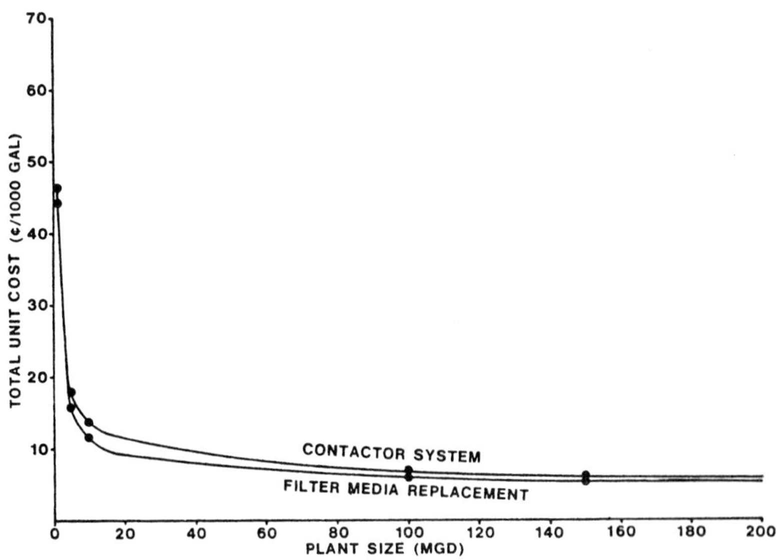

Replacing sand by carbon represents a short-term possibility for water treatment plants with low capital investment but high operating costs. A separate contactor system represents a longer term and permanent solution with higher capital investment requirements but with lower operating costs, as shown in Table 8.9 on the following page. These capital investment requirements are discussed in the following section.

Capital Investment: All of the cost data presented to this point in the analysis have been in terms of unit costs (including amortized capital and O&M costs).

It is important to present a different perspective by examining total investment costs. Separate contactor systems, as can be seen from Table 8.9, are more capital intensive than replacing the filter media by activated carbon. To build these systems utilities must raise considerable amounts of initial capital.

TABLE 8.9: AMORTIZED CAPITAL AND O&M COSTS FOR CONTACTOR VERSUS FILTER MEDIA REPLACEMENT (¢/1,000 gal)

System	1 mgd	5 mgd	10 mgd	100 mgd	150 mgd
			Amortized Capital Costs		
Media Replacement	19.5	5	3.5	1.5	1.1
Contactors	30.2	10.2	8.2	4.3	4.1
			Operating and Maintenance Costs		
Media Replacement	21.5	10.5	8.2	4.5	4.0
Contactors	16.1	7.3	5.4	2.4	2.2
			Total Cost		
Media Replacement	41.1	15.5	11.7	6.0	5.1
Contactors	46.3	17.5	13.6	7.3	6.3

A major part of the capital investment for an on-site reactivation system is the furnace. Table 8.10 summarizes the estimated cost of these types of furnaces.

TABLE 8.10: ESTIMATED CONSTRUCTION COST OF GRANULAR ACTIVATED CARBON REACTIVATION FURNACES

Furnace Type	Capacity (lb/hr)	Estimated Total Cost ($ x 10^6)
Multiple-hearth	5,000	4.2
Infrared*	5,000	0.8
Rotary-kiln	5,000	**
Fluidized-bed*	5,000	1.2

*Because furnaces of this size have not been manufactured these estimates are very preliminary.

**Insufficient information is available to estimate a cost for this type of furnace.

TYPES OF REACTIVATION SYSTEMS

The material in this section is taken from Appendix A of the *Interim Guide for Organic Contaminants* (see bibliography). It assumes that separate contactors of GAC will be used for water purification and evaluates the types of possible GAC reactivation systems.

Introduction

Although the internal pore structure providing a large surface area per unit weight is developed during initial manufacture (thermal activation) of granular activated carbon, the surface area is finite and eventually becomes covered with adsorbate, and adsorption ceases. To continue with effective adsorption the granular activated carbon is processed to remove these adsorbed materials (reactivation).

The most common technique for processing granular activated carbon to renew the adsorption capacity of its surfaces is to drive off and oxidize the adsorbed organic compounds in environment containing steam and little oxygen at high temperature (approaching 1000°C). The problem is to design a facility and choose reactivation conditions, such that the maximum amount of adsorbed materials is removed with a minimum of change in the properties of the granular activated carbon.

Basic Furnace Types

The four basic types of furnaces in use for the reactivation of granular activated carbon are: the multiple-hearth, the rotary-kiln, the infrared-tunnel, and the fluidized-bed furnace. Of the five currently operating on-site reactivation facilities in Europe, two are multiple-hearth furnaces and three are various designs of the fluidized-bed furnace.

Figure 8.10 is a cross section of a multiple hearth furnace; Figure 8.11 is a cross section of a one-bed fluidized-bed furnace; Figure 8.12 is a diagram of the infrared-tunnel furnace; and Figure 8.13 shows a rotary-kiln furnace. The Water Supply Research Division currently has three projects under way for the evaluation of reactivation facilities.

FIGURE 8.10: CROSS SECTION OF MULTIPLE-HEARTH FURNACE

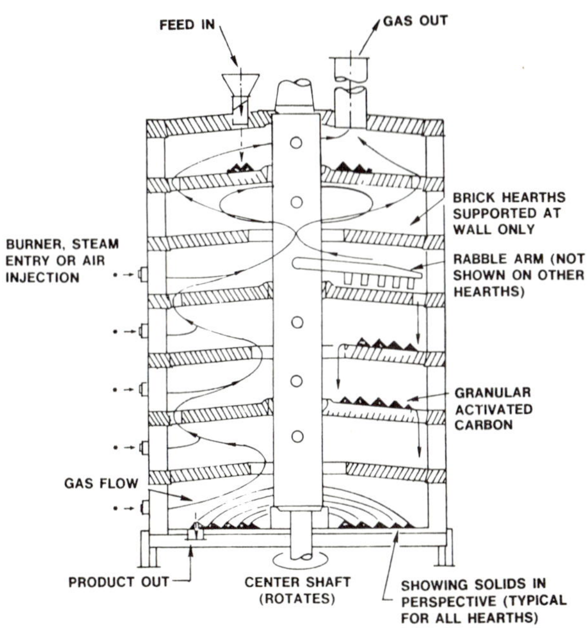

FIGURE 8.11: CROSS SECTION OF FLUIDIZED-BED FURNACE

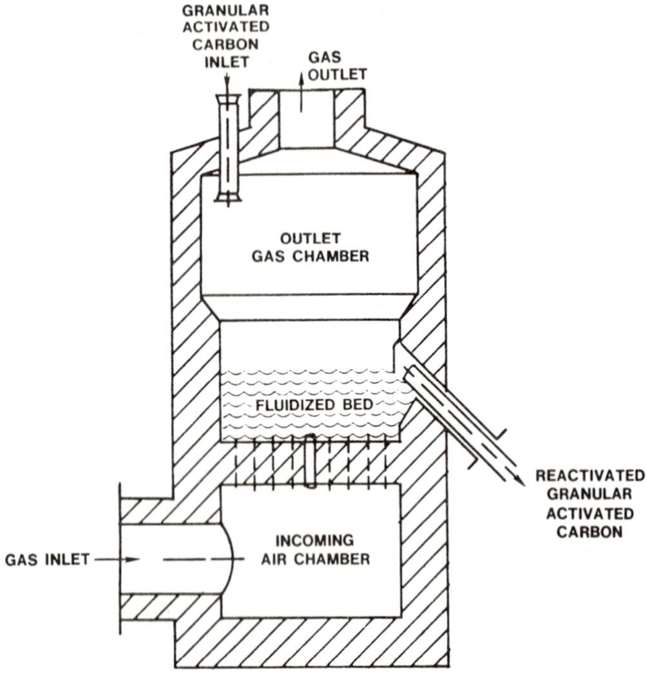

FIGURE 8.12: CROSS SECTION OF INFRARED-TUNNEL FURNACE

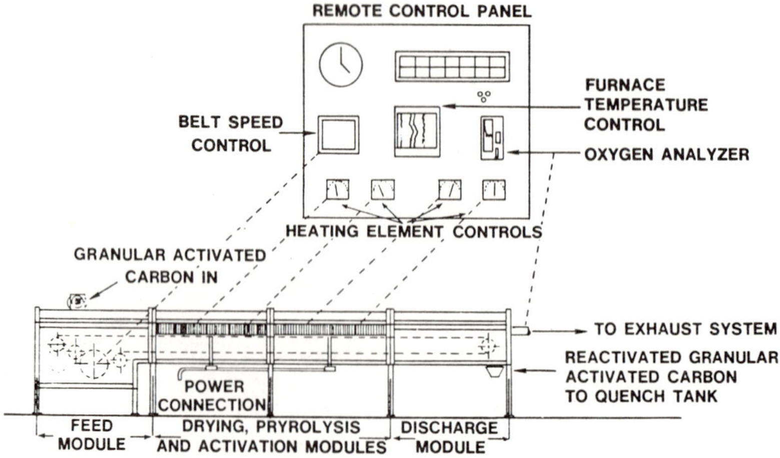

FIGURE 8.13: CROSS SECTION OF ROTARY-KILN FURNACE

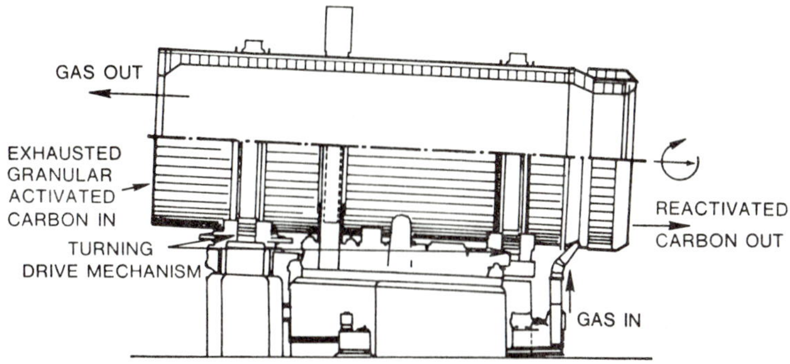

Proper design and operation of a thermal reactivation facility is necessary to avoid any unwanted change in the properties of granular activated carbon. For example, Juhola (1) reported a change in pore size distribution upon several cycles of reactivation, as shown in Table 8.11.

TABLE 8.11: CHANGES IN GRANULAR ACTIVATED CARBON PROPERTIES ON SUCCESSIVE REACTIVATIONS

Cycle	Ash Content, %	Iodine Number*	Molasses Number*	Bulk Density
Initial	5.7	1090	250	0.469
1	7.6	1040	310	0.468
2	8.6	935	290	0.469
3	9.5	940	350	0.473

*These numbers are based on the adsorption of the respective materials under a standard set of conditions. The iodine number is the measure of pores between 1 and 2.5 nm; the molasses number is the measure of pores greater than 2.8 nm.

This change in pore structure was then reflected in a decrease in performance as measured by a decrease in initial percent chemical oxygen demand reduction from 55 to 40% over eight reactivation cycles.

Smith (2) agreed, pointing out that, based on Figure 8.14, practically no change occurred in pore size distribution in the 1 to 1.9 nm diameter range, although a marked change occurred in the pores between 1.9 and 2.5 nm in diameter. He also noted that from 2.5 nm diameter and above, again very little change occurred in the pore structure.

Finally, DeJohn and Hutchins (3) also reported that the effect of thermal reactivation may be to reduce the surface area in the micropores of granular activated carbon. They further state that the properties of granular activated carbon derived from lignite are not changed as much as the properties of granular activated carbon derived from another source, bituminous coal. Their data showed a decline in specific surface area of bituminous coal based activated carbon from

1,100 m²/gm for virgin material to 700 m²/gm (36%) after four reactivations, although lignite-based activated carbon only declined from 650 m²/gm for virgin material (initially lower) to 500 m²/gm (23%) after five reactivations.

FIGURE 8.14: CUMULATIVE MICROPORE VOLUMES AS A FUNCTION OF MICROPORE DIAMETER

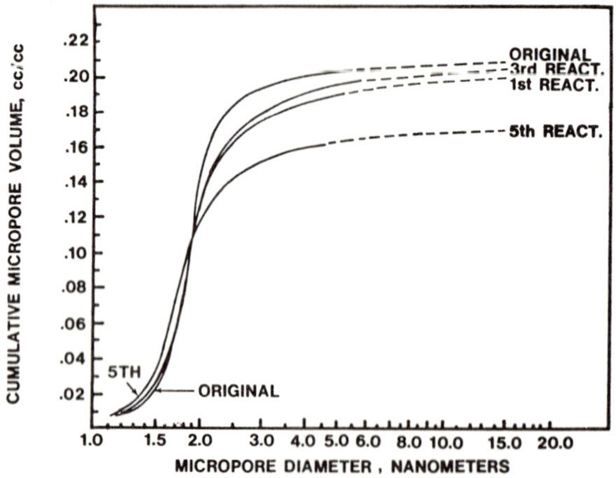

In contrast to these reports, Juntgen (4) noted that, based on Figure 8.15, an appropriate time-temperature relationship (800°C-20 minutes) during reactivation can be developed that will not alter activated carbon significantly from its virgin state.

FIGURE 8.15: EFFECT OF RESIDENCE TIME ON ACTIVITY AND CHANGE OF WEIGHT OF ACTIVATED CARBON

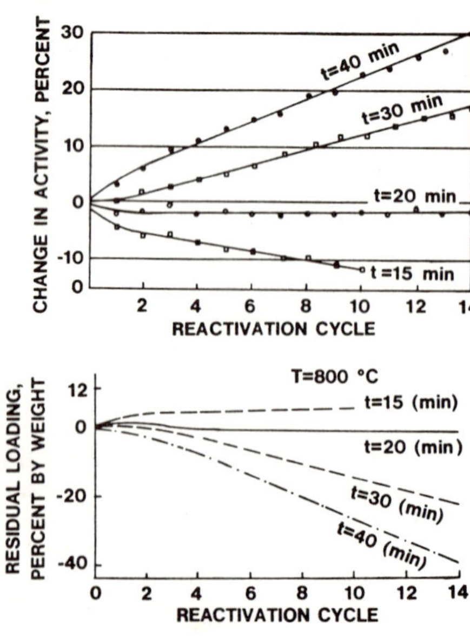

In support of this, Weissenhorn (5) collected data on the decline of granular activated carbon performance as measured by ultraviolet absorbance of the effluent with relation to the volume of water treated per unit volume of granular activated carbon and showed, according to Figure 8.16, that virgin activated and reactivated carbon were nearly equivalent.

FIGURE 8.16: INFLUENCE OF REACTIVATION ON GRANULAR ACTIVATED CARBON PERFORMANCE

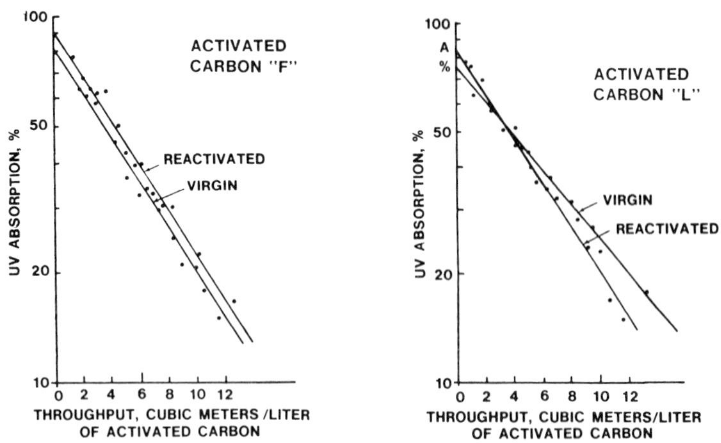

Finally, the Shirco Company of Dallas, Texas, gives the data shown in Figure 8.17 and Table 8.12.

FIGURE 8.17: INFLUENCE OF REACTIVATION ON IODINE NUMBER

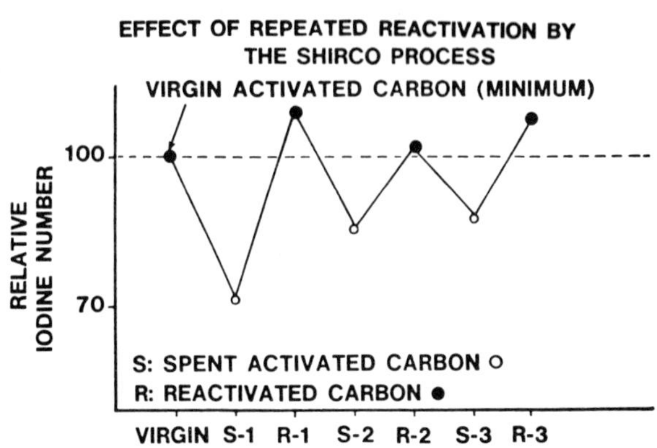

TABLE 8.12: TYPICAL RESULTS OF CARBON REACTIVATION IN THE SHIRCO ELECTRIC INFRARED FURNACE

Activated Carbon	Apparent Density (g/cc)	Iodine No.	Molasses No.
ICI Hydrodarco 3000			
Virgin	0.352	550	333
Spent	0.458	413	302
Reactivated	0.397	596	380
Calgon Filtrasorb 400			
Virgin	0.48	1167	355
Spent	0.62	480	250
Reactivated	0.48	1122	376
Westvaco Nuchar WV-L			
Virgin	0.594	834	371
Spent	0.678	588	178
Reactivated	0.593	868	380
Carborundum			
Virgin	0.525	950	220
Spent	0.667	320	137
Reactivated	0.523	1132	230

References

(1) Juhola, A.J., "Regeneration of Activated Carbon," in: Papers and Proceedings of a Water Research Association Conference at the University of Reading, United Kingdom, Paper 9, 177-202, (Feb. 1974), 398 pp.

(2) Smith, S.B., "Techniques of Activated Carbon Regeneration," in: Proceedings of the Physical-Chemical Treatment, Activated Carbon Adsorption in Pollution Control Technology Transfer Seminar, Sponsored by Environment Canada and Pollution Control Association of Canada, 1975 Report M 1000.01, Chemical Division, Westvaco Corporation, Covington, Virginia, (1975), Mimeo, 50 pp.

(3) DeJohn, P.B. and Hutchins, R.A., "Treatment of Dye Wastes with Granular Activated Carbon," *Textile Chemist and Colorist, 8,* No. 4, 34-38, (April 1976).

(4) Juntgen, H., "Phenomena of Activated-Carbon Regeneration," in: Translation of Reports on Special Problems of Water Technology—Volume 9—Adsorption," Conference held in Karlsruhe, Federal Republic of Germany, 1975, p. 269-283, EPA-600/9-76-030, U.S. Environmental Protection Agency, Cincinnati, Ohio (December 1976), 454 pp.

(5) Weissenhorn, F.J., "Testing of Activated Carbon Filters in Waterworks," in: Translation of Reports on Special Problems of Water Technology—Volume 9—Adsorption," Conference held in Karlsruhe, Federal Republic of Germany, 1975, pp. 238-250, EPA-600/9-76-030, U.S. Environmental Protection Agency, Cincinnati, Ohio (Dec. 1976), 454 pp.

Economic Analysis of Proposed Amendments

The material in this chapter is taken from a two-part report made by Temple, Barker & Sloane, Inc. of Wellesley Hills, Massachusetts for the U.S. Environmental Protection Agency, Office of Water Supply. The first section of the report, published in August 1977, is entitled *Economic Impact Analysis of a Trihalomethane Regulation for Drinking Water,* and the second part is called *Economic Analysis of Proposed Regulations on Organic Contaminants in Drinking Water,* and was published in December 1977.

INTRODUCTION

During the past year the Office of Water Supply at the Environmental Protection Agency has focused special attention developing regulations on organic contaminants in drinking water. The result of the Agency's investigations is a proposed amendment to the Interim Primary Drinking Water Regulations which will be the first phase of a program for organics control.

The proposed new regulations are of two types: a maximum contaminant level on trihalomethanes in drinking water; and a treatment requirement for granular activated carbon or equivalent treatment on systems susceptible to contamination from synthetic organic compounds. Both regulations will apply only to large community water systems, those serving resident populations of over 75,000 at this time. It is expected that they will be extended to smaller systems at some future date.

The purpose of this report is to present the economic analysis for the proposed regulations. The economic analysis of the proposed trihalomethane regulation will be given first. Then the economic evaluation of the proposed treatment requirement for the control of synthetic organics will be given, after which the combined economic impacts of the two regulations will be reviewed. Since some community water systems are expected to be affected by both regulations, the combined impacts of both are less than the sum of the individual assessments.

THE TRIHALOMETHANE REGULATION

Analytic Structure

Regulatory Criteria: Naturally-occurring organics have become a regulatory concern primarily because of the evidence that chlorine combines with precursor matter in water to form chloroform, and other related compounds, which are suspected carcinogens. The regulation to reduce the level of these contaminants in drinking water contains the following parameters:

(1) a maximum contaminant level (MCL) for trihalomethanes (THM) of 0.10 mg/l; and
(2) a lower boundary for the size of water systems to be covered by the regulation which has been set at 75,000 persons served. In addition, systems serving between 10,000 and 75,000 will be required to monitor for THM.

An MCL of THM at 0.10 mg/l is the level which the Office of Water Supply has selected for the proposed regulatory action. This level allows some flexibility in the type of treatment which can be used as a remedy. It does not, therefore, necessarily force systems to add any particular treatment.

The lower boundary for systems covered by the regulation (75,000 people) is the point at which over 50% of the population served by community water systems is covered. The number of systems of that size and larger is relatively small, 390 compared to 2,685 for all systems serving populations of over 10,000, and nearly 35,000 community water systems in total.

In addition to the extent of the coverage offered by the proposed regulation, there are additional reasons for limiting the size of the systems included. The majority of the water systems in the regulated category use surface water which is more likely to have organic contamination. As the size of the systems decreases, the water used is more likely to be drawn from a groundwater source which, in turn, is less likely to produce trihalomethane concentration at or above the MCL.

Further, it is more likely among larger systems that skilled scientists and operators will be available to develop and manage modified treatment practices which assure no reduction of the water's microbiological quality. Nevertheless, as water quality problems and the feasibility of implementing appropriate treatments among small systems become more certain, the regulation may be extended to cover smaller systems. The first step will be to require monitoring for THM among systems serving between 10,000 and 75,000 people.

Number of Community Water Systems: This analysis addresses primarily the costs which large systems will incur as a result of the regulation. It is, therefore, important to illustrate how large a portion of the population is served by these large systems. At the same time, it should be noted that 92% of the community water systems in operation serve under 16% of the population. Figure 9.1 illustrates the percentage of water systems in each size category and the related portion of the population which they serve. The numbers of systems in those categories under 75,000 are referred to in a later section (Sensitivity Analysis) which summarizes the costs which would result under different regulatory configurations.

FIGURE 9.1: CUMULATIVE PERCENT OF POPULATION SERVED BY COMMUNITY WATER SYSTEMS

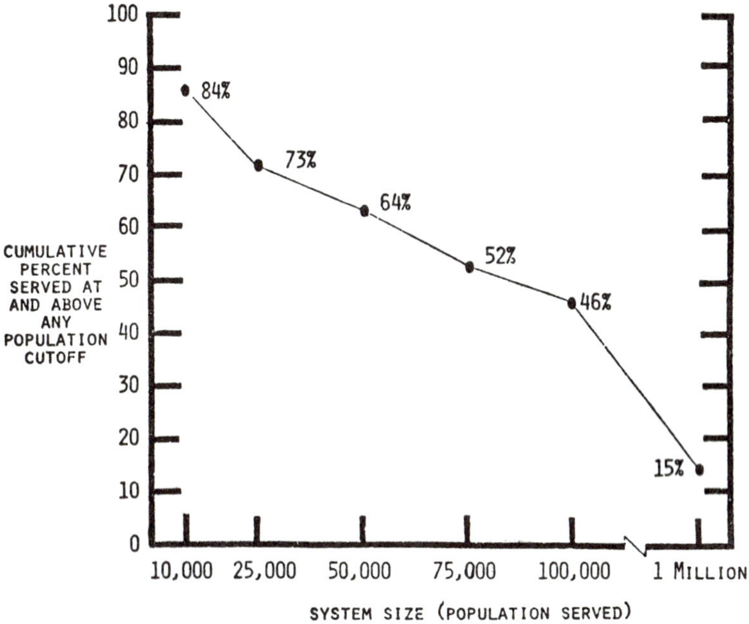

Comparative Distribution of Number of Systems and Population Served by Size of System

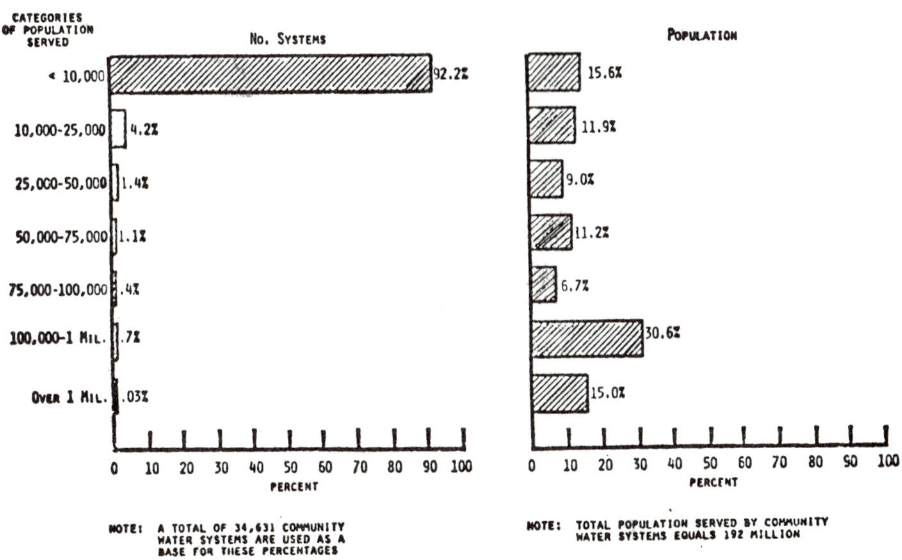

The systems serving over 75,000 have been subdivided into three size groupings: 75,000 to 100,000; 100,000 to 1 million; and over 1 million. These size categories permit the cost analysis to reflect such differences among systems as the economies of scale associated with the sizing of equipment for new treatments.

Having established that over 50% of the population is served by the 390 systems covered by the regulation, the following discussion summarizes the treatments which are expected to be adopted by those large systems which exceed the MCL.

Available Treatment Alternatives

There are three general categories of treatment possibilities. The selection of the appropriate category for a specific water system depends in part upon the magnitude of the system's THM level and the system's existing treatment facility. Systems, of course, will tend to select the alternative which is least disruptive to their current practices and still complies with the regulation.

The major treatment options which are available to meet a THM regulation are described below:

(1) The first alternative consists of minor modifications to current procedures. These modifications include moving the point of disinfection, adjusting the chlorine dosage, or improving existing conventional coagulation and sedimentation practices. This approach would enable systems where the MCL is exceeded by a small amount to comply at minimal cost.

(2) Changing disinfectants is the second category of treatments. Since it is the use of chlorine which causes part of the organic problems, some systems may choose to use other chemicals for disinfection. The available alternatives are chloramines, ozone, and chlorine dioxide.

(3) Using a tertiary adsorbent is the most complex and costly alternative. Systems with the most serious organic contamination may select treatment techniques which require the use of granular activated carbon (GAC), resins, or an equivalent. This analysis has used the costs of installing GAC in contactors following conventional filtration as those which represent the most likely treatment technique in this category.

Profile of Systems' Response to MCL Regulations

In order to complete the basis for estimating the total costs of the regulation, the number of systems which will select each of the treatments above must be established. These estimates were arrived at by first estimating the number of systems which will exceed the MCL and then allocating these systems according to the treatments they are most likely to select.

Of the 390 community water systems serving populations of 75,000 or more, about 86 have been estimated to exceed an MCL of THM at 0.10 mg/l, based upon the water quality data currently available.

The number of systems which will select each of the treatments available has been estimated on the basis of the THM level, current treatment practices, and the economics of the treatment options. The distribution by treatment category presumes that 45% of the affected systems as likely to change disinfectants and 30% likely to use an adsorbent. The remaining 25% would modify existing disinfection or other procedures. As the table shows, approximately 24 million people are served by the systems which would be likely to exceed the standard prior to any corrective measures.

TABLE 9.1: MOST PROBABLE TREATMENT SELECTION BY WATER SYSTEMS AFFECTED BY MCL REGULATION OF THM AT 0.10 mg/l

	Move Point of Disinfectant or Adjust Dosage	Change Disinfectant	Use Adsorbent	Total
Number of Systems	21	39	26	86
Percent of total affected	25%	45%	30%	100%
Population affected	2,260,000	14,470,000	7,160,000	23,890,000

The distribution of selected treatment was arrived at by the following method.

Regulatory Compliance Strategies

Since there is no empirical method for predetermining the choice which will be made by each affected water system, a different approach was necessary. The approach chosen is a step-by-step procedure which can be tracked easily and modified as new information becomes available. A logical sequence of decision points was designed to distribute the systems covered by the regulation according to the most likely path they would follow. The decision made at each point is consistent with certain criteria. The criteria are based upon:

(1) the treatments currently used: If a system does not chlorinate it will not be affected by a THM regulation, and therefore will require no new treatment.

(2) water source used: If a system uses surface water as its primary source, it is more likely to exceed a given level of THM contamination. Hence, the number of water systems using water from ground or surface sources affects the number of systems which will exceed a given level and therefore require treatment.

(3) degree to which water quality exceeds MCL: If the presence of THM is only slightly in excess of the standard, then minimal modifications to procedures will be adequate for compliance. As the level of contamination increases, a system must consider more significant (and costly) treatment techniques.

(4) economic considerations: The presumption is that systems will adopt the least cost treatment strategy which satisfies the regulation.

(5) treatment effectiveness: The presence of THM above certain levels can probably be controlled only by the use of adsorbents. This is because of the likelihood that high disinfectant demand water cannot be adequately disinfected without generating a considerable amount of by-products of unknown hazards. Consequently, those few systems with a very high level of THM are likely to require the addition of the most costly treatment.

The estimates presented below are the result of considering these criteria. The attached decision tree structure (Figure 9.2) illustrates the paths expected to be followed for compliance with the THM at 0.10 mg/l regulation by each of the 390 water systems which serve more than 75,000 people.

FIGURE 9.2: REGULATORY COMPLIANCE STRATEGIES FOR MCL OF TRIHALOMETHANES AT 0.10 mg/l

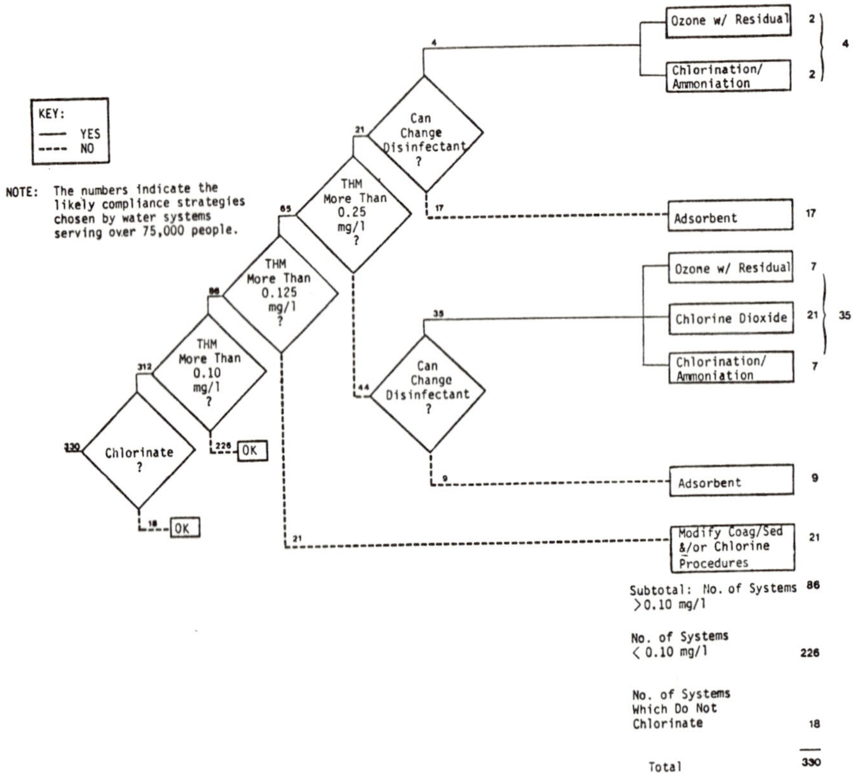

Sixty of these systems purchase the majority of their water from other systems which are presumed to provide the treatment. Thus, a total of 330 systems may have to add treatments, though 18 of these are excluded since they do not presently chlorinate. Of the remaining 312, some 86 are estimated to have THM levels above 0.10 mg/l and hence would require treatment. The 21 systems which are estimated to exceed the 0.10 mg level by less than 25% are assumed to be able to comply by the least costly method–minor modification of existing chlorination or other procedures.

The remaining 65 systems are split into those above and below a THM level of 0.25 mg/l. Of those estimated to be over the 0.25 level, about 20% are assumed to be able to comply by changing disinfectant and the remainder (80%) would be required to use an adsorbent to achieve compliance; in these cases, changing disinfectants would not bring the system into compliance with the regulation. As noted previously, the disinfectant demand would be excessive in those instances where high levels of total organic carbon are present. Eighty percent of those below the 0.25 level are assumed to change disinfectant and the remaining 20% to use an adsorbent. The results of these treatment selections are that 39 systems would change disinfectants and 26 would use adsorbents as a compliance strategy.

COST OF THE THM REGULATION

The cost analysis of a THM regulation combines the assumptions above on the number of systems likely to be affected by the regulation, and the treatments which those systems will tend to select. The results are presented first in terms of the national costs for all large systems requiring treatment and, second, in terms of the costs to individual systems.

National Cost Estimates

The economic implications of a THM regulation at 0.10 mg/l, covering systems of 75,000 people or more, are summarized below in terms of five key measures:

(1) Capital expenditures requirements during the 1976-1981 period are projected to be $154.4 million (1976 dollars).
(2) External financing requirements during the same period for those capital expenditures are projected to be $145.2 million under the regulation.
(3) Annual operating and maintenance (O&M) expenses in 1981 for the required treatments are estimated at approximately $25.9 million.
(4) Annual revenue requirements in 1981, reflecting the amortization of capital expenditures and the O&M expenses, are expected to increase by a total of $36.0 million for the 65 systems which are likely to have cost impacts.
(5) Per capita costs, simply in terms of total revenue impacts divided by the population served by systems with cost impacts, are projected to be $2.07 per year in 1981.

As Table 9.2 shows, over 87% of the aggregate costs of this regulation is expected to be borne by systems serving 100,000 people or more.

TABLE 9.2: SUMMARY OF TOTAL COSTS FOR AN MCL REGULATION OF THM AT 0.10 mg/l (MILLIONS OF 1976 DOLLARS)

	Systems Serving Populations of.		
	75,000 to 99,000	100,000 and above	Total
Capital Expenditures, 1976-1981	20.5	133.9	154.4
External Financing, 1976-1981	19.3	125.9	145.2
Operating and Maintenance Expenses, 1981	2.6	23.3	25.9
Revenue Requirements, 1981	4.0	32.0	36.0
Annual per Capita Cost*, 1981 (dollars)	2.30	2.04	2.07

*Revenue requirements divided by population served by cost-impacted systems.

These cost figures include the expenses of all 86 systems adding or altering treatment practices. Table 9.3 breaks down these costs into those attributable to each treatment category. About 89% of the capital costs are due to the 26 systems which are anticipated to add adsorbents, though these systems are only 30% of the number affected by the regulation.

TABLE 9.3: SUMMARY OF COSTS BY TREATMENT CATEGORY FOR AN MCL REGULATION OF THM AT 0.10 mg/l (MILLIONS OF 1976 DOLLARS)

	No. of Systems	Capital Expenditures	Annual Revenue Requirements
Change Disinfectant	39	17.3	7.8
Use Adsorbent	26	137.1	28.2
Move Point of Disinfectant	21	0.0	0.0
Total	86	154.4	36.0

Costs to a Typical System: [Note: Further cost figures for alternate disinfectants are given in Chapter 4.] The costs for the four types of treatments–ozonation, chlorine dioxide, chlorination/ammoniation and tertiary adsorbent–can best be compared on the basis of additional per capita costs for a typical water system. They are as follows:

(1) Ozonation (plus residual disinfectant) is the most capital intensive of the three disinfectant treatments. Systems serving over 1 million people would need capital expenditures of about $6 million each. Annual per capita costs range from 39 to 93 cents.
(2) Chlorine dioxide treatment requires only minor investment but considerable expense for the purchase of sodium chlorite. Per capita costs range from 68 to 81 cents per year.
(3) Chlorination/ammoniation is the least expensive treatment with annual per capita costs in the 28 to 47 cent range.
(4) Adding a tertiary adsorbent is the most expensive treatment and involves substantial capital expenditures (approximately $19 million for a typical system serving over 1 million people) as well as continuing operating expenses for reactivation. Per capita costs range from $3.30 to $6.11.

The capital expenditures, annual revenue requirements, and per capita costs are shown in Table 9.4 for each treatment for each of the three size categories over 75,000.

TABLE 9.4: COMPLIANCE COSTS FOR A TYPICAL WATER SYSTEM UNDER AN MCL REGULATION OF THM AT 0.10 mg/l (1976 DOLLARS)

	75,000-100,000	100,000-1 Million	Over 1 Million
Average population served per system	85,000	188,000	1,560,000
Ozone			
Capital expenditures	720,000	1,275,000	5,900,000

(continued)

TABLE 9.4: (continued)

	75,000-100,000	100,000-1 Million	Over 1 Million
Ozone			
Revenue requirements/year	79,443	122,331	605,364
Annual per capita cost	.93	.65	.39
Chlorine dioxide			
Capital expenditures	20,300	20,800	37,800
Revenue requirements/year	68,800	153,500	1,068,000
Annual per capita cost	.81	.81	.68
Chlorination/ammoniation			
Capital expenditures	39,000	46,000	74,000
Revenue requirements/year	40,300	74,800	439,600
Annual per capita cost	.47	.40	.28
Tertiary adsorbent			
Capital Expenditures	2,500,000	4,300,000	18,500,000
Revenue requirements/year	519,000	808,000	4,671,000
Annual per capita cost	6.11	4.30	3.30

It is clear that the range of costs is broad across treatments and size categories. The use of a tertiary adsorbent is considerably more expensive than any of the alternate disinfectants for all size categories. However, the use of a tertiary adsorbent has the ancillary benefit of generally reducing organic chemicals in addition to THM. Its use may also result in reduced disinfectant demand.

Among the disinfectants, chlorine plus ammonia is always the least expensive. In the case of ozone compared to chlorine dioxide, there are distinct economies of scale for the use of ozone. This implies that most systems serving over one million people would select ozone over chlorine dioxide on the basis of cost alone. As can be seen in the table, the economies of scale for chlorine dioxide are less because the operating costs increase more directly with production than is the case for ozonation.

The number of systems selecting each of the treatments will have a significant impact on the total costs of the proposed amendment. This is clearly illustrated by the differences in the costs to individual systems of each of the treatments. The next selection, therefore, includes an alternative distribution of treatment selection as one of the three sensitivity analyses included for comparison.

Monitoring Costs

In addition to the treatment costs which the 65 water systems covering over 75,000 will incur, there are specific monitoring requirements included in the regulation. The costs for monitoring which would be incurred by systems exceeding the MCL and adding treatments have already been included in the cost estimates presented above. In addition, two other categories of water systems will be required to conduct monitoring for THM.

First, all those systems serving over 75,000 people which are not expected to alter their current treatment practices will be required to continue monitoring at a minimum frequency of five samples per quarter. Second, all water systems serving between 10,000 and 75,000 people will be required to monitor for THM at a minimum frequency of two samples per quarter.

The annual national monitoring costs for the 307 systems serving over 75,000 which will probably not need to alter current treatment practices amounts to about $153,500. This estimate is based on a $25 per sample cost, assuming the five samples per quarter minimum, or $500 a year per system. The annual monitoring costs for the second category, the approximately 2,300 systems serving between 10,000 and 75,000 are estimated at $460,000. This estimate is based upon two samples per quarter at $25 each, or $200 a year per system.

Monitoring costs were computed based upon a survey of contract analytical laboratories currently performing THM analyses. The cost of equipping an existing laboratory with an appropriate gas chromatograph is dependent upon which analytical procedure is selected and the type of instrument. The basic instrumentation for the "liquid-liquid" extraction method consists of a gas chromatograph with an "Electron Capture" detector and recorder; the base cost is approximately $5,000. The basic instrumentation for the "purge and trap" method consists of a gas chromatograph, a "Hall" detector, purge and trap sample concentrator, and recorder; the basic cost is approximately $10,000. In either case, some additional expenditures for accessories would be added.

Per sample costs ranged from $25 to $100. After these regulations have been promulgated, the increased volume of business and competitive factors would be expected to reduce the analytical costs to well below $25 per sample.

EPA expects that a number of community water systems will choose to purchase the equipment and monitor for THM on-site more frequently than the minimum, for operational control as well as for compliance purposes. An additional benefit from purchase and on-site analytical capability, is that the gas chromatograph is versatile and can be used to monitor for the presence of many other organic chemical contaminants besides THMs.

COST OF A REGULATION ON SYNTHETIC ORGANICS

The proposed regulation for the control of synthetic organics in drinking water is a treatment requirement. It specifies the use of granular activated carbon or an equivalent treatment for community water supplies which are susceptible to contamination by synthetic organic compounds. In its proposed form it will only apply to systems serving resident populations of over 75,000.

Cost at an Individual System Level

It is expected that community water systems affected by this regulation will incorporate granular activated carbon into their treatment trains by construction of postfiltration contactors for the carbon. Such contactors are assumed in this analysis to be circular, each approximately 15 to 30 feet in diameter and 10 feet in height.

The number of contactors needed at each plant will depend upon plant capacity and the length of time which water must be in contact with the carbon in order to achieve adequate organic removal efficiency ("contact time"). The required contact time will vary from site to site and is expected to range from nine minutes to perhaps as high as eighteen minutes. A typical plant serving 280,000 people in 1981 (producing 50 mgd) would have 10 to 20 contactors depending upon its required contact time. Carbon movement to and from the contactors for regeneration is assumed to be via automated slurry systems.

On-site regeneration of the spent carbon is anticipated through multiple-hearth furnaces constructed for this purpose. Regeneration frequencies will vary from site to site depending upon raw water quality and other factors, but should generally be between 2 and 6 months. The furnaces will vary in size depending upon volume of granular activated carbon utilized and frequency of regeneration. A typical installation for a system serving 280,000 people in 1981 (producing 50 mgd) might have 6 hearths with an outside dimension of approximately 12 ft in diameter and 20 ft in height. These will generally be fueled by distillate oil or natural gas.

The capital expenditure requirements for the contactors, furnaces, initial carbon fill, and other capital items are shown in Table 9.5 for an individual water system. The capital expenditures vary significantly with different contact times and less significantly with different regeneration frequencies. Contact time affects virtually all the capital expenditure items almost linearly—a doubling requires twice as many contactors, twice the initial carbon fill, and twice the furnace capacity for regeneration. Different regeneration frequencies, however, affect only the furnace capacity, not other capital improvements.

TABLE 9.5: CAPITAL EXPENDITURES FOR GAC* INDIVIDUAL WATER SYSTEMS (MILLIONS OF 1976 DOLLARS)

	Regeneration Frequency			
	6 Months		2 Months	
Contact Time	**9 min**	**18 min**	**9 min**	**18 min**
Population Served				
75,000-100,000	2.1	3.6	2.9	5.1
100,000-1 million	4.9	9.0	6.1	11.3
Over 1 million	12.4	22.6	15.1	27.6

*Granular activated carbon, including contactors, initial carbon fill, plant modifications, and regeneration furnaces.

The capital requirements for individual systems under the most favorable assumptions would be \$2.1 million for an average system serving 75,000 to 100,000 people, \$4.9 million for an average system serving 100,000 to one million people, and \$12.4 million for an average system serving over one million people. The capital requirements would increase to almost 2½ times those levels under the least favorable conditions.

Operations and maintenance (O&M) expenses, in contrast, change more over the range of regeneration frequencies shown than over the range of contact times. These expenses consist primarily of labor and furnace operating costs and the cost of makeup carbon to compensate for losses of approximately 7% experienced in each regeneration cycle. These expenses, therefore, vary with both the volume of carbon used (a function of contact time) and the number of times it is regenerated each year. The estimated O&M costs for alternatives considered are presented in Table 9.6 on the following page.

TABLE 9.6: ANNUAL OPERATIONS AND MAINTENANCE EXPENSES FOR GAC* INDIVIDUAL WATER SYSTEMS (MILLIONS OF 1976 DOLLARS)

	Regeneration Frequency			
	6 Months		2 Months	
Contact Time	**9 min**	**18 min**	**9 min**	**18 min**
Population Served				
75,000-100,000	0.2	0.3	0.3	0.6
100,000-1 million	0.4	0.7	0.7	1.3
Over 1 million	1.6	3.0	3.1	5.6

*Including labor, fuel and makeup carbon during regeneration.

The annual O&M expenses for granular activated carbon treatment for the control of synthetic organics would range from \$0.2 million to \$0.6 million for an average system serving 75,000 to 100,000 people depending upon contact time and regeneration frequency. O&M costs for the larger system sizes would vary similarly, from minimum cost levels of approximately \$0.4 million and \$1.6 million, respectively, under the most favorable conditions.

Annual revenue requirements for this application of granular activated carbon have been computed from the capital expenditure and O&M figures above by assuming amortization of the capital expenses over a 40-year period at an average interest rate of approximately 8% for new debt financing. The resulting figures are shown in Table 9.7 below.

TABLE 9.7: ANNUAL REVENUE REQUIREMENTS FOR GAC INDIVIDUAL WATER SYSTEMS (MILLIONS OF 1976 DOLLARS)

	Regeneration Frequency				1981
	6 Months		2 Months		Baseline
Contact Time	**9 min**	**18 min**	**9 min**	**18 min**	**Revenues**
Population Served					
75,000-100,000	0.4	0.6	0.6	1.0	2.2
100,000-1 million	0.8	1.5	1.3	2.3	6.1
Over 1 million	2.7	4.9	4.3	7.9	23.6

The effect of GAC utilization on these water systems will be approximately a 10 to 45% increase in annual revenue requirements by 1981 depending upon design specification. The significance of that effect is probably best examined in terms of cost per capita and cost per residential customer as shown in Table 9.8 on the following page.

The cost per capita shown in Table 9.8 is simply the total annual revenue requirement divided by the population served by the water system. It provides a measure of the cost of this form of health protection and is an upper bound on the possible cost to individual consumers if no costs were allocated to nonresidential customers. By this measure the utilization of GAC treatment will result in costs of approximately \$2 to \$4 per capita under the most favorable conditions and \$7 to \$11 per capita under the least favorable conditions.

TABLE 9.8: ANNUAL PER CAPITA AND CUSTOMER COSTS OF GAC, 1981
INDIVIDUAL WATER SYSTEMS
(1976 DOLLARS)

	Regeneration Frequency			
	6 Months		2 Months	
Contact Time	9 min	18 min	9 min	18 min
Annual Cost per Capita				
Population Served				
75,000-100,000	3.80	6.70	6.30	11.00
100,000-1 million	3.10	5.60	4.80	8.70
Over 1 million	2.20	4.00	3.60	6.50
Annual Residential Customer Bill Impact*				
Population Served				
75,000-100,000	5.70	10.00	9.50	16.50
100,000-1 million	4.70	8.40	7.20	13.10
Over 1 million	2.90	5.30	4.80	8.60

*For a family of three.
Note: See text for explanation of those calculations.

Actually the increase in water rates will usually be less than this per capita cost because some of the costs will be borne by nonresidential customers of the water system. The other set of figures in the table is an estimate of the likely increase in annual water rates for an average family of three assuming that nonresidential customers pick up the same proportion of GAC costs that they do of other system costs. On this basis the GAC treatments will result in annual increased water bills of approximately $3 to $6 per family under the best conditions shown and $9 to $17 per family under the worst conditions shown.

Costs at the National Level

A major factor in assessing the national economic effects of the proposed treatment requirement is the number of systems which may ultimately be required to install and operate the treatment. EPA has reviewed its monitoring data and its inventory of community water systems and estimated that approximately 50 systems would be impacted by this proposed regulation. The exact number, of course, will depend upon site-specific water quality and other factors. Higher or lower numbers of systems affected would result in proportionately higher or lower national costs in the following tables.

As Table 9.9 shows, there are almost 400 water systems in the country which serve populations of over 75,000 people and which therefore would be subject to this regulation. It is generally thought that systems which rely primarily upon surface water sources for their water will be the most likely to be affected. Those account for just over 200 systems in this size range.

The 50 systems which are expected to be affected include some in regions or river basins which have been identified as likely to need treatment and others which are statistically estimated and cannot be specifically identified at this time. The 50 systems represent approximately one out of every eight systems in this size range and one of every four which relies primarily upon surface water. The national costs of applying granular activated carbon to these 50 systems are shown in Table 9.10 based upon the individual system costs presented earlier. [Note: Costs of regeneration of GAC are discussed at length in Chapter 8.]

TABLE 9.9: WATER SYSTEMS POTENTIALLY AFFECTED BY PROPOSED TREATMENT REQUIREMENT

	Population Served Category			
	75,000-100,000	100,000-1 million	Over 1 million	Total
Total systems	147	232	11	390
Systems with primary surface water*	59	137	10	206
Expected to be affected**	15	32	3	50

*TBS estimate, *Survey of Operating and Financial Characteristics of Community Water Systems,* April 1977.
**EPA estimate.

TABLE 9.10: NATIONAL COSTS OF GAC TREATMENT REQUIREMENT (MILLIONS OF 1976 DOLLARS)

	Regeneration Frequency			
	6 Months		2 Months	
Contact Time	**9 min**	**18 min**	**9 min**	**18 min.**
Capital Expenditures				
Population Served				
75,000-100,000	31	54	44	76
100,000-1 million	157	288	197	360
Over 1 million	37	68	45	83
Total	225	410	286	519
Annual O/M Expenses				
Population Served				
75,000-100,000	3	5	5	9
100,000-1 million	12	23	24	43
Over 1 million	5	9	9	17
Total	20	37	38	69
Annual Revenue Requirements				
Population Served				
75,000-100,000	5	9	9	15
100,000-1 million	26	47	40	74
Over 1 million	8	15	13	24
Total	39	71	62	113

The aggregate national costs for the proposed treatment requirement range as follows:

(1) Capital expenditures will be in the range of $225 to $286 million if an average nine-minute contact time will be adequate to achieve the desired organics removal levels. If contact times have to increase to the point where they average 18 minutes nationally,

then the capital expenditures total would be $410 to $519 million.

(2) Operations and maintenance expenses are more dependent upon regeneration frequency than contact time. The national range of O&M expenses will be $20 to $37 million if an average regeneration frequency of six months will be adequate to achieve the desired performance. However, if the average has to be as low as two months then the O&M expenses would be $38 to $69 million per year.

(3) Annual revenue requirements reflect both sets of uncertainties and fall in three general positions under the range of assumptions shown. Under the best conditions they could total $39 million per year for the 50 systems. Under the least favorable assumptions they could be as high as $113 million. However, if either one of key operating characteristics, contact time or regeneration frequency, turned out favorably and the other did not, then revenues would total $62 to $71 million per year.

COMBINED COSTS OF TRIHALOMETHANE AND SYNTHETIC ORGANICS REGULATIONS

The two proposed regulations have each been evaluated separately in terms of economic impacts on the country, individual systems and customers. The majority of the systems is expected to be affected by only one of these regulations. For those systems and their customers, the costs of compliance will simply be the costs which are described in the appropriate individual analysis. A smaller number of systems, approximately 15, are expected to be affected by both of the regulations. Since the costs for these systems have been included in both of the separate cost analyses, the combined costs of the two regulations are less than the sum of the individual costs.

For those systems affected by both restricted regulations, the choice of compliance method will be to the use of granular activated carbon, and the operating parameters of contact time and regeneration frequency will have to be designed to achieve both sets of requirements. Accordingly, some individual systems may be forced to operate with longer contact times or shorter regeneration frequencies than would be possible if affected by only one regulation, though all are expected to fall within the ranges described in the previous section.

In combination the two regulations are expected to require that between 61 and 78 large water systems utilize granular activated carbon or an equivalent treatment. Of that number, 11 to 28 systems will install carbon treatment solely to comply with the trihalomethane regulation, 35 will install it solely to meet the treatment requirement and 15, as noted above, will use it to meet both requirements. (The range associated with the trihalomethane regulation shown in this and later tables reflects uncertainty on the proportion of affected systems which will utilize carbon, ranging from a low of 30% to a high of 50%.)

The national costs of the combined regulations are shown in Table 9.11 on the following page.

TABLE 9.11: NATIONAL COMBINED COSTS THM AND TREATMENT REQUIREMENT REGULATIONS (MILLIONS OF 1976 DOLLARS)

	. . . . Regulation	. . . .		
	THM Only	Treatment Requirement Only	Both	Total Costs
Low Cost Assumptions				
Capital Expenditures	66	155	70	291
O/M Expenses	14	14	6	34
Annual Revenues	18	27	12	57
Mid-Cost Assumptions				
Capital Expenditures	66-166	198	88	352-452
O/M Expenses	14-23	26	12	52-61
Annual Revenues	18-32	43	19	80-94
High Cost Assumptions				
Capital Expenditures	166	359	160	685
O/M Expenses	23	48	21	92
Annual Revenues	32	78	35	145

The national costs may vary significantly depending upon the local conditions which prevail at the systems affected by the regulations. The table presents costs for three different sets of assumptions:

(1) Low-cost assumptions, which project that only 11 systems would add GAC treatment under the trihalomethane regulation alone and that the treatment requirement could be satisfied with 9-minute contact time and 6-month carbon regeneration.

(2) Mid-cost assumptions, which show a range of 11 to 28 systems which will add carbon under the trihalomethane regulation alone and which assume that the regeneration frequency would be shortened to 2 months.

(3) High-cost assumptions, which project that 28 systems would add carbon under the trihalomethane regulation alone and that the regeneration frequency would be reduced to 2 months and the contact time would be increased to 18 minutes.

DEMAND ON SUPPLYING INDUSTRIES

The treatments required by both the THM and the synthetic organics regulations will result in large increases in the demand for GAC and for regeneration furnaces. The demand for GAC as initial fill will be determined by the number of systems adding GAC treatment and the contact time required. The annual replacement of GAC lost in the regeneration process will be set by the frequency of regeneration.

Each water system adding GAC treatment is likely to purchase a regeneration furnace, the size of which will vary according to the quantity of carbon the system plans to regenerate. For the largest systems, generally those serving more than one million people, the purchase of more than one furnace would be required. Systems with multiple treatment plants would also generally be expected to purchase two or more furnaces.

Table 9.12 below estimates the quantity of carbon, both as initial fill and annual replacement, which will likely result from the combined THM and treatment regulations. The three rows of this table represent the low-cost, mid-cost and high-cost assumptions discussed previously.

TABLE 9.12: DEMAND ON SUPPLYING INDUSTRIES OF THE COMBINED REGULATIONS

	Number of Systems Selecting GAC	Demand for GAC ..... (million lb)...... Initial Fill*	Annual Replacement**	Expected Demand for Furnaces
Low Cost Assumptions	61	112	26	64-80
Mid-Cost Assumptions	61-78	112-134	47-56	64-97
High Cost Assumptions	78	211	89	81-100

*Assumes 22,600 lb of GAC per MGD for 9 minte contact time and 45,200 lb of GAC for 18 minute contact time.
**Based on 7% losses per regeneration cycle.

The demand for both GAC and regeneration furnaces generated by the proposed regulations is within the capacity of the respective industries. The activated carbon industry was reported to have an unutilized annual production capacity of some 100 million pounds of GAC in mid-1976. Since that time, one firm has announced plans to add new capacity for GAC production. The industry's expanded capacity appears likely to exceed anticipated demand (excluding new uses for drinking water treatment) by the same 100 million pound level until 1980. Hence, the industry is capable of supplying the 112 to 211 million pounds of GAC needed by the affected water utilities over the next three years.

It should be noted, however, that the highest estimate of 211 million pounds corresponding to the high-cost assumptions would require full use of the industry's capacity for more than two years and could only be supplied within that time frame with advance planning and contacting and in the absence of large competing demands. In the longer perspective, beyond the initial requirement of this regulation in its first two or three years, the carbon industry has indicated an intent to expand its capacity to whatever level is required to supply the recurring annual volumes of carbon needed by the water supply industry.

The regeneration furnaces needed by the utilities adding GAC treatment could be supplied by the furnace manufacturers over a three-year time period. Multi-hearth furnace producers have estimated that their collective capacities could be increased by as many as 100 furnaces annually. Rotary kilns, infrared electric furnaces and fluidized bed furnaces also could be made available to water utility customers. The design and installation of reactivation furnaces, particularly multihearth furnaces, requires some 18 months of lead time, though this time frame might be reduced by the availability of preengineered designs in the future.

Assessment of an Interim Requirement for GAC in Existing Filters

Since the design and installation of contactors for granular activated carbon is expected to require a significant lead time, perhaps two to four years, EPA has included an interim measure in the proposed regulation. The measure requires systems which are susceptible to synthetic organic contamination to replace the sand or other filter media in their existing filter beds with granular activated carbon while the full treatment is being designed and constructed. Such systems would not be expected to make major capital modifications to their filters in response to this interim requirement. Nor would they necessarily be required to adopt the same operating practices which they will eventually employ when contactors are used—that is, they might have shorter contact times and longer regeneration cycles in the interim period.

Two approaches to interim treatment are possible. One involves no on-site regeneration of carbon, and depends upon the carbon manufacturers to take back spent carbon and perform the regeneration service; the other assumes the water system would purchase its own furnace and perform the carbon regeneration on-site. For the eventual level of treatment required under the proposed regulations it is expected that all of the large systems affected would take the latter approach. However, to meet an interim requirement some might wish to postpone capital commitments and select the former approach.

The costs during the interim period would be less than those anticipated with the final design. The initial capital investment would consist of the purchase of GAC to fill the beds and, for those systems taking the second approach, a regeneration furnace. Operating costs relate primarily to the replacement or regeneration of the spent carbon.

The incremental costs of the potential interim treatment are comprised of three factors: (1) the capital investment in carbon, furnace and plant modifications which are not recoverable for use in the final treatment technique; (2) those operating expenses for fuel, labor, and carbon replacement during the interim period which have no carry-over to the final treatment; and (3) the financing costs for the capital equipment for the additional years prior to ultimate compliance.

Per capita interim treatment costs for an average system serving 100,000 to one million people at a six-month regeneration cycle, would be \$3.30 per year without on-site regeneration and \$1.50 per year with it. If the regeneration frequency were extended to twelve months that per capita cost would decline to \$1.70 and \$1.10 per year, respectively.

Removal of Inorganic Contaminants

The material in this section is from an article by Thomas J. Sorg of the USEPA Water Supply Research Division, Cincinnati, Ohio, it was in the *Journal of the American Water Works Association,* February 1978. This article first presents a summary of the USEPA Water Supply Research Division's overall project to evaluate removal techniques for the ten inorganic contaminants for which MCLs were promulgated on June 24, 1977. It then gives specific details on the technology for nitrate and fluoride removal. References for this section start on p 300.

INTRODUCTION

The general intention of Congress in passing the 1974 Safe Drinking Water Act (1) was to upgrade the quality of U.S. drinking water. One of the first major steps toward this goal was the promulgation of the EPA National Interim Primary Drinking Water Regulations (2) (NIPDWR) on December 24, 1975. These regulations established the maximum contaminant levels (MCL) for ten inorganic contaminants (Tables 10.1 and 10.2). Six months later, EPA promulgated a second set of regulations for radionuclides (3), which set maximum levels for radium-226 and radium-228, gross α-particles, β-particles, and photon radioactivity (Table 10.1). Both sets of regulations took effect on June 24, 1977.

The MCLs for inorganic contaminants are very similar to the 1962 Public Health Service (PHS) Standards (4), which were health related and constituted the basis for rejection of a water supply. The only revisions made to the PHS list were the elimination of cyanide, the addition of mercury, and the mandatory requirement for nitrate. The real major difference between the PHS standards and NIPDWR is the number of systems regulated. Whereas the PHS standards applied to only interstate carrier water-supply systems (approximately 700), the new regulations affect most of the approximately 40,000 community systems. Also, some regulations, such as nitrate, apply to all public (community and noncommunity) water supplies, which are estimated to number in the hundreds of thousands.

TABLE 10.1: NATIONAL INTERIM PRIMARY DRINKING WATER REGULATIONS FOR NINE INORGANIC CONTAMINANTS AND RADIONUCLIDES

Contaminant	Maximum Contaminant Level (mg/l)
Arsenic	0.05
Barium	1
Cadmium	0.010
Chromium	0.05
Lead	0.05
Mercury	0.002
Nitrate (as N)	10
Selenium	0.01
Silver	0.05
Radium-226 and -228 combined	5**
Gross α-particle activity	15**
Gross β-particle activity*	4***

*Based on a 2 liter/day drinking water intake, except for tritium and strontium-90. Average annual concentrations of tritium and strontium-90 assumed to produce a dose of 4 mrem/year are 20,000 and 8 pCi/l, respectively.

**pCi/liter.

***mrem/year

TABLE 10.2: NATIONAL INTERIM PRIMARY DRINKING WATER REGULATIONS FOR FLUORIDE

Contaminant	Temperature °F	Temperature °C	Maximum Contaminant Level (mg/l)
Fluoride*	≤53.7	≤12.0	2.4
	53.8-58.3	12.1-14.6	2.2
	58.4-63.8	14.7-17.6	2.0
	63.9-70.6	17.7-21.4	1.8
	70.7-79.2	21.5-26.2	1.6
	79.2-90.5	26.3-32.5	1.4

*The MCL for fluoride is determined by the annual average of the maximum daily air temperature for the location in which the community water system is situated.

The monitoring requirements stipulated in the NIPDWR are leading to the sampling of many water supplies for inorganic contaminants for the first time. No doubt exists that these monitored data will reveal many new problems, particularly for small communities. Communities discovering that their water supply does not meet the regulations will be confronted with decisions to upgrade the quality of their drinking water. Many small communities that do not treat their water will very likely seek an alternate source as their first course of action. If an alternate source is not available, treatment of an existing supply must be considered. Because of the general lack of data on the removal of inorganic substances from water, much of the information presented in this series of articles on the eight heavy metals will be drawn from the research efforts of the Water Supply Research Div., USEPA, Cincinnati, Ohio.

The basic philosophy of the research program is to first evaluate the effectiveness of conventional coagulation and lime softening processes. If they are found to be ineffective, then ion exchange and reverse osmosis are studied.

The WSRD research program consists of two phases: (1) a series of laboratory jar tests to determine the critical variables, such as pH, coagulant type and dose, and contaminant concentration, affecting removal of contaminants from water and (2) pilot plant tests to verify laboratory results. All of the laboratory studies and a significant portion of pilot plant tests have now been completed.

Overview of Contaminant Removal

No single treatment technique has been found to be effective for the removal of all ten inorganic contaminants listed in the NIPDWR. Chemical coagulation and lime softening are effective methods for removing most of the eight heavy metals, but are not useful for either nitrate or fluoride. If a water has several contaminants that cannot be effectively removed by one treatment technique, several unit processes may be required. It is hoped that most communities will not be faced with this problem and will be able to utilize a single treatment method to provide a good-quality drinking water at a reasonable cost.

The removability of substances from water by a specific treatment technique is determined by a number of factors. The most significant are: (1) the form, anion or cation; (2) the chemical oxidation state (valence); and (3) the solubility of the various complexes formed in water. Some metals, such as lead and cadmium, are found as cations and will form hydroxide or carbonate complexes, or both, that have low solubility limits in water. Others, such as selenium and arsenic, are found as anions. Table 10.3 lists the principal ionic forms of the contaminants found in aqueous solutions.

Several metals have more than one oxidation state, such as arsenic(III) and (V), chromium(III) and (VI), and selenium(IV) and (VI). The valence of the contaminant found in the water is important because studies have shown that removals by chemical coagulation methods are significantly different for the two valence states.

Identification of the valence of the contaminant, therefore, should be made prior to selecting the proper treatment technique. In general, the reduced form is found in groundwaters and the oxidized form in surface waters. In some cases, both forms may be found, particularly if each form is rather stable, as in the case of selenite (+4) and selenate (+6). The common valences of the ten contaminants are also shown in Table 10.3.

Knowledge of the geology and the industry in a locality may provide information about possible problems that will occur. Some contaminants are more naturally occurring, for example, fluoride, arsenic, selenium, and radium-226, and are more likely to be found in groundwaters than in surface waters. Others, such as chromium, lead, and cadmium, are likely to result from an industrial discharge and, therefore, are more commonly found in surface waters. Nitrates are more likely to be found in agricultural regions and in places where sanitary sewers are not available.

TABLE 10.3: PRINCIPAL FORM AND SOURCES OF INORGANIC CONTAMINANTS

Inorganic Contaminant	Valence	Principal Forms in Water	Principal Source	Probable Occurrence
Arsenic	+3 (arsenite)	anion $H_2AsO_2^{-1}$	mineral	groundwater
	+5 (arsenate)	anion $H_2AsO_4^{-1}$	mineral	surface water
		anion $HAsO_4^{-2}$	industry	
Barium	+2	cation Ba^{+2}	mineral	groundwater
Cadmium	+2	cation Cd^{+2}	industry	surface water
Chromium	+3	cation Cr^{+3}	industry	surface water
	+5 (chromate)	anion $HCrO_4^{-2}$	industry	surface water
	+5 (dichromate)	anion $HCr_2O_7^{-3}$	industry	surface water
Fluoride	-1	anion F^{-1}	mineral	groundwater
Lead	+3	cation PB^{+2}	industry	surface water
Mercury	+2 (inorganic)	cation Hg^{+2}	industry	surface water
	+2 (organic)	cation CH_3Hg^+ (methyl mercury)	industry	surface water
Nitrate	+5	anion NO_3^{-1}	agricultural	groundwater
Radium-226	+2	cation Ra^{+2}	mineral	groundwater
Selenium	+4 (selenite)	anion $HSeO_3^{-1}$	mineral	groundwater
		anion SeO_3^{-2}		
	+6 (selenate)	anion SeO_4^{-2}	mineral	groundwater surface water
Silver	+1	cation Ag^+	industry	surface water

The best estimates of most water-supply experts are that the most common problems will be occurrences of nitrate, fluoride, and arsenic. Areas of the country also exist where selenium, barium, and Ra-226 will be common problems. All of these contaminants are the more naturally occurring ones and will most likely be found in small water supplies using groundwater sources.

The effectiveness of chemical coagulation and lime softening for metal removal will be highlighted in the discussions on each contaminant. The removal mechanisms may be either one or a combination of several processes: precipitation, coprecipitation, and adsorption. As previously mentioned, many metals form insoluble hydroxides and carbonates and can be precipitated out of solution. For the most part, the pH of the processes must be closely controlled because the solubilities of these complexes are normally pH dependent.

Some metals will coprecipitate with either iron or aluminum hydroxide. Chemical coagulation studies have shown that higher removals are generally obtained with iron coagulants than with alum. Iron salts develop a stronger and heavier floc than alum, but coprecipitation may also be a significant factor. Many metals form iron-metal complexes in the natural environment, and similar types of complexes may be formed in chemical coagulation treatment.

Some metals, e.g., mercury, lead, and silver, adsorb quite readily onto the natural turbidity in the water. Removal of metals by adsorption is dependent on the amount of turbidity in the raw water, the quality of the floc developed, and the degree of turbidity removal.

Reverse osmosis, electrodialysis, and distillation are other potential metal-removal processes, but these techniques are rather expensive and not commonly used. These processes, in particular reverse osmosis (RO), do have some advantages for special applications. Thus, RO is being utilized in Florida for treating small water supplies that are high in total dissolved solids and Radium-226. Even with its drawbacks of high costs and disposal of sizeable amounts of reject water, the advantages of good removability for all contaminants, compactness, simplicity of operation, and low maintenance and manpower requirements make RO an attractive solution for some small communities with special problems.

NITRATE REMOVAL FROM WATER

Health Effects

Nitrate in drinking water is associated with a blood disorder in infants called methemoglobinemia. Development of this illness comes about by bacterial conversion of the relatively innocuous nitrate ion to nitrite in the gastrointestinal tract. The nitrite then converts the hemoglobin, an oxygen-carrying substance in the blood, to methemoglobin that cannot transport oxygen. The result is oxygen deprivation or suffocation. Development of the illness is generally confined to infants less than 3 months old, although it has been known to affect children up to the age of six. The regulation was established for nitrate rather than nitrite rarely occurs in water in significant amounts except in highly polluted waters.

The 1969 Community Water Supply Study (5) on 955 water utilities showed that finished water from 26 utilities exceeded the then 45 mg/l nitrate limit (now, 10 mg/l NO_3 as N). It is anticipated that nitrate excess will be one of the most frequently reported violations, certainly a major problem for small communities in agricultural areas utilizing groundwater.

Treatment Methods

Several techniques have been studied for the removal of nitrate from drinking water: (1) chemical reduction, (2) biological denitrification, and (3) ion exchange. Of these three processes, only the latter one, ion exchange, is considered to be feasible and practical for full-scale use by water utilities. Reverse osmosis and electrodialysis are two other treatment methods that are effective for nitrate removal and will be discussed.

Chemical Reduction: The chemical reduction of nitrate to a nitrogen gas has been investigated by Gunderloy et al (6)(7) and O'Brien (8). After a literature survey, Gunderloy selected eight reducing agents and two deammination agents for testing. Results of this investigation showed that ferrous iron, iron powder, and hydrazine, or its salts, had appreciable reducing power; glucose had a small amount of reducing power; and formaldehyde, carbon, sulfur dioxide, and carbon monoxide were inactive. Studies with deammination agents, sulfamic acid, and urea, each in combination with reducing agents, showed no nitrogen loss. Separate studies also showed that sulfamate could only produce deammination under acid conditions and that urea was ineffective. Of the three reducing agents that passed the initial screening tests, only ferrous iron was determined to be economically attractive. Additional laboratory studies for ferrous iron revealed the need for a catalyst, cupric ion, for denitrification and for the reaction to take place in an alkaline solution. The results of this project also showed that the reduction process is time dependent, the composition of the water can affect the products produced, and only about 70% of the nitrate is reduced.

The chemical-reduction process is rather involved and cannot be considered either a practical or economical method for water-supply treatment. Even if the process could be refined and be demonstrated successfully on a pilot- or full-scale basis, the method still would pose serious operational problems for small systems. Moreover, because about eight mols of iron per mol of nitrate are required for reduction, a voluminous amount of sludge would be produced for disposal.

Biological Denitrification: The biological denitrification process has been studied for the purification of both wastewater and drinking water (9)-(12). The process consists of using denitrifying organisms in a filter bed to reduce the nitrate to nitrogen gas. The biological filter is followed by a conventional sand filter to remove any carry-over bacterial growth. The organisms require an organic energy source. An organic substrate, therefore, must be provided for treating drinking water because groundwater supplies are essentially free of organic material. The most frequently used organic source is methanol because it is effective and economical.

Biological denitrification has not been enthusiastically received by water utilities because of several undesirable features. First, the process requires careful control to prevent bacterial and methanol breakthrough. Methanol is a hazardous chemical, and even though low concentrations are required, there is always the potential for a system failure. Furthermore, methanol is an organic material that is being added to a water supply that is essentially free of organics. Second, several days are required to establish the bacterial population in the filter. If the biological mass is lost for some reason, the system will be out of service for the time required to reestablish the biological filter. For these reasons, the process is not very practical and, as long as other alternatives exist, it is doubtful that this process will be given any serious consideration for water-supply treatment.

Ion Exchange: Currently, ion exchange is the only treatment technique used on a full-scale level to remove nitrate from drinking water. In 1974, a full-scale plant was constructed by the Garden City Park Water Dist., Nassau County, N.Y., to treat groundwater containing 20 to 30 mg/l of nitrate-nitrogen (13). A continuously regenerated ion-exchange process is used, the resin moving through a closed loop. The plant was designed to treat 0.076 m^3/s (1,200 gpm) and to lower the nitrate-nitrogen level to less than 2 mg/l.

The design of the Garden City Park plant was based upon several laboratory tests conducted with a prototype ion-exchange plant in 1970 by the Chemical Separations Corp. in Oak Ridge, Tenn. (14). The tests were conducted using Oak Ridge water with chemical additives and nitrate-laden groundwater from several locations on Long Island, N.Y. Results indicated that the nitrate-nitrogen could be consistently lowered from as high as 50 mg/l to 5 mg/l and less. Tests also were run to determine the amount of salt needed to regenerate the ion-exchange resin and to estimate the operating costs. Although the Garden City Park nitrate removal plant has been in existence for several years, it has not been operating continuously for any extended period of time for various reasons, including failure of the required continuous-nitrate-monitoring system. However, when the plant is operating, it is effective. The Garden City Park plant is the only plant of its kind in the U.S. and little information is available on its operation and costs. The system, therefore, has provided little guidance for the design of other new facilities. The plant also has several atypical operational features so that operating and cost data do not have widespread application through the U.S.

First, the ion-exchange system is somewhat unique. The resin material moves countercurrent through a closed loop and is regenerated continuously. The major advantage of this system is that the waste brine solution is produced continuously and slug discharges are avoided. Although the system has been used successfully here and in other industrial applications, it is doubtful that a small community could afford such an elaborate system or provide the trained manpower necessary to operate the system properly. The operational and cost data, therefore, would not apply to the more conventional and, probably, less expensive fixed-bed system.

Second, the well water treated by the Garden City Park plant is very low in total dissolved solids (TDS ~ 200 mg/l). Thus, the removal efficiency, chemical usage, and so forth, cannot necessarily be projected to other locations with a significantly different quality of water. The selective preference of a monovalent ion, such as nitrate, over a divalent ion, such as sulfate, and vice versa, is a function of the ionic concentration of the water. Buelow et al (15) and Clifford and Weber (16) conducted studies on the use of various strong-base and weak-base ion-exchange resins for nitrate selectivity. Buelow confirmed that nitrate takes preference over sulfate in relatively high ionic-concentration water when ion-exchange resins, such as those employed by the Chemical Separation Corp., are used.

Clifford and Weber recently investigated the selectivities of all nineteen strong-base and thirteen weak-base resins commercially available from U.S. manufacturers for sulfate, nitrate, chloride, and bicarbonate. These studies were part of an investigation to compare a two-bed ion-exchange system (strong acid, weak base) with the single-bed strong-base system for nitrate removal.

The investigators concluded from their studies that sulfate is always preferred over nitrate and that at less than 3,000 mg/l TDS, the selectivity sequence for all commercially available resins is sulfate > nitrate > chloride >> bicarbonate. Somewhat surprisingly, they also concluded that strong- and weak-base resins used for nitrate removal should be selected on the basis of high-nitrate-chloride selectivity and not on the basis of low-sulfate-nitrate selectivity. The reason for this is that high-sulfate-nitrate selectivity actually improves the economics of nitrate removal in column operations because the sulfate is taken up so quickly at the top portion of the bed that the majority of the bed is available for nitrate and the other less selective species.

Clifford and Weber also stated that other significant design factors are the nitrate and sulfate concentration, bed depth, and flow rate. Finally, they concluded that, although the spent regenerant solution from the two-bed system may be easier to dispose of, this system is about one and a half to two times as expensive to construct and operate as the single-bed process. The studies conducted by Buelow (15), Chemical Separations Corp. (14), and Clifford and Weber (16) all support a recommendation that to evaluate the nitrate-removal efficiencies of ion exchange accurately for a specific water, trial runs should be made with the water to be treated rather than with simulated water.

Electrodialysis and Reverse Osmosis: A literature survey indicates few data are available on the removal of nitrate from water by these methods. Electrodialysis studies on irrigation water (17) containing 26 mg/l nitrate-nitrogen and secondary wastewater effluent (18) with 12.4 mg/l of nitrate-nitrogen produced nitrate removals of 46 and 39%, respectively, by single-stage systems. Reverse-osmosis studies on aqueous solutions spiked with 2.5 to 250 mg/l of nitrate-nitrogen achieved slightly higher removals of 68 to 73% by single-stage systems (19). Higher removals should be achievable by multistage systems.

Reverse-osmosis equipment literature published by various manufacturers generally lists a rather wide removal range (60 to 95%) for nitrate. Ion rejection by reverse osmosis is directly related to the size and valence of the ion or molecule. Monovalent nitrate is, therefore, not as effectively removed from water as the divalent ions, such as sulfate. In addition, such other factors as the number of stages, type of membrane, and operating pressure also will affect nitrate-removal efficiency.

Neither electrodialysis nor reverse-osmosis systems have been installed solely for the removal of nitrate from water. Consequently, no removal data are available for full-scale systems. Laboratory data and equipment-manufacturers' information indicate that both processes can remove nitrate but that the amount of removal is dependent on system design.

Nitrate Summary

Chemical reduction and biological denitrification are not practical methods for nitrate removal from drinking water. Ion exchange is the only treatment technique that has been used on a full-scale basis specifically for nitrate removal and has been demonstrated to lower nitrate-nitrogen levels from 20 to 30 mg/l to less than 2 mg/l. The one existing system is atypical, however, and no operating or cost data are available, particularly as they apply to small-system application.

Reverse osmosis and electrodialysis are other effective methods that can provide 40 to 95% removal of nitrate. These methods are more expensive than ion exchange; however, they deserve consideration, particularly if the water supply is high in total dissolved solids or the supply has a multiple-contaminant problem that cannot be solved with a single-bed, ion-exchange system.

Because of the lack of definitive information on treatment techniques for nitrate removal and the potential magnitude of the problem, the Water Supply Research Lab., EPA, has given nitrate a high priority for future research. In August 1977, a research grant was funded to the Charlotte Harbor Water Assn., Charlotte Harbor, Fla., to study the removal of nitrate and the other inorganic contaminants from groundwater by high- [2.8×10^6 Pa (400 psi)] and low- [1.4×10^6 Pa (200 psi)] pressure reverse-osmosis treatment systems. This project will determine the removal effectiveness of each system on at least two different quality groundwaters spiked with varying concentrations of nitrate and other inorganic contaminants. The operating costs and energy requirements also will be determined for each system.

In addition to the Charlotte Harbor project, WSRD expects to fund several other research grants during fiscal year 1978 to evaluate and compare ion-exchange and reverse-osmosis systems for treating small community water supplies. These projects are only in the development stage, and it will be some time before any useful information will be available.

FLUORIDE REMOVAL FROM WATER

Health Effects

Fluoride is a normal constituent of all diets, and it is considered an essential nutrient (20). Fluoride is also added to many drinking waters in small quantities to prevent dental caries. Above an optimum range, however, fluoride will cause the mottling of teeth, particularly in young children during the years of tooth calcification. This disorder, known as dental fluorosis, alters the structure and appearance of the enamel, producing a dark discoloration of the permanent teeth. The degree of discoloration increases as the fluoride concentration increases.

Although the problem of dental fluorosis was known for many years, it was not until 1926 that McKay (21) related it to drinking water, and several more years before it was shown to be specifically caused by fluoride in drinking water [Churchill (22) and Smith (23)]. The intake of excessively large amounts of fluoride also can be expected to produce other health effects: (1) bone changes from long-term consumption of water containing 8 to 20 mg/l of fluoride (24); (2) crippling fluorosis from a total intake of 20 mg of fluoride per day for 20 or more years (25); and (3) death from a single dose of 2,250 to 4,500 mg (24). Fluoride is a substance, therefore, that can provide beneficial or detrimental effects depending on concentration and total amount ingested.

The MCL established for fluoride varies with the average maximum air temperature because the amount of water (and consequently the amount of fluoride) consumed is influenced by the air temperature (Table 10.2). In 1969, a USPHS report on community water supplies (26) reported that finished water from 1,200 water utilities contained fluoride concentrations exceeding 2.0 mg/l. Based upon these data and more recent reports from some of the southeastern states, many small community water supplies are thought to have fluoride concentrations exceeding the MCL.

Treatment Methods

During the years immediately following the discovery of fluoride as the cause of dental fluorosis, many investigations were conducted on various methods and materials for fluoride removal. Reviews of these studies have been reported by Maier (27)-(29), and Savinelli and Black (30). Consequently, only the salient points will be presented in this discussion.

Boruff (31) was the first to investigate a variety of materials for fluoride removal. The materials studied included aluminum sulfate, sodium silicate, ferric fluoride, sodium aluminate, zeolites, bauxite, silica gel, and lime. With the possible exception of lime and aluminum sulfate, none of these materials was found to be very practical.

Alum Coagulation: In addition to the studies on alum coagulation by Boruff (31) (32), this technique has also been investigated by Kempf (33), Scott el al (34), and Culp and Stoltenberg (35). Culp and Stoltenberg showed that 250 mg/l of alum were required to reduce the fluoride concentration in a groundwater from 3.6 to 1.5 mg/l and 350 mg/l was needed to reduce it to 1.0 mg/l. Results among the various researchers varied considerably, probably because of variation in such test conditions as the raw water, mixing, and pH. All results, however were consistent in that large doses of alum were required to remove small amounts of fluoride.

Because the alum doses required for fluoride removal are much greater than those commonly used for turbidity and color removal, the technique has not been considered a very practical solution. Culp and Stoltenberg, however, concluded from their investigation that the treatment technique could remove fluoride and that the method should not be overlooked for treatment of soft, high-fluoride water supplies.

Activated Carbon: McKee and Johnston (36) investigated the use of powdered activated carbon for fluoride removal and showed it to achieve excellent removals. Unfortunately, the process is pH dependent, with good results obtainable only at pH 3.0 or less. At pH 8.0, little fluoride removal was achieved. From a practical standpoint, therefore, the use of this material would be expensive because of the need to lower and raise the pH of the water supply.

Lime Softening: Boruff (31) showed that lime softening could remove fluoride from water, and several years later Scott et al (34) showed that the decrease was a function of the amount of magnesium removed during the softening process. The fluoride-removal mechanism was determined to be coprecipitation with magnesium hydroxide.

If sufficient amounts of magnesium are not present in the water, a magnesium salt must be added in quantities necessary to accomplish the desired level of fluoride removal. This treatment method is, therefore, similar to alum coagulation in that it has limited application. Because of the large quantities of chemical that would normally be required, the method is adaptable to only low-fluoride, high-magnesium water requiring softening.

Ion Exchange: Because of the limited application of either alum coagulation, activated carbon, or lime softening, the most frequently employed treatment method for fluoride removal has been ion exchange, using either activated alumina, bone char, or granular tricalcium phosphate.

Bone was one of the first materials suggested for use as an ion-exchange media because of its known affinity for fluoride. The main ingredient of bone is a carbonate apatite, $Ca_9\ (PO_4)_6 \cdot CaCO_3$. Studies by Smith and Smith (38) on a degreased, caustic-and-acid treated bone material showed that it could reduce the fluoride concentration very effectively from 3.5 to less than 0.2 mg/l. The removal mechanism suggested was an exchange of the carbonate radical with fluoride. Because of the high cost of bone, the material has not been widely used.

Bone char, another bone material, has also been investigated by Burwell et al (39) and has been utilized in several full-scale defluoridation plants. Bone char is ground animal bones charred to remove all organics and consists essentially of tricalcium phosphate and carbon. This material was developed for decoloring cane syrups in sugar manufacturing and is more economical than bone. The suggested removal mechanism for fluoride also was the replacement of the carbonate radical. When material capacity for fluoride is exhausted, the bone char is regenerated with a caustic soda. One of the first full-scale defluoridation plants to use this material was the USPHS plant in Britton, S.D. This plant operated from about 1953 to 1971 with bone char and the data developed indicated an exchange capacity of 102 g fluoride/m^3 bone char when removing 5 mg/l of fluoride.

A potential drawback for the use of bone char was discovered by Bellack (40). During an investigation on the removal of fluoride with bone char on an arsenic-bearing water, Bellack observed that arsenic could also be removed by this material but that the arsenic-removal process was irreversible. Because arsenic competes with fluoride and because it cannot be stripped off by the normal caustic regeneration process, the fluoride capacity of bone char decreases with use and eventually the bone char has to be replaced. Bone char, therefore, would not be very practical for fluoride waters that also contain arsenic.

Bone char is soluble in acid, and the pH of the raw water and acid rinse must be carefully controlled to minimize losses of the media. Fluoride removal increases with decreasing pH, but because of the problem of solubility and for reasons of consumption and distribution, a pH of 7.0 or greater is recommended.

Synthetic bone material, tricalcium phosphate, also has been produced by reacting phosphoric acid with solutions of lime. Both powder and coarse granular forms were developed and investigated for fluoride removal (41)-(44). Granular tricalcium phosphate is similar to other ion-exchange media that require regeneration. The material is regenerated with a caustic solution, usually sodium hydroxide, and then rinsed with a dilute acid solution. Granular tricalcium phosphate material was the first commercially available medium used in defluoridation plants—at Climax, Colo. (45) in 1937 and Scoba, Miss. in 1940. By 1949, both plants had been abandoned.

Granular tricalcium phosphate was also the first medium chosen from the Britton, S.D. defluoridation plant (46). It was used from 1948 to 1951, when it was replaced by bone char. It was found to be efficient but also very friable, and the plant experienced attrition losses of about 42% per year during the 4 years used. The fluoride capacity was rated at 685 g fluoride/m^3. Its capacity, however, decreased about 3% for each 100 mg/l of sulfate in the raw water. Reports on its use in the Britton plant indicated that in addition to its high attrition losses, the medium was not as effective as bone char and required more frequent regeneration. Although the material also has been reported in use in several small-scale defluoridation systems in California (47), large-scale experience has not shown it to be a good fluoride-removal medium.

Activated alumina (Al_2O_3) is calcined granules of hydrated alumina. The material was first studied for fluoride removal by Boruff (31), and thereafter Fink and Lindsay (48) and Swope and Hess (49) successfully demonstrated its use on a small scale. During the 1930s and early 1940s, several other investigations were conducted on patented materials that were based upon the use of aluminum oxide.

The most well-known and longest-operating defluoridation plant in the U.S. was constructed in Bartlett, Tex. by USPHS in 1952 (27). This plant utilized activated alumina as the fluoride-exchange medium and continued to operate with essentially the original medium until this past year when it was closed down. Activated alumina is produced by the major aluminum manufacturers and is available in various mesh sizes. The size in use in full-scale defluoridation plants is

28 to 48 mesh. The advantages of activated alumina for fluoride removal are several. First, the medium is somewhat specific for fluoride and, therefore, has a relatively high fluoride-exchange capacity. Maier (29) reported that the Bartlett plant showed an exchange capacity of 1,590 g/m^3 when the raw water contained 8 mg/l of fluoride and 920 g/m^3 at 3 mg/l of fluoride. Studies by Savinelli and Black (30) showed that the capacity is a function of the fluoride concentration, pH of the treated water, and the amount of regenerant used—the greater the fluoride concentration, the greater the fluoride capacity; the lower the pH, the greater the capacity; and the larger the regenerant solution, the greater the fluoride capacity.

Another advantage of alumina is that the capacity is not affected by the sulfate or chloride concentration of the water: Savinelli and Black showed this for concentrations of up to 1,000 mg/l for each anion. These results tend to confirm the results of the work of Kubli (50), who found the anion-alumina adsorption series in order of decreasing preference to be the following: OH^-, $PO_4^{\equiv}$, $Cr_2O^=$, F^-, $SO_3^=$, $CrO_4^=$, NO_2^-, Cl^-, NO_3^-, $MnO_4^=$, $SO_4^=$. This affinity preference offers a particular advantage over synthetic strong-base resins that remove many anions and whose selectivity series is commonly reported as follows: I^-, HSO_4^-, NO_2^-, Br^-, Cl^-, OH^-, HCO_3^-, $H_2PO_4^-$, F^-.

Another significant advantage of activated alumina is its relatively low cost—about \$425 to \$500/m^3 (\$12 to \$14/ft^3), about 10% the cost of a synthetic anion resin and about 25% the cost of bone char (Table 10.4).

TABLE 10.4: FLUORIDE REMOVAL MEDIA

Material	. . . . Density		Average Fluoride . . . Capacity. . . .		. . . . . . Cost.	
	1,000 kg/m^3	lb/ft^3	g/m^3	g/ft^3	\$/m^3	\$/ft^3
Activated alumina	0.88	55	020-1,480	26-45	435-530	12-15
Bone char	0.85	53	1,030	29	1,350-1,700	38-45
Strong base resin	0.68-0.72	42-45	*	*	4,400-5,300	125-150

*Variable depending on ratio of fluoride in total anions.

The regeneration process for activated alumina is similar to those prescribed for bone char. A caustic solution, usually 1% sodium hydroxide, is used to strip the fluoride from the alumina. Following this, the medium is neutralized with a dilute acid rinse, normally 0.05 N sulfuric acid, and a water rinse. Savinelli and Black reported on the use of alum as a regenerant, but a review of the literature does not indicate that alum actually has been used for regeneration at a full-scale plant. The use of the caustic and acid solutions is probably the major disadvantage to the use of alumina and bone char because the reagents require careful control and handling and, therefore, the need for well-trained operators.

With the closing of the Bartlett, Tex. plant, only two large-scale activated alumina defluoridation plants are operating—at Desert Center, Calif., and at the X-9 Ranch near Tucson, Ariz. Within the near future, a new plant will be constructed and operating at Gila Bend, Ariz. The Desert Center plant, constructed in 1970, has a rated capacity of 5,680 m^3/day (1.5 mgd) and is lowering the fluoride concentration from about 8 mg/l to less than 1 mg/l. The X-9 Ranch plant, which has a rated capacity of 2,650 m^3/day (0.7 mgd), has been operating for 3 years and has been successfully reducing the fluoride from near 5 mg/l to below 1 mg/l. The Gila Bend plant will have a design capacity of 3,840 m^3/day (0.75 mgd) and will be treating 5 mg/l fluoride water.

No data or information have been published on the design or operation of these three plants. During site visits by the author to the two operating facilities, however, plant personnel at both reported the fluoride-exchange capacity (4,520 g/m^3) to be about four times the capacity obtained at the Bartlett, Tex. plant. The operating costs, excluding amortization, were estimated at \$0.57–0.76/m^3 (\$0.15–0.20 per 1,000 gal).

Thompson and McGarvey (51) report on the use of strong-base anion-exchange resins for fluoride removal when these resins are used in the chloride form. Because other anions in the raw water will also be removed, the fluoride-exchange capacity depends on the ratio of fluoride to total anions. The capacity reported for Amberlite XE-75 was approximately 88 g/m^3 of resin when the ratio of fluoride to total anions was 0.05.

The capacity of resins increases rapidly with higher fluoride total-anion ratios, but most fluoride waters have a very low ratio. Because of their relatively low capacity and high cost compared with other media, such as activated alumina, synthetic anion resins usually are not considered for fluoride removal. If, however, other anions need to be removed along with fluoride, these resins have advantages over alumina and bone char.

Reverse Osmosis: Limited data are available on the removal of fluoride by reverse osmosis. Hindin (19) conducted laboratory studies on aqueous solutions containing only fluoride using a very small 7-cm (2.75-in) reverse-osmosis cell containing a commercially available cellulose acetate membrane. The results of this investigation showed for the test conditions imposed that the fluoride concentration could be lowered from 58.5 to 1.0 mg/l (93.3% removal).

Literature published by the various reverse-osmosis manufacturers indicates a wide removal range, 40 to 96%. One company, DuPont, states that fluoride removal is pH dependent. Its data indicate that removals will range from approximately 45 to 90% as the pH increases from 5.5 to 7. It also states that these data apply to brackish water supplies containing 110 mg/l of fluoride and do not necessarily apply to waste streams with higher fluoride concentration.

Although reverse osmosis is not normally used on low TDS water and has not been used specifically for fluoride removal, the process should be an effective method. In spite of its high operating costs, this process may be practical for treating a small community water supply and should be considered as an alternative.

Fluoride Summary

Alum coagulation and lime softening can remove fluoride from water, but these treatment techniques have very limited application and are not considered practical. Reverse osmosis can also remove fluoride, but the process has not been demonstrated on a full-scale basis strictly for fluoride removal. This relatively new treatment method may be practical for fluoride removal in special circumstances. Ion exchange is the only treatment method that has been utilized on a full-scale basis.

Many materials have been investigated for their ability to remove fluoride, but only activated alumina, bone char, and tricalcium phosphate have been tried in full-scale operations. Of these three materials, only activated alumina and bone char have been demonstrably successful and of these two materials, activated alumina seems to be most accepted and proved for fluoride removal from drinking water.

Fluoride treatment technology research has been given high priority, along with nitrate removal, by WSRD during FY-78. The Charlotte Harbor, Fla. reverse-osmosis project will also investigate the removal of fluoride from groundwater. In addition, at least one other research project during the year is expected to be funded on the removal of fluoride by activated alumina for small community water supplies.

REFERENCES

(1) Public Law 93-523, Safe Drinking Water Act, 93rd Cong. S. 433 (December 16, 1974).
(2) National Interim Primary Drinking Water Regulations, EPA, Wtr. Prog., *Fed. Reg.*, 40:248 (December 24, 1975).
(3) Drinking Water Regulations, Radionuclides, EPA, *Fed. Reg.*, 41:133 (July 9, 1976).
(4) Drinking Water Standards, PHS, DHEW, PHS Publ. No. 956 (1962).
(5) *Community Water Supply Study: Analyses of National Survey Findings*, Bu. Wtr. Hygiene, Envir. Health Service, PHS, HEW (July 1970).
(6) Gunderloy, F.C. et al, "Development of a Chemical Denitrification Process," Project No. 17010 EEX, *Wtr. Pol. Control Res. Ser.,* EPA (October 1970).
(7) Gunderloy, F.C. et al, "Dilute Solution Reactions of the Nitrate Ion as Applied to Water Reclamation," Rpt. No. TWRC-1, FWPCA, Cincinnati, Ohio (October 1968).
(8) O'Brien, W.J., "Chemical Removal of Nitrate from Potable Water Supplies," Contribution No. 23, Kansas Water Resources Res. Instit., Manhattan, Kan., NTIS PB-193023 (June 1968).
(9) McCarty, P.L., Beck, L., and St. Amant, P.P., "Biological Denitrification of Waste Waters by Addition of Organic Materials," *Proc. 24th Ind. Waste Conf.*, Purdue Univ., Lafayette, Ind. (May 1969).
(10) Smith, J.M. et al, "Nitrogen Removal from Municipal Waste Water by Columnar Denitrification," *Envir. Sci. Tech.*, 6:3:260 (March 1972).
(11) Jeris, J.S. et al, "Biological Fluidized-Bed Treatment for BOD and Nitrogen Removal," *Jour. WPCF*, 49:5:816 (May 1977).
(12) St. Amant, P.P. and McCarty, P.L., "Treatment of High Nitrate Waters," *Jour. AWWA*, 61:12:659 (December 1969).
(13) Gregg, J.D., "Nitrate Removal at Water Treatment Plant," *Civ. Engrg.*, 43:4:45 (April 1973).
(14) Higgen, I.R. and Ferner, J.D., "Demonstration of Nitrate Removal from Water Supplies," Prepared for Garden City Pk. Wtr. Dist. and Bethpage Wtr. Dist. by Chemical Separations Corp., Oak Ridge, Tenn. (August 1970).
(15) Buelow, R.W. et al, "Nitrate Removal by Anion Exchange Resins," *Jour. AWWA*, 67:9:528 (September 1975).
(16) Clifford, D.A. and Weber, W.J., "Nitrate Removal from Water Supplies by Ion Exchange," EPA Draft Rpt., Cincinnati, Ohio (1977).
(17) Sword, B.R., "Desalination of Irrigated Return Waters," Am. Geophys. Union Nat. Fall Meet., Hydrology Section, San Francisco, Calif. (December 1969).
(18) Takenaka, H.H., Chen, C.L., and Miele, R.P., "Demineralization of Wastewater by Electrodialysis," Off. Res./Devel. EPA, Cincinnati, Ohio (October 1975).

(19) Hindin, E., Dunstan, G.H., and Bennett, R.J., "Water Reclamation by Reverse Osmosis," Bull. 310, Tech. Ext. Service, Washington State Univ., Pullman, Wash. (August 1968).

(20) *Food Nutrition Board, Recommended Daily Allowance*, National Res. Coun. Publ. 1964, Nat. Acad. Sci., Washington, D.C., 7th ed. (1968).

(21) McKay, F.S., "Mottled Enamel: A Fundamental Problem in Dentistry," *Dental Cosmos*, 67:9:847 (September 1925).

(22) Churchill, H.V., "Occurrence of Fluorides in Some Waters of the United States," *Ind. Engrg. Chem.*, 23:9:996 (September 1931).

(23) Smith, M.C., Lantz, E.M., and Smith, H.V., Univ. Arizona Expt. Sta. Tech. Bull. 32 (1931).

(24) Hodge, H.C. and Smith, F.A., "Some Public Health Aspects of Water Fluoridation," *Fluoridation as a Public Measure*, A.A.A.S., J.H. Shaw, editor, Publ. No. 38, Washington, D.C. (1954).

(25) Roholm, K., *Fluoride Intoxication, A Clinical-Hygiene Study*, H.K. Kewis & Co. Ltd., London (1937).

(26) "Natural Fluoride Content of Community Water Supplies," DHEW Public Health Service, N.I.H., Bethesda, Md. (1969).

(27) Maier, F.J., "Methods of Removing Fluoride from Water," *Am. Jour. Public Health*, 37:12:1559 (December 1947).

(28) Maier, F.J., "Defluoridation of Municipal Water Supplies," *Jour. AWWA*, 45:8:879 (August 1953).

(29) Maier, F.J., "Partial Defluoridation of Water," *Public Works*, 91:11:90 (November 1960).

(30) Savinelli, E.A. and Black, A.P., "Defluoridation of Water with Activated Alumina, *Jour. AWWA*, 50:1:33 (January 1958).

(31) Boruff, C.S., "Removal of Fluorides from Drinking Waters," *Ind. Engrg. Chem.* 26:1:69 (January 1934).

(32) Boruff, C.S., Buswell, A.M., and Upton, W.V., "Adsorption of Fluorides from Salts by Alum Floc.," *Ind. Engrg. Chem.*, 29:10:1154 (October 1937).

(33) Kempf, C.A. et al, "Studies on the Removal of Fluoride from Drinking Water in the State of Iowa," *Iowa Acad. of Sci.*, 43:191 (1936).

(34) Scott, R.D. et al, "Fluoride in Ohio Water Supplies–Its Effect, Occurrence, and Reduction," *Jour. AWWA*, 29:1:9 (January 1937).

(35) Culp, R.L. and Stoltenberg, H.A., "Fluoride Reduction at LaCrosse, Kansas," *Jour. AWWA*, 50:3:427 (March 1958).

(36) McKee, R.H. and Johnston, W.S., "Removal of Fluorides from Drinking Water," *Ind. Engrg. Chem.*, 26:8:849 (August 1934).

(37) Finkbeiner, C.S., "Fluoride Reduction Plant Installed at Village of Bloomdale, Ohio," *Wtr. Works Engrg.*, 91:7:990 (July 1938).

(38) Smith, H.V. and Smith, M.C., "Bone Contact Removes Fluorine," *Wtr. Works Engrg.*, 90:1:600 (January 1937).

(39) Burwell, A.L., Case, L.C., and Goodnight, C.H., "Fluoride Removal from Drinking Water," Okla. Geo. Survey Cir. No. 25 (1945).

(40) Bellack, E., "Arsenic Removal from Potable Water," *Jour. AWWA*, 63:7:454 (July 1971).

(41) Adler, H., Klein, G., and Lindsay, F.K., "Removal of Fluorides from Potable Water by Tricalcium Phosphate," *Ind. Engrg. Chem.*, 30:2:163 (February 1938).

(42) Behrman, A.S. and Gustafson, H., "Removal of Fluoride from Water," *Ind. Engrg. Chem.*, 20:9:1011 (September 1938).

(43) Goodwin, R.C. and Litton, J.B., "Removal of Fluorides from Public Water Supplies," *Ind. Engrg. Chem.*, 33:8:1046 (August 1941).

(44) MacIntire, W.H. and Hammond, J.W., "Removal of Fluorides from Natural Waters by Calcium Phosphate," *Ind. Engrg. Chem.*, 30:2:160 (February 1938).

(45) Wamsley, R. and Jones, W.E., "Fluoride Removal," *Wtr./Sew. Works*, 94:8:272 (August 1947).

(46) Maier, F.J., "Water Defluoridation at Britton: End of an Era," *Public Works*, 102:6:70 (June 1971).

(47) Harmon, J.A. and Halechman, S.G., "Defluoridation of Drinking Water in Southern California," *Jour. AWWA*, 57:2:245 (February 1965).

(48) Fink, G.J. and Lindsay, F.K., "Activated Alumina for Removing Fluoride from Drinking Water," *Ind. Engrg. Chem.*, 28:8:947 (August 1936).
(49) Swope, H.G. and Hess, R.H., "Removal of Fluorides from Natural Waters by Defluorite," *Ind. Engrg. Chem.*, 29:4:424 (April 1937).
(50) Kubli, H.A., "Contribution to the Knowledge of Anion Separation by Means of Adsorption on Alumina," *Helv. Chim. Acta,* Switzerland, 3:453 (1947).
(51) Thompson, J. and McGarvey, F.Y., "Ion-Exchange Treatment of Water Supplies," *Jour. AWWA*, 45:2:145 (February 1953).

The material in this section is from an article in the *Journal of the American Water Works Association,* July 1978, by Thomas J. Sorg and Gary S. Logsdon. It concerns the specific treatment technology for the removal of arsenic and selenium from drinking water. References for this section start on p 317.

ARSENIC REMOVAL FROM WATER

Health Effects

Of the ten inorganic contaminants listed in the National Interim Primary Drinking Water Regulations (NIPDWR), arsenic (0.05 mg/l MCL) is certainly most well-known by the average person as a hazard to human health. Although some organic forms have been used for medical treatment, arsenic is not known to be an essential element for humans. Arsenic can occur in four oxidation states in water, but generally is found in only the trivalent and pentavalent states. The element forms both inorganic and organic compounds; the inorganic compounds are more toxic than the organic compounds, and the trivalent form is more toxic than the pentavalent form. The toxicity is also dependent on the chemical form, and the route, duration, and rate of arsenic exposure (1).

No adverse health effects have been reported from the ingestion of water containing arsenic levels of 0.1 mg/l (1), but chronic or subacute toxicity has been observed with the ingestion of a few milligrams per day for 2 weeks or longer (2). The initial symptoms reported are skin erythrema, edema, and pigmentation, and gastrointestinal and neurological disturbances. Other conditions reported to be caused by high levels of arsenic are neuropathy, increased heart attacks, and vascular injury leading to gangrene and "black foot disease" (1)(2).

The maximum no-observed-adverse-effect for arsenic in water is less than 100 μg/l. The intake from 2 liters of water per day containing 20 μg/l is, therefore, slightly greater than 10% of the median total intake, reported to be 137 to 300 μg/l, from all sources in the U.S. From these basic assumptions, the NAS concludes that the "current mandatory U.S. drinking water limit of 50 μg/l provides a meager margin of safety." Although the NAS does not specifically recommend that the existing 0.05 mg/l limit for arsenic be lower, it definitely does not support raising the MCL.

The 1969 Community Water Supply Study (3) on 969 water utilities reported that finished water from less than 1% of the utilities exceeded the 0.05 mg/l limit. Although this study indicated that arsenic was not a serious problem in finished waters, other information suggests otherwise. In 1970, Kopp and Kroner (4) reported that 7% of 1,500 water samples collected from 150 rivers in the U.S. contained arsenic. A year later, Durham et al (5) found 21% of 727 samples from rivers and lakes contained arsenic.

Because arsenic is a common mineral in many western states, it is probably present in many groundwater supplies serving small communities. Because the western states agencies have many small community water supplies to sample, the magnitude of the arsenic problem will not be determined for several years.

Arsenic Chemistry Overview

Arsenic can occur in aquatic systems in four oxidation states; however, it is normally found as an anion with acid characteristics in only the trivalent (arsenite) and pentavalent (arsenate) forms. In natural waters, the valence and the species are dependent on oxidation-reduction conditions and the pH of the water.

In a system including oxygen and sulfur, Ferguson and Gavis (6) showed that the soluble arsenite species are H_3AsO_3, $H_2AsO_3^-$ and $HAsO_3^=$. The soluble arsenate species are H_3AsO_4, $H_2AsO_4^-$, H_2AsO_4, and $AsO_4^{\equiv}$. In the pH range 4–10, the predominant arsenite form is H_3AsO_3 and the predominant arsenate species are $H_2AsO_4^-$ and $HAsO_4^=$. The rate of oxidation of arsenic(III) to arsenic(V) with oxygen was found to be very slow at neutral pH values, but proceeds measurably in several days in strong alkaline or acid solutions.

Based upon a general review of the literature, arsenite is more likely to be found in anaerobic groundwaters and arsenate in aerobic surface waters. Because information on conventional coagulation and lime softening treatment shows that arsenic removal is valence dependent, steps should be taken to identify the arsenic form causing an arsenic problem before evaluating possible solutions.

Very little work has been done to identify the specific forms of arsenic found in drinking water supplies because the commonly used analytical techniques such as atomic absorption do not distinguish between valence states. During fiscal year 1977, EPA awarded a research grant to Texas A&M Univ. to develop a reliable arsenic speciation analytical technique. Problems of sample preservation resulted in the project being expanded to develop suitable methods of arsenic sample preservation. Progress reports have indicated that neither task has a simple solution.

Geochemistry and chemistry investigations on the removal of arsenic from solution indicate that arsenate should be easier to remove from water than arsenite by conventional coagulation treatment. LaPeintre (7) showed that arsenate species coprecipitate with, or adsorb onto, hydrous iron oxides; Osishi and Sanell (8) reported that arsenate can also be absorbed by aluminum hydroxide and clays. Studies on arsenite, on the other hand, indicate that it may not be easily removable.

Mackenthun's measurements (9) of arsenic concentrations in sodium arsenite-treated lakes showed that removal of arsenic from solution and its accumulation in sediments occurred only occasionally. Frear (10) and Gilderhus (11) also reported little removal of sodium arsenite from arsenic-rich soils or sediments. Based upon the results of these studies, conventional coagulation treatment should be more effective for arsenate removal than for arsenite removal.

Treatment Methods

The only treatment methods that have been thoroughly studied for arsenic removal are conventional coagulation and lime softening. Very few investigations have studied the use of ion exchange, powdered or granular activated carbon, reverse osmosis, or electrodialysis.

Conventional Coagulation: Investigations on arsenic removal from water by conventional coagulation treatment began as early as 1934, with work by Buswell et al (12), and more recently by Shen (13), Gulledge and O'Connor (14), and Logsdon et al (15). All of these investigations indicate that removal of arsenic by coagulation treatment is a function of the oxidation state of the arsenic, type and dose of coagulant, the pH of the treated water, and the initial arsenic concentration.

In the early studies of Buswell et al, jar and pilot-plant tests on water were spiked with large amounts of sodium arsenate [As(V)] and sodium arsenite [As(III)]. Ferric sulfate and alum were used as coagulants. Ferric sulfate was found to be more effective than alum, and arsenic(III) removals similar to arsenic(V) removals could be achieved by first oxidizing the arsenic(III) to (V) with calcium hypochlorite. For example, in pilot-plant tests, a 25 mg/l concentration of arsenic(V) could be reduced to 5 mg/l with 136 mg/l of ferric sulfate. With oxidized arsenic(III), the same dose of ferric sulfate was slightly less effective; arsenic(III) concentration was reduced from 25 mg/l to about 7 mg/l.

More recent investigations used jar-test studies with water spiked with arsenic concentrations of 0.05 to 10.0 mg/l. Gulledge and O'Connor studied the effect of the type and dose of coagulant and of pH on the removal of 0.05 mg/l of arsenic(V). Logsdon et al investigated the effect of type and dose of coagulant, pH, and the initial arsenic concentration on removal of both arsenic(III) and (V) from well water.

The effects of pH and coagulant on arsenic(V) found by Gulledge and O'Connor showed that, in the pH range 5 to 7, 30 mg/l doses of ferric sulfate and alum achieved greater than 90% removal of arsenic(V). Above pH 7.0, alum removal ability decreased with increasing pH to 47% at pH 8, whereas ferric sulfate removal ability remained above 95%.

The data of Logsdon et al on the effect of pH and coagulant on arsenic(III) (0.3 mg/l) in well water showed that ferric sulfate coagulation achieved higher arsenic(III) removals (40 to 60%) than alum coagulation (5 to 15%), but the overall removals of arsenic(III) were considerably less than those obtained on arsenic(V) by Gulledge and O'Connor. When Logsdon et al repeated the jar tests following the oxidation of the spiked well water with 2 mg/l of chlorine, the arsenic(III) removals by both ferric sulfate and alum were almost identical to the arsenic(V) removals.

The differences in removals between arsenic(V) and nonoxidized arsenic(III), particularly with alum coagulation, clearly demonstrate the significance of identifying the form of the arsenic contaminant before selecting a treatment method or changing the operation of an existing treatment facility. Furthermore, if alum is the preferred coagulant, the significance of pH control is demonstrated by these studies. Initial arsenic concentration also affects removal. Because removal data are commonly reported on a percentage basis, the significance of percent of removal in achieving the MCL is heavily dependent on the initial concentration. The jar-study results of Logsdon et al on the effect of initial concentration showed that 30 mg/l of either alum or ferric sulfate applied to spiked river water in the optimum pH range achieved removals of 95% or more on arsenic(V) concentrations to about 1.0 mg/l. Above that level, the percentage of removal decreases with increasing concentrations. With these data converted into a form that shows the amount of arsenic(V) remaining in the treated water, it was indicated that the arsenic MCL can be achieved with coagulation doses of around 30 mg/l, with arsenic(V) concentrations of 1.5 mg/l or less.

Similar data for the removal of arsenic(III) from well water for ferric sulfate and alum coagulation respectively showed that the MCL is not attainable when the arsenic(III) concentration is 0.1 mg/l or higher unless arsenic(III) is oxidized to arsenic(V) prior to treatment. A comparison showed ferric sulfate to be more effective than alum on low turbidity well water. With preoxidation and 30 mg/l of ferric sulfate, the MCL can be achieved with arsenic(III) concentrations of 0.8 mg/l or lower. With preoxidation, the MCL is attainable with 30 mg/l of alum when the arsenic(III) concentration is less than 0.2 mg/l.

If the arsenic MCL (0.05 mg/l) cannot be achieved with coagulant doses of 30 mg/l, removals can be improved by increasing the coagulant dose. Results of studies of Logsdon et al on determining the effect of increased coagulant dose on arsenic(III) and arsenic(V) removal showed that, in the optimum pH range of 5.7 to 8.0, ferric sulfate can achieve greater increases in removals than alum at the same incremental dosage increase. The benefits derived by increasing the dose were also shown to be greater with higher turbidity river water than with low turbidity well water, indicating that turbidity may also affect removals.

In addition to the jar-test studies done by Logsdon et al at WSRD, pilot-plant studies were made on arsenic removal. The WSRD pilot plant is capable of treating two parallel streams, each 7.6 l/m (2 gpm), of water. The parallel treatment trains were designed to be operated in a conventional manner, but with the flexibility to bypass any of the treatment steps and to be operated in series.

Each treatment train has two rapid-mix basins, a flocculation basin, a sedimentation tank and four parallel filters. The test water is pumped from a 28.4 m^3 (7,500 gal) stainless steel storage tank to the first rapid mix tank. From here, the water flows by gravity through coagulation, flocculation, and sedimentation steps to the filter columns.

The filtration rate of the filters is controlled by pumps on the effluent side of each filter. Although the system can employ four parallel filters, each with a diameter of 10.8 cm (4.25 in), at a loading rate of 2.5×10^{-3} m/s (4 gpm/sq ft), the actual number used varied from one to four. The filter materials were changed periodically and the following different filter media were used:

1. Dual-media–51 cm (20 in) anthracite with an ES of 1.2 and UC of 1.5 over 30 cm (12 in) of sand with an ES of 0.4 and UC of 1.7,
2. GAC–76 cm (30 in),
3. GAC–76 cm (30 in),
4. Exhausted GAC–76 cm (30 in), and
5. GAC–76 cm (30 in).

The test waters for the pilot-plant studies were Cincinnati tap water; Ohio river water; Glendale, Ohio well water; and a low turbidity, gravel-pit water from the Cincinnati area. Except for Cincinnati tap water, all test waters were trucked to the plant on an as-needed basis in a 19 m^3 (5,000 gal) tank trailer.

A normal pilot-plant test ran 90 to 100 hours of continuous treatment from Monday morning to Friday afternoon. Because none of the test waters contained arsenic, they were spiked with either sodium arsenite or sodium arsenate. One liter grab samples of the raw, settled, and filtered water were normally collected at 12-hour intervals. Samples were preserved with concentrated nitric acid and analyzed for arsenic by atomic absorption spectrophotometer methods. A summary of the arsenic removal results is given in Table 10.5. Each removal percentage is an average of 8 to 10 grab samples.

The dual-media filter results were plotted for comparison with jar-test results. Although the test conditions and waters were not always identical, the pilot-plant removal results generally agreed with jar-test results. They confirmed that: (1) arsenic(V) is more effectively removed than arsenic(III) by alum and ferric sulfate coagulation; (2) ferric sulfate coagulation achieves higher arsenic(III) removals than alum; and (3) removals of arsenic(V) by alum coagulation are very dependent on the pH of the treated waters. No pilot-plant studies were conducted on the removal of oxidized arsenic(III).

The significance of the oxidation state and the type of coagulant on arsenic removal was also shown in an investigation by Shen (13) on Taiwan groundwater containing naturally occurring arsenic. Oxidation and coagulation studies with varying doses of potassium permanganate and chlorine as oxidizing agents indicated that chlorine was the better oxidizing agent, although both oxidants significantly increased iron coagulation removals. Aeration alone, and in combination with coagulation, had no apparent effect on arsenic removals, indicating that aeration cannot oxidize arsenite to arsenate in a short time. After a number of laboratory and pilot-plant studies, Shen concluded that the only treatment technique continuously capable of achieving high arsenic removal was ferric chloride coagulation preceded by chlorine oxidation and followed by sedimentation and filtration. Based on Shen's laboratory and pilot-plant results, a full-scale arsenic removal plant was constructed in Taiwan in 1969 to treat 150 m^3 (0.04 mgd). During the first 59 days of operation, when the raw water arsenic concentration was 0.6 to 0.94 mg/l, plant operation data showed arsenic removals from 82-100% (below detectable limit).

Lime Softening: Arsenic can be removed very effectively from hard water by lime and excess lime softening treatment, with the amount of removal dependent upon the pH and the valence of the arsenic contaminant. Jar-test study results reported by Logsdon et al (15) showed that arsenic(V) was more readily removed from hard water than arsenic(III) and removals of both forms increased with increasing pH.

The results of these lime softening studies on well water, having an initial hardness of around 300 mg/l as $CaCO_3$, show that, at pH 10.5 and above, almost 100% removal was obtained on water spiked with 0.4 mg/l of arsenic(V). The best removal of arsenic(III) was about 80% at pH 11.1 and above. Below the optimum pH range, arsenic(V) and (III) removals decreased with decreasing pH, with arsenic(V) removals always higher than arsenic(III) removals. As with conventional coagulation treatment, arsenic(III) removals almost identical to arsenic(V) removals can be achieved by oxidizing arsenic(III) to arsenic(V) prior to treatment.

TABLE 10.5: RESULTS OF CONVENTIONAL COAGULATION PILOT-PLANT TESTS FOR ARSENIC REMOVAL

Test No.	Date Started	Length of Test (hr)	Test Water	Coagulant Type	Coagulant Dose (mg/l)	pH of Treated Water	Arsenic Contaminant Form	Arsenic Contaminant Raw Water Concentration (mg/l)	Percentage Arsenic Removed: Settled Water	Filtered Water: Dual-Media Filter	Filtered Water: Granular Activated Carbon Filters: Virgin GAC*	Exhausted GAC*	Virgin GAC**
1	11/11/74	101	Cincinnati tap	alum	26	7.4	As V	0.41	56	79	–	–	–
2	11/18/74	102	Cincinnati tap	alum	26	8.3	As V	0.39	31	44	–	–	–
3	12/9/74	102	Cincinnati tap	ferric sulfate	19	8.4	As V	0.50	75	89	–	–	–
4	12/16/74	102	Cincinnati tap	ferric sulfate	15	7.2	As V	0.30	48	78	–	–	–
5	4/28/75	96	gravel pit	ferric sulfate	19	7.6	As III	0.43	25	47	–	–	–
6	5/5/75	99	gravel pit	ferric sulfate	18	8.5	As III	0.48	38	58	–	–	–
7	5/12/75	99	gravel pit	ferric sulfate	18	6.5	As III	0.45	20	36	–	–	–
8	5/19/75	100	gravel pit	ferric sulfate	15	6.5	As III	0.25	6	29	–	–	–
9	5/26/75	99	gravel pit	ferric sulfate	15	8.5	As III	0.27	27	52	–	–	–
10	6/2/75	100	gravel pit	alum	27	8.5	As III	0.39	1	1	–	–	–
11	6/9/75	102	gravel pit	alum	20	7.5	As III	0.47	3	6	–	–	–
12	6/16/75	100	gravel pit	alum	17	7.5	As III	0.54	9	11	–	–	–
13	6/23/75	100	gravel pit	alum	15	6.5	As III	0.46	8	13	–	–	–
14	11/10/75	95	Ohio River	alum	24	8.9	As V	0.37	14	14	11	–	–
15	11/17/75	100	Ohio River	ferric sulfate	25	6.7	As V	0.39	81	96	98	–	–
16	12/1/75	98	Ohio River	ferric sulfate	26	8.8	As V	0.33	46	62	61	–	–
17	8/30/76	100	Ohio River	ferric sulfate	29	7.0	As III	0.12	36	64	90	91	–
18	9/20/76	92	Ohio River	ferric sulfate	28	8.3	As III	0.12	61	82	80	84	–
19	1/17/77	92	Ohio River	ferric sulfate	30	8.0	As V	0.26	80	95	97	95	–
20	5/23/77	93	Ohio River	alum	29	7.0	As III	0.16	14	21	–	35	44
21	7/11/77	95	Ohio River	alum	27	8.1	As V	0.17	17	23	–	37	24
22	8/8/77	89	Ohio River	ferrous sulfate	33	7.0	As III	0.23	40	76	–	83	84
23	8/15/77	93	Ohio River	ferrous sulfate	28	8.5	As III	0.20	52	73	–	68	64
24	8/29/77	90	Ohio River	ferrous sulfate	30	8.6	As V	0.81	62	70	–	75	76
25	10/31/77	88	Ohio River	alum	25	7.3	As V	0.16	74	95	–	96	94
26	11/7/77	91	Ohio River	alum	27	7.7	As V	0.22	71	79	–	82	83
27	11/14/77	92	Ohio River	alum	48	6.4	As III	0.62	18	18	–	29	36
28	1/9/78	76	well water	alum	28	7.6	As III	0.21	10	19	–	27	29
29	2/6/78	70	well water	ferric sulfate	33	7.5	As III	0.22	48	62	–	72	67
30	2/13/78	90	well water	ferric sulfate	30	6.4	As III	0.19	30	47	–	71	58

*Filtrasorb 200, Calgon Corp. **Hydrodarco 1030, ICI United States, Inc.

The capability of achieving the arsenic MCL by lime softening is dependent on the initial concentration and percentage of removal. Logsdon et al (15) studied the effect of initial concentration on arsenic(V) and (III) removal. In the optimum pH range, the data indicate that the arsenic MCL can be achieved if the concentration of arsenic(V) or oxidized arsenic(III) in the raw water is 0.35 mg/l or lower. If arsenic(III) levels are less than 0.1 mg/l, preoxidation is not necessary to achieve the MCL, if softening is done at pH 10.7 or above.

The WSRD conducted several lime-softening pilot-plant studies on removing arsenic from well water. Two test runs were completed at pH 9.5 and 11.3 for each arsenic form. The pilot-plant system was operated in the same manner as conventional coagulation for softening at pH 9.5. At pH 11.6, modifications provided for recarbonation and additional settling; the settled water was pumped to the settling tank of the parallel treatment system, where the water was recarbonated to lower the pH to 9.6. The recarbonated water flowed through the second settling tank before entering the filters.

To determine if additional settling occurred, extra grab samples of the effluent water were collected from the second settling basin. The results of the four lime-softening pilot-plant test runs are listed in Table 10.6. The arsenic(V) removal results are identical to those in the jar test studies and the arsenic(III) removal results were comparable. No pilot-plant test runs were done with oxidized arsenic(III).

TABLE 10.6: RESULTS OF LIME-SOFTENING PILOT-PLANT TESTS FOR ARSENIC REMOVAL FROM WELL WATER

				Arsenic Contaminant		Percent Arsenic Removed			
						Settled Water		Filtered Water	
Test No.	Date Started	Length of Test (hr)	pH of Treated Water	Form	Raw Water Concentration (mg/l)	First Stage	Second Stage	Dual-Media	GAC*
1	2/3/75	100	11.3	As III	0.34	63	69	71	72
2	2/10/75	100	9.5	As III	0.48	10	–	24	26
3	3/3/75	59	9.5	As V	0.42	49	–	53	53
4	3/24/75	99	11.3	As V	0.58	98	98	99	99

*Filtrasorb 100, Calgon Corp.

Ion Exchange: Very few ion exchange studies on the removal of arsenic from water have been reported in the literature. The limited data from these studies show, however, that arsenic can be removed with anion exchange resins but not with cation resins. Because arsenic occurs as an anion and cannot be removed with cation resins, neither the common household water softener nor the conventional zeolite softening system can remove arsenic from water.

The practicability of removing arsenic by anion ion exchange is questionable because the effect of competitive anions in the water is not known. The arsenic ion exchange studies reported in the literature were basically short screening tests to determine if arsenic could be removed by this treatment technique. None of the studies evaluated the effect of other anions on the efficiency of arsenic removal.

Bellack (17) accidentally discovered that activated alumina and bone char could remove arsenic from water. Because arsenic sorption on bone char caused an irreversible change in the chemical structure of the char, Bellack concentrated his investigation on the use of activated alumina. Column studies using a synthetic water containing 0.06 mg/l of arsenic and 4.8 mg/l of fluoride showed that alumina could lower the arsenic concentration from 0.06 mg/l to around 0.003 mg/l. (Fluoride concentration was also reduced to 1.0 mg/l or less.)

Bellack also did column studies on a well water containing more than 0.1 mg/l of naturally occurring arsenic, which showed that arsenic removal was dependent on flow rate and pH of the raw water. The optimum flow rate was established at 1.5 to 2 x 10^{-3} m/s (2.5 to 3 gpm/sq ft) and removal increased as the pH decreased from 10 to 7. Under optimum conditions, the initial arsenic content in the effluent was below the detectable limit (0.003 mg/l). The studies also showed that the method of regeneration is significant in maximizing the efficiency of the process.

Because bone char cannot be regenerated for arsenic removal, its use is limited to applications in which it can be discarded when exhausted. Activated alumina can be regenerated and may be economical for arsenic treatment, particularly on waters that are also high in fluoride. The advantages of activated alumina over anion exchange resins include its lower cost and its preference for arsenic in the presence of sulfate and chloride anions.

Until 1977, information on the removal of arsenic with activated alumina was limited to the 1970 laboratory study by Bellack. During 1977, Rubel and Alcoa (18) conducted a pilot study on an Arizona groundwater containing about 1.0 mg/l of naturally occurring arsenic, which showed that activated alumina achieved removals below the detectable limit of 0.005 mg/l. Although this information indicates that activated alumina may be a very promising treatment method, pilot-plant studies are recommended before this technique is selected because of the limited amount of laboratory pilot-plant data and the lack of full-scale treatment experience.

Activated Carbon: Jar-test studies by Logsdon et al (15) showed that powdered activated carbon (PAC) is not effective for the removal of either arsenic(III) or (V). Granular activated carbon (GAC) is very effective for removing organics from water and wastewater. A general review of the technical literature indicates that GAC will also remove some metal salts such as mercury chloride (19). Because of the widespread interest in GAC for water treatment, the WSRD pilot-plant studies included one or two GAC-filter columns used in parallel with a dual-media filter, in some test runs.

The virgin GAC material was replaced only occasionally when the plant was shut down for maintenance or repairs. Six months was the longest period that any GAC filter was used, with the usual period being two months. Total organic carbon (TOC) removal was not monitored; thus no data were available to show whether any GAC filters became exhausted because of TOC removal. (None of the virgin GAC was believed to have been exhausted during the arsenic studies, however. Data from other WSRD pilot-plant studies showed 40% TOC removal at ten months of continuous treatment of settled coagulated Ohio River water with virgin GAC material.) Exhausted GAC material was obtained from a full-scale water treatment plant in Virginia. This material was considered to be exhausted for TOC removal for all tests and was generally replaced when the other filter materials were replaced.

The arsenic removal data in Tables 10.5 and 10.6 show that, except for test runs 17 and 20 of Table 10.5, little difference in removal exists between the GAC filters and the dual-media filter, thus indicating that the GAC did have arsenic adsorption capability. In tests 17 and 20, arsenic removals with the exhausted and virgin GAC media were similar, but both were slightly higher than those achieved by the dual-media filter. The reason suggested for this difference is turbidity removal, not adsorption, because the GAC filters usually had shorter filter runs and slightly better turbidity removal.

Reverse Osmosis and Electrodialysis: A general review of the literature including equipment manufacturer information revealed no data on the removal of arsenic from drinking water by either reverse osmosis or electrodialysis. One reverse osmosis equipment specification bulletin did list an estimated arsenic rejection range of 90 to 95%. Both treatment methods are very effective for the removal of most dissolved solids and, therefore, should be effective for arsenic removal.

Arsenic Treatment Summary

Conventional coagulations using alum or iron can achieve more than 90% removal of environmental levels of arsenic(V) at pH 7 or below. Above pH 7, iron coagulation can also obtain high arsenic(V) removal, but alum coagulation removals decrease with increasing pH. Removal can be increased by increasing coagulant dose.

Lime softening is also very effective for arsenic(V) removal, achieving about 100% reductions at pH 10.6 and above. Below this pH level, the removal decreases with decreasing pH. Coagulation and lime softening treatment are poor to moderately effective for the removal of arsenic(III).

Iron coagulation can achieve 50 to 60% removal in the normal pH operating range. Lime softening removals are pH dependent, with best removals (70 to 80%) occurring above pH 11. Below pH 11, removal decreases with decreasing pH to as low as 15 to 20% at pH 10.2. The least effective method for arsenic(III) removal is alum coagulation with removal below 20% achieved throughout the common operating pH range.

Arsenic(III) removals identical to those of arsenic(V) can be achieved by conventional coagulation and lime softening if the arsenic(III) is oxidized to arsenic(V) prior to treatment. Chlorine and potassium permanganate easily oxidize arsenic(III) to arsenic(V) whereas aeration was not found effective. Ozone and other strong oxidants should also be successful. Preoxidation with chlorine may not be always advisable because of the potential formation of chloroform and other trihalomethanes by the reaction of chlorine with organic material in the water. This potential should be investigated thoroughly before a preoxidation step is chosen and an oxidant selected, because the solution to one problem could create a more serious one.

Ion exchange treatment using an anion exchange resin, activated alumina, or bone char can remove both forms of arsenic. Little information is available, however, on the competitive effect of other anions on the efficiency of these materials.

Bone char cannot be regenerated and, therefore would have application only on a throw-away basis. Pilot-plant studies conducted on the actual water to be treated would be an absolute necessity prior to the design of an ion exchange plant.

Because reverse osmosis and electrodialysis are very effective on the removal of most dissolved solids, both treatment methods should also be effective for arsenic removal.

The WSRD research plan for fiscal year 1978 for the removal of arsenic is to study the effectiveness of reverse osmosis and ion exchange with activated alumina. The Charlotte Harbor, Fla. reverse osmosis project, described in the nitrate removal section, will provide reverse osmosis rejection data for various arsenic concentrations in natural waters. Laboratory studies on the removal of arsenic from water by reverse osmosis and activated alumina will also continue at the Environmental Research Center in Cincinnati, Ohio.

SELENIUM REMOVAL FROM WATER

Health Effects

Prior to 1962, the USPHS limit for selenium in drinking water that constituted grounds for rejecting a water supply was 0.05 mg/l. Because of concern over the possible carcinogenic properties of selenium, the USPHS lowered the limit to 0.01 mg/l in the 1962 revised PHS standards. The limit was maintained by EPA in establishing the 1976 NIPDWR.

Controversy continues to exist over the MCL for selenium since data supporting carcinogenicity have not been forthcoming or confirmed in properly conducted studies. Moreover, selenium is considered an essential nutrient and some evidence suggests that it may provide some protection against certain types of cancer (2). On the other hand, chronic exposure to excess selenium results in dermatitis and central nervous system and gastrointestinal disturbances, and very large doses can cause death (2). Selenium therefore, is similar to fluoride and some other elements in that it has both adverse and beneficial effects depending on the amount ingested. Some studies also indicate that the beneficial or adverse effects of selenium are dependent on the mode of intake, the form of selenium, the presence of other trace metals, and other factors (1).

The EPA supporting background information for the NIPDWR states that the 0.01 mg/l limit provides a safety factor of about three, considering the lower end of the range of selenium intakes associated with minor toxic effects in man (1). This factor was based upon an estimated average selenium intake of an American person as 200 μg per day from sources other than water, and on the assumption that an individual consumes two liters of water per day.

The National Academy of Sciences (NAS) report (2) on the NIPDWR draws no firm conclusion on the adequacy of the selenium MCL. It does state, however, "that most evidence indicates that there is a greater overall potential for selenium deficiency than for selenium toxicity at current levels of selenium intake." The report further states that "the no-observed-adverse-health effect level for selenium in water is at least 100 μg/l and appears to be as great as 500 μg/l." NAS recommended a nine-point research program that included studies on developing baseline data on selenium levels in humans in health and disease.

Chemistry Overview

Selenium is similar to arsenic in several ways. Selenium has several oxidation states, but only two are predominant in water: selenium(IV) (selenite) and selenium(VI) (selenate). Like arsenic, selenium occurs as an anion in water and thus has acid characteristics.

Hingston (25), in studying the effect of pH on selenium(IV) removal, showed that a maximum amount of adsorption occurs at any one pH and that this amount decreases with increasing pH. Howard compared the use of two types of clays to ferric hydroxide and found that ferric hydroxide was much more effective than the clays for selenium(IV) removal.

Another significant finding of selenium chemistry that has application to its removal from water is that the transformation of selenite to selenate is very slow, and that both forms in the same solution are very stable (22)(26). Neither form can be oxidized easily or reduced to its other valence state. These studies indicate that conventional coagulation treatment should remove selenium(IV) much more effectively than selenium(VI) and that removals are dependent on various operating conditions.

Data from the 1969 Community Water Supply study showed that only 16 of 952 finished water samples exceeded the 0.01 mg/l selenium limit. The maximum level reported was 0.065 mg/l; the minimum level was 0.002 mg/l; and the mean 0.003 mg/l. Thus, there were very few problems in the finished water of the treatment systems studied.

Little information is available on which to estimate the magnitude of the selenium problem. Because selenium is not a common industrial pollutant, it is more likely to be found in groundwaters than in surface waters. Several surveys have shown that selenium problems may be numerous in certain localities.

Engberg (29) reported that the 0.01 mg/l selenium limit was exceeded in 40.3% of 139 groundwaters surveyed in Nebraska. The two highest concentrations were 0.48 mg/l and 0.103 mg/l. Thirty-nine surface waters were also sampled and 25% of these had maximum levels of more than 0.01 mg/l. The highest level reported was 0.02 mg/l, twice the MCL.

Selenium is associated with uranium mining and has been reported in groundwaters in mining areas of Colorado, Arizona, and New Mexico. An EPA study (30) of the water quality, in the Grants Mineral Belt, N.M., showed selenium levels of 0.05 to 0.11 mg/l in industry potable water supplies.

Studies on the occurrence of selenium in ground and surface waters have reported results as total selenium and have not identified the specific species found. For general monitoring purposes, total selenium concentration measurements are certainly sufficient. However, because the effectiveness of certain treatment techniques, such as lime softening and conventional coagulation, depends on the oxidation state of the selenium, identification of the oxidation state may be necessary before a selection of treatment. No simplified analytical techniques to identify the form of selenium exist. However, ferric hydroxide precipitation of selenium(IV) has been suggested to separate selenium(IV) from selenium(VI). In lieu of more refined analytical techniques, a standard jar-test, using an iron coagulant in the pH range 6 to 7, could determine the valence of selenium in raw water. If the iron coagulant lowers the selenium level of the raw water significantly, selenium(IV) is present. If the raw water contains a mixture of the two valence forms, the technique will be less useful.

Treatment Methods

Very little work has been done on removing selenium from drinking water. The only in-depth studies have been conducted by the WSRD, and were reported in part by Logsdon et al (15) and Sorg and Logsdon (31). These studies were primarily coagulation and lime-softening laboratory jar-test studies. Since these reports were published, pilot-plant studies have been done and are described here.

A review of the wastewater treatment literature showed a number of studies done on the removal of trace metals from wastewater, but rarely was selenium included among the list of contaminants. The most pertinent wastewater studies were reported by Hannah et al (32) and Linstedt et al (33).

Conventional Coagulation: Jar-test and pilot-plant tests conducted by the WSRD on the removal of selenium from ground and surface waters by conventional coagulation showed that selenium removal is dependent on the oxidation state and the initial concentration of selenium, the pH of the treated water, and the type and doses of coagulant. Under optimum operating conditions, the highest removals of selenium(IV) were about 80%. This treatment technique was ineffective for removing selenium(VI), with 10% or less removal obtained.

The removal curves indicate that ferric sulfate achieved much higher selenium(IV) removals than alum, and that ferric sulfate coagulation removals are pH dependent. The highest removals (70 to 80%) were achieved with iron coagulation in the 6 to 7 pH range with Ohio River water. Alum coagulation removals of selenium(IV) were significantly less (20%) than those of iron coagulation with both types of water. Moreover, the pH had little effect on removals with the 25 mg/l dose used in the experiments. If higher selenium(IV) removals are required, removals can be increased by using larger coagulant doses.

The jar-test studies with selenium(VI) showed less than 10% removal by both ferric sulfate and alum coagulation through the pH range 6 to 8. When larger coagulant doses were used, no significant increases in removals were obtained with either coagulant.

Following completion of the jar-test experiments, the WSRD ran pilot-plant studies. Alum and ferric sulfate coagulation tests done with the variety of spiked test waters used in the arsenic tests produced the results shown in Table 10.7.

The pilot-plant results were almost identical to those from jar-test studies. Ferric sulfate and alum coagulation achieved less than 10% removal of selenium(VI). Selenium(IV) removals varied from near zero to about 70% depending on the coagulant used and the pH of the water. The pilot-plant results confirmed the conclusion drawn from the jar-test studies, that although ferric sulfate is much more effective than alum, even ferric sulfate coagulation is not a very good treatment method for removing selenium(IV).

TABLE 10.7: RESULTS OF CONVENTIONAL COAGULATION PILOT-PLANT TESTS FOR SELENIUM REMOVAL

Test No.	Date Started	Length of Test (hr)	Test Water	Coagulant Type	Coagulant Dose (mg/l)	pH of Treated Water	Selenium Contaminant Form	Selenium Contaminant Raw Water Concentration (mg/l)	Percentage Selenium Removed: Settled Water	Filtered Water: Dual-Media Filter	Filtered Water: GAC Filters Virgin GAC*	Filtered Water: GAC Filters Exhausted GAC*	Filtered Water: GAC Filters Virgin GAC**
1	5/19/75	100	gravel pit	ferric sulfate	15	6.5	Se IV	0.040	20	58	–	–	–
2	5/26/75	99	gravel pit	ferric sulfate	15	8.5	Se IV	0.050	0	0	–	–	–
3	6/2/75	100	gravel pit	alum	27	8.5	Se IV	0.061	5	8	–	–	–
4	6/9/75	102	gravel pit	alum	20	7.5	Se IV	0.066	23	20	–	–	–
5	6/19/75	100	gravel pit	alum	17	7.5	Se IV	0.050	4	10	–	–	–
6	6/23/75	100	gravel pit	alum	15	6.5	Se IV	0.033	3	6	–	–	–
7	9/20/76	92	Ohio River	ferric sulfate	28	8.3	Se IV	0.017	26	30	41	36	–
8	1/31/77	94	Ohio River	ferric sulfate	25	7.8	Se IV	0.025	64	78	80	80	–
9	2/28/77	100	Ohio River	ferric sulfate	23	6.9	Se IV	0.027	81	81	81	81	–
10	3/7/77	99	Ohio River	ferric sulfate	32	6.5	Se VI	0.097	11	14	23	28	–
11	4/11/77	100	Ohio River	alum	31	6.6	Se IV	0.019	12	20	–	17	17
12	4/25/77	100	Ohio River	alum	30	8.1	Se IV	0.030	28	36	–	30	29
13	5/2/77	100	Ohio River	alum	29	8.1	Se VI	0.076	6	8	–	6	10
14	6/6/77	100	Ohio River	alum	30	8.3	Se IV	0.029	29	20	–	34	27
15	6/13/77	100	Ohio River	alum	30	8.3	Se VI	0.055	0	5	–	20	14
16	7/18/77	82	Ohio River	alum	34	7.9	Se IV	0.022	14	18	–	21	14
17	7/25/77	93	Ohio River	ferric sulfate	29	7.9	Se IV	0.024	47	58	–	63	60
18	9/12/77	91	Ohio River	ferrous sulfate	29	8.1	Se VI	0.042	0	0	–	0	0
19	9/19/77	72	Ohio River	ferrous sulfate	32	8.3	Se VI	0.083	1	2	–	0	5
20	10/31/77	88	Ohio River	alum	25	6.8	Se VI	0.028	18	29	–	0	0
21	1/9/78	76	well water	alum	28	7.6	Se IV	0.056	18	29	–	36	39
22	2/6/78	70	well water	ferric sulfate	33	7.5	Se IV	0.079	69	74	–	74	73
23	2/13/78	90	well water	ferric sulfate	30	6.4	Se IV	0.047	3	79	–	77	77

*Filtrasorb 200, Calgon Corp.

**Hydrodarco 1030, United States, Inc.

Lime Softening: Jar-test and pilot-plant studies on the removal of selenium by lime softening showed that this treatment technique is not very effective for the removal of either selenium(IV) or selenium(VI). Selenium(VI) removals were similar to those of conventional coagulation treatment; less than 10% removal through the pH range 9.0 to 11.5. Removals of selenium(IV) were only slightly higher. The maximum removal, obtained with hard well water spiked with 0.1 mg/l selenium(IV), was about 50% at pH 11.5. Below this pH, removals decreased with decreasing pH. When the test was repeated with only 0.03 mg/l of selenium(IV), the percentage of removals was about the same.

Pilot-plant tests were conducted on the removal of selenium(IV) at pH 9.5 and 11.3 with the same well water used in the jar-test studies. Although the pilot-plant results were slightly higher than those in jar-test studies, they were not significantly higher, indicating that lime softening is not a good method for selenium removal.

Ion Exchange: Because conventional coagulation and lime softening are ineffective for selenium(VI) removal and only moderately effective for selenium(IV) removal, the WSRD investigated the use of ion exchange. Column tests were run using distilled water spiked with 0.1 mg/l of either selenium(IV) or (VI). Six liters of the test solution were passed through two ion-exchange columns in series. The first column contained a cation resin and the second one, an anion resin. For all samples, 12.5% removal of selenium(IV) was effected by the cation resin and more than 98% by the two resins in series. The selenium(VI) test results were almost identical to those of selenium(IV). The cation effluent samples showed a 7.4% removal, with cumulative removals by both resins above 98% for all samples. Because selenium occurs as an anion, the low removal in the cation column was expected.

In a study by Linstedt et al (33) investigating the ion-exchange process for selenium(IV) removal as part of a wastewater pilot-plant study, the results were similar to the WSRD results with distilled water. A cation ion-exchange resin removed only 0.9%, whereas cation and anion resins in series removed over 99%. In these tests, essentially all of the selenium(IV) removal occurred in the anion exchange resin column.

Because activated alumina has been shown to remove arsenic from water, and because of the similarities between arsenic and selenium, the WSRD recently began to investigate activated alumina for selenium removal, using column tests with activated alumina. Separate experiments were conducted for the removal of selenium(IV) and (VI) with distilled water and a hard groundwater from Wyoming, Ohio. The studies with spiked distilled water (0.3 mg/l) showed very high removals (95% and above) of both selenium(IV) and (VI). When the experiments were repeated with the groundwater with total dissolved solids (TDS) of 608 mg/l, however, selenium(IV) removals ranged from 62 to 88% and essentially no selenium(VI) was removed (34).

Variables such as pH of the influent water, flow rate, and regeneration procedures were tested, but data were insufficient to evaluate the effect of these variables on selenium(IV) and (VI) removal. The only conclusion that can be drawn at this time is that activated alumina may be effective for selenium(IV) removal and ineffective for selenium(VI) removal. Further research is required to evaluate the variables that affect selenium removal and to determine operating conditions to optimize removal. If selenium removal is ultimately found to be valence dependent, this would be a shortcoming in the use of activated alumina.

Activated Carbon: The WSRD conducted jar-test and pilot-plant studies to determine the removal of selenium(IV) and (VI) from water by powdered activated carbon (PAC), which showed that less than 4% of either selenium species could be removed with PAC doses up to 100 mg/l on well water spiked with selenium concentrations of 0.03 and 0.1 mg/l. Tests done at several pH values showed that pH had no effect on removals. Results indicated that granular activated carbon (GAC) does not adsorb selenium.

Reverse Osmosis and Electrodialysis: The WSRD conducted two brief experiments to determine selenium removal by reverse osmosis (RO). The RO system was a small portable unit containing a cellulose acetate membrane. During a 60-minute test run with Cincinnati tap water (360 mg/l TDS) spiked with 0.1 mg/l of selenium(IV), seven product water samples showed selenium removals above 97%. The RO operating pressure was 1,300 kPa (190 psi), the flow rate was 7.3.

For the selenium(VI) experiments on Cincinnati tap water spiked with 0.1 mg/l of selenium(VI), the TDS content was raised to almost 3,000 mg/l with sodium chloride and calcium sulfate. The RO flow rate was 1.5 l/min, the operating pressure was 1,295 kPa (185 psi), and the raw water pH was 6.3. Fifteen samples collected during a 140-min test run all had selenium(VI) removals above 97%.

No information was found on the removal of either selenium(IV) or (VI) by electrodialysis. Electrodialysis, like reverse osmosis, has good rejection characteristics for most ions and, therefore, should be effective for selenium removal.

Selenium Treatment Summary

Lime softening and conventional coagulation treatment with either alum or iron coagulants are ineffective for removing selenium(VI) from water. The only treatment methods that can achieve high removals (90% or above) of selenium(VI) are ion exchange, reverse osmosis, and electrodialysis. None of these techniques, however, has been demonstrated on either a small or large scale, and little research has been done on the proper operational procedures for ion exchange or on the effect of competing anions on this process.

Moderate removals of 40 to 80% of selenium(IV) can be achieved by iron coagulation treatment. The percentage of removal is dependent on the pH of the treated water, the iron coagulant dose, and the initial selenium concentration. Alum coagulation is not a good method for selenium(IV) removal; only 15 to 20% removal can be obtained with alum doses of 20 to 30 mg/l. The effectiveness of lime softening for selenium(IV) removal lies between that of iron and alum coagulation. Lime softening removals are pH dependent but even at optimum pH, removals are only 35 to 45%.

The most effective treatment methods for selenium(IV) removals are the same as those given for selenium(VI) removal: ion exchange, reverse osmosis, and electrodialysis. None of these techniques has been demonstrated on even a pilot scale. WSDR is planning further research on selenium removal by activated alumina and RO.

REFERENCES

(1) "National Interim Primary Drinking Water Regulations," EPA-570/9-76-003, Office of Wtr. Sply., Washington, D.C. (1976).

(2) "Drinking Water and Health, Recommendations of the National Academy of Sciences," *Fed. Reg.*, 42:132:35764 (July 11, 1977).

(3) McCabe, L.J. et al, "Survey of Community Water Sypply Systems," *Jour. AWWA*, 62:11:670 (November 1970).

(4) Kopp, J.F. and Kroner, R.C., *Trace Metals in the Water of the United States*, Fed. Wtr. Qual. Admin., Cincinnati, Ohio (1970).

(5) Durham, W.H., Hem, J.D., and Heidel, S.G., "Reconnaissance of Selected Minor Elements in Surface Waters of the United States," Geol. Surv. Cir. 643, U.S. Dept. Interior, Washington, D.C. (1970).

(6) Ferguson, J.F. and Gavis, J.A., "A Review of the Arsenic Cycle in Natural Waters," *Wtr. Res.*, 6:11:1259 (November 1972).

(7) LaPeintre, M., "Solubilization par les eaux naturelles de l'arsenic lie au fer dans les roches sedimentaires," *C.R. Acad, Sci.*, 239:359 (1954).

(8) Osishi, H. and Sanell, E.B., "Geochemistry of Arsenic," *Geochim. Cosmochim. Acta.*, 7:1 (1955).

(9) Mackenthun, K.M., "Aquatic Weed Control with Sodium Arsenite," *Sew. and Ind. Wastes*, 22:8:1062 (August 1950).

(10) Frear, O.E., *Agricultural Chemistry*, 2:532, Van Nostrand, Toronto (1952).

(11) Gilderhus, P.A., "Some Effects of Sublethan Concentrations of Sodium Arsenite on Bluegills and the Aquatic Environment," *Trans. Am. Fish. Soc.*, 95:289 (1966).

(12) Buswell, A.M. et al, "War Problems in Analysis and Treatment," *Jour. AWWA*, 35:10:1303 (October 1943).

(13) Shen, Y.S., "Study of Arsenic Removal from Drinking Water," *Jour. AWWA*, 65:8543 (August 1973).

(14) Gulledge, J.H. and O'Connor, J.T., "Removal of As(V) from Water by Adsorption on Aluminum and Ferric Hydroxide," *Jour. AWWA*, 65:8:548 (August 1973).

(15) Logsdon, G.S., Sorg, T.J., and Symons, J.M., "Removal of Heavy Metals by Conventional Treatment," *Proc. 16th Water Quality Conference–Trace Metals in Water Supplies: Occurrence, Significance and Control*, Univ. Bull., No. 71, p. 111 (1974).

(16) Calmon, C., "Comments," *Jour. AWWA*, 65:8:586 (August 1973).

(17) Bellack, E., "Arsenic Removal from Potable Water," *Jour. AWWA*, 63:7:454 (July 1971).

(18) Rubel, F., Jr., personal communication, (January 31, 1978).

(19) "Symposium on Activated Carbon D-114," Atlas Chem., ICI America Inc., Wilmington, Del. (1972).

(20) "Pure Drinking Water, Polymetrics Reverse Osmosis Equipment," Bull. 950–Rev. 1, Polymetrics, Inc., Santa Clara, Calif. (1974).

(21) "Selenium," *Medical and Biologic Effects of Environmental Pollutants Series*, Nat. Res. Council, Nat. Acad. of Sciences, Washington, D.C. (1976).

(22) Plotnikov, V.I., "Coprecipitation of Small Quantities of Selenium," *Zhur. Neorg. Khim.*, 3:8:1761 (1958).

(23) Plotnikov, V.I., "Coprecipitation of Selenium and Tellurium with Metal Hydroxides," *Russian Jour. of Inorg. Chem.*, 5:3:351 (March 1960).

(24) Plotnikov, V.I., "Coprecipitation of Small Amounts of Selenium and Zinc with Metal Hydroxides," *Russian Jour. of Inorg. Chem.*, 9:2:245 (February 1964).

(25) Hingston, F.J., Posner, A.M., and Quirk, J.P., "Adsorption of Selenite by Geothite," *Advances in Chemistry Series* 79:82 (1968).

(26) Geering, H.R. et al, "Solubility and Redox Criteria for the Possible Forms of Selenium in Soils," *Soil Sci. Soc. Amer. Proc.*, 32:35 (1968).

(27) Howard, J.H. III, "Control of Geochemical Behavior of Selenium in Natural Waters by Adsorption on Hydrous Ferric Oxides," *Trace Substances in Environmental Health–V*, (D.D. Hemphill, editor), Univ. of Missouri, Columbia (1972).

(28) Olson, O.E. and Jensen, C.W., "The Absorption of Selenate and Selenite Selenium by Colloidal Ferric Hydroxide," *Proc. S. Dak. Acad. Sci.*, 115:21 (1940).

(29) Engberg, R.A., "Selenium in Nebraska's Groundwater and Streams," Neb. Wtr. Surv. Paper No. 35, Univ. of Nebraska, Conservation and Surv. Div., Lincoln, Nebraska (December 1973).
(30) "Impact of Uranium Mining and Milling on Water Quality in the Grants Mineral Belt," New Mexico, USEPA, Region VI, Dallas, Tex., EPA 906/9-75-002 (September 1975).
(31) Sorg, T.J. and Logsdon, G.S., "Removal of Selenium from Water–State of the Art," *Proc. Symposium on Selenium-Tellurium in the Environment*, Industr. Health Foundation, Inc., Pittsburgh, Pa. (May 1976).
(32) Hannah, S.A., Jelus, M. and Cohen, J.M., "Removal of Uncommon Trace Metals by Physical and Chemical Treatment Processes," *Jour. WPCF*, 49:11:2297 (November 1977).
(33) Linstedt, K.D., Houch, C.P. and O'Connor, J.T., "Trace Element Removals in Advanced Wastewater Treatment Processes," *Jour. WPCF*, 43:7:1507 (July 1971).
(34) Ball, R., "Removal of Selenium from Drinking Water Using Activated Alumina," Unpublished senior thesis, Dept. of Civ. Engrg., Univ. of Cincinnati, Cincinnati, Ohio (September 1977).
(35) Kubli, H.A., "Contribution to the Knowledge on Anion Separation by Means of Adsorption on Alumina," *Helv. Chim. Acta, Switzerland,* 3:453 (1947).
(36) Thiem, L. and Connor, J.F., "Removal of Barium, Cadmium, Mercury, and Selenium from Drinking Water Supplies," Dept. of Civ. Engrg., Univ. of Missouri, Columbia, Mo. (1977).

REMOVAL OF THE OTHER INORGANIC CONTAMINANTS AND RADIONUCLIDES ACCORDING TO THE NIPDWR

USEPA Water Supply Research Division will publish details on the treatment technology for removal of the other inorganic contaminants specified in the National Interim Primary Drinking Water Regulations in the *JAWWA* at intervals of 4 to 6 months. These contaminants are barium, cadmium, chromium, lead, mercury, silver, and the radionuclides–radium-226 and radium-228, gross alpha particles, beta particles and photon radioactivity.

REMOVAL OF OTHER SELECTED INORGANIC CONTAMINANTS

> The material in this last section of the chapter is excerpted from a report prepared for USEPA Water Supply Division by Ralph Stone and Company, Inc. of Los Angeles, California (PB-258 271; see bibliography for details).
>
> The report contains information on treatment methods for removal of various organic compounds, including bis-ethers, various organic insecticides, etc., as well as the inorganic contaminants. None of the material on organic contaminants is being used here since the first chapters of this book have already dealt with the removal of such toxic organic compounds with material which is more up-to-date.

Table 10.8 is a summary table, showing the contaminants which will be discussed in this section, their sources, uses, toxicology, and the treatment methods which have been used to get rid of them.

Table 10.9 is a summary table of the effectiveness of the treatment methods discussed.

TABLE 10.8: SUMMARY OF PERTINENT CONTAMINANT PROPERTIES

Contaminant	Natural Sources	Industrial Uses	Toxicity (LD_{50})	Treatment Methods	Notes
Antimony (Sb)	Stibnite Sb_2S_3 (California, Utah, Idaho, Nevada)	Alloys Batteries Paints Glass Medicines	2 mg/kg body weight	Lime coagulation H_2S precipitation Ion exchange Distillation Electrodialysis	
Beryllium (Be)	Beryl $Be_3Al_2(SiO_3)_6$ (New England, Pennsylvania, Virginia, North Carolina, Alabama, S. Dakota, Colorado, California)	Alloys Nuclear reactor parts Rocket fuels Computer parts	86 mg/kg	Ion exchange Distillation Electrodialysis	More toxic as a vapor than in aqueous solution
Cobalt (Co)	Cobaltite CoAsS Skutterudite (Co, Ni, Fe)As_3 (Ontario, Idaho)	Alloys Pigments Medicines	180 mg/kg	Magnetite adsorption Coagulation/precipitation Distillation Electrodialysis	An essential nutrient in trace amounts
Lithium (Li)	Lepidolite $(F,OH)_2KLiAl_2Si_3O_{10}$ Amblygonite LiAl(F,OH) Spodumene $LiAl(SiO_3)_2$ (New England, California)	Alloys Catalysts Organic synthesis	757 mg/kg	Ion exchange Distillation Electrodialysis	Ubiquitous in nature in trace amounts LiCl used as a salt substitute in low sodium diets

(continued)

TABLE 10.8: (continued)

Contaminant	Natural Sources	Industrial Uses	Toxicity (LD_{50})	Treatment Methods	Notes
Molybdenum (Mo)	Molybdenite MoS_2 (New England, New York, Pennsylvania, Colorado, New Mexico, Washington)	Alloys Glass Ceramics Inks	125 mg/kg	Lime Coagulation Ion exchange Chitosan precipitation Distillation Electrodialysis	Essential trace element for plants
Nickel (Ni)	Nicolite NiAs Pentlandite $(Fe,Ni)_9S_8$ (Ontario, New Jersey, Tennessee, Colorado, California)	Alloys Plating Batteries Pigments Organic synthesis	1,600 mg/kg	Reverse osmosis Coagulation/precipitation Ion exchange Distillation Electrodialysis	
Tungsten (W)	Wolframite $(Fe,Mn)WO_4$ Scheelite $CaWO_4$ (Connecticut, North Carolina, Missouri, S. Dakota, Colorado, Nevada, Arizona, Calif.)	Alloys Lamp filaments	1,190 mg/kg	Ion exchange Coagulation/precipitation Distillation Electrodialysis	
Vanadium (V)	Carnotite $K_2(UO_2)_2(VO_4)_2 \cdot 3H_2O$ Vanadinite $Pb_5(VO_4)_3Cl$ (Arizona, New Mexico, Utah, Colorado, Pennsylvania)	Alloys Glassware Dyes Inks Catalysts Pesticides	160 mg/kg	Coagulation/precipitation Ion exchange Distillation Electrodialysis	More toxic as vapor or dust than in solution

TABLE 10.9: SUMMARY OF TREATMENT METHOD EFFECTIVENESS

Treatment Method	Contaminant	Influent Concentration	Effluent Concentration	Percent Removal
Adsorption,	Cobalt	–	–	86
magnetite	Nickel	–	–	81
Lime coagulation	Antimony	–	–	90
	Cobalt	12 mg/l	1.4 mg/l	–
	Molybdenum	–	–	10
	Nickel	160 mg/l	0.08 mg/l	–
	Nickel	2,935 mg/l	<1 mg/l	–
Alum coagulation	Nickel	16 mg/l	11.8 mg/l	–
	Vanadium	249 mg/l	1 mg/l	–
Chitosan coagulation	Cobalt	50 mg/l	–	62
	Molybdenum	2,500 mg/l	36 mg/l	–
	Nickel	50 mg/l	45 mg/l	–
	Tungsten	500 mg/l	170 mg/l	–
Hydrogen sulfide precipitation	Antimony	–	<2 mg/l	–
Distillation	Metals	–	–	~100
Electrodialysis	Ionic contaminants	–	–	≤95
Ion exchange				
Amberlite CG 400	Antimony	13 μg	~0	–
Dowex 50	Beryllium	8.4-10.5 mg	–	93-100
Wofatit MC 50	Cobalt	11-12 mg	–	~100
	Nickel	11-12 mg	–	~100
DEAE cellulose	Molybdenum	0.1-5 mg/l	–	≥96
	Tungsten	0.004-2 mg/l	–	≥98
ISM-1	Lithium	–	–	~100
De-acidite FF	Vanadium	5 mg/l	–	96
Dowex-1-X8	Vanadium	5.58 μg/l	–	97
Reverse osmosis	Nickel	93.6 μg	–	99.6

Antimony

Occurrence: Antimony is a heavy metal used in the manufacture of bearings, alloys (such as Babbitt metal and hard lead), storage battery grids, pewter, printing type, lead electrodes, rubber, textiles, paint, and glass. Its presence in water supplies can usually be traced to waste discharges from one of these industries. Antimony in solution exists as the Sb^{+3} or Sb^{+5} cation, or as the antimonite [SbO_3^- or $Sb(OH)_6^-$] anion.

Toxicology: The toxicity of antimony is similar to arsenic's, although less acute. Antimony poisoning can result in dermatitis, keratitis, conjunctivitis, nasal septal ulceration, and liver damage. The industrial threshold limit value (TLV) is 0.5 mg/m^3. Antimony's median lethal dose (LD_{50}) varies from 100 mg/kg as antimony metal to 4,000 mg/kg as SbO_5. A nonofficial recommended drinking water limit not to be exceeded is 0.1 mg/l.

Treatment Methods:

(1) Lime Coagulation – Removal efficiencies of better than 90% have been demonstrated.
(2) Hydrogen Sulfide Precipitation – Antimony sulfide has a solubility of less than 2 mg/l. Removal should be accomplished at least to that limit provided that stoichiometric or better quantities of H_2S are used.
(3) Ion Exchange – Laboratory tests with Amberlite CG-400 have demonstrated that at least 13 μg can be removed with 10 g of resin.
(4) Distillation.
(5) Electrodialysis.

Beryllium

Occurrence: Natural weathering of beryllium-containing minerals can produce beryllium concentrations in water of up to 0.31 μg/l. Higher concentrations are usually the result of industrial waste discharges. Industrial uses and products include nuclear reactor components, solid rocket fuels, filaments in neon and fluorescent tubes, and some alloys. In aqueous solutions, the ion normally exists as Be^{+2}.

Toxicology: Airborne beryllium is extremely toxic. Waterborne beryllium is not nearly as toxic as a rule, except for certain of its salts. Exposure to relatively low concentrations of the metal or one of these salts may cause death. Sublethal doses may cause dermatitis, chemical conjunctivitis, ulcerations, rickets, blood or bone tumors, kidney stones, and liver damage. The industrial 8-hr TLV is 2 μg/m^3 with a 30-min peak limit of 25 μg/m^3. Tests on minnows have yielded 96-hr median tolerance limits (TL_m) of 0.15 mg/l in soft water and 15 mg/l in hard water.

Treatment Methods:

(1) Ion Exchange – Experiments on solutions containing 8.4 to 10.5 mg beryllium treated with Dowex 50 have yielded removals of 93 to 100%.

(2) Distillation.
(3) Electrodialysis.

Cobalt

Occurrence: Cobalt is commonly used in alloy manufacture, nuclear technology, blue pigments for glass and china, and as a binder in the tungsten carbide tool industry. Cobalt in water can result from waste discharges from these industries or from weathering of cobalt-containing minerals. Natural concentrations can range to as much as 0.019 mg/l, with an average of 0.002 mg/l. In solution, it exists primarily as Co^{+1}.

Toxicology: Cobalt is reported to be beneficial or nontoxic in concentrations up to 7 μg/day. In excess it can cause polycythemia and respiratory ailments. The TLV is 0.1 mg/m^3. No official or unofficial drinking water limits have been established for cobalt.

Treatment Methods:

(1) Sorption on Magnetite – Removal efficiencies up to 94% have been achieved in bench-scale water treatment experiments.
(2) Carbon Adsorption – At a pH of 2.8, the cobalt-APDC (ammonium pyrrolidine dithiocarbamate) complex can be quantitatively extracted into methylisobutyl ketone (MIBK). Stoichiometric quantities of APDC are required. The APDC complex can be adsorbed onto activated carbon.
(3) Precipitation – $Ca(OH)_2$ reduced cobalt from 12 to 1.4 mg/l. Chitosan polymer treatment reduced cobalt concentrations by 62% in 50 mg/l cobalt test solutions.
(4) Ion Exchange – Wofatit MC50 chelating resin has demonstrated virtually complete removal of cobalt from influent solutions containing 11 to 12 mg cobalt.
(5) Distillation.
(6) Electrodialysis.

Lithium

Occurrence: Although lithium in water usually results from natural causes, industrial discharges are at times responsible for large or widely fluctuating concentrations. Lithium is used industrially in certain alloys, as a getter in vacuum tubes, in lithium-hardened bearing metals, in aircraft fuels, and in the synthesis of catalysts for various chemical industries. Lithium salts are used in porcelain enamels, air conditioning, and multipurpose greases. In solution, it is found as the Li^+ ion.

Toxicology: Lithium is not known to be toxic at the concentrations most likely to be encountered. Lithium chloride (LiCl) is used as a salt substitute in low-sodium diets.

Treatment Methods:

(1) Ion Exchange – The ion sieve cation exchange resin ISM-1 is extremely selective for lithium and has demonstrated virtually complete removals at low concentrations.

(2) Distillation.
(3) Electrodialysis.

Molybdenum

Occurrence: Molybdenum can occur naturally in drinking water supplies at concentrations up to 20 μg/l due to weathering of minerals. It can also be found in water as a result of waste discharges from industries related to the manufacture of glassware, ceramics, printing inks, electrical equipment, and certain steel alloys. In solution, it can exist as the cation Mo^{+2}, Mo^{+4}, Mo^{+5}, or as the molybdate anion (MoO_4^{-2}).

Toxicology: Molybdenum is mildly toxic to man. In low concentrations, it is an essential mineral nutrient acting as a catalyst in the protein xanthine oxidase, a metabolic growth factor. Animal tests have demonstrated that an excess of molybdenum can cause liver and kidney damage. Molybdenum can concentrate in plant tissues up to 15 times the exposure concentration. The TLV's for soluble and insoluble compounds are 5 and 10 mg/l, respectively. Experiments on minnows have yielded 96-hour LD_{50}s in soft and hard water of 70 and 370 mg/l, respectively.

Treatment Methods:

(1) Lime Coagulation – At a pH of 8.2, molybdenum concentrations were reduced by only 10%.
(2) Ion Exchange – Several resins have shown a capability of removing 96+% of molybdenum from solutions of 0.1 to 5 mg/l, including De-Acidite FF and diethylaminoethyl (DEAE) cellulose in the SCN form.
(3) Precipitation – Chitosan polymer is reported to have reduced molybdenum concentrations from 2,500 to 36 mg/l.
(4) Distillation.
(5) Electrodialysis.

Nickel

Occurrence: Nickel occurs naturally at concentrations up to 0.072 mg/l with an average concentration of about 0.005 mg/l. Wastewater discharges from industries associated with nickel plating, nickel alloying, storage battery manufacturing, organic hydrogenation, and the manufacture of nickel-chrome resistance wire can contribute to its presence in water sources. In solution, it exists primarily as Ni^{+2}.

Toxicology: Nickel-contaminated water can cause dermatitis in sensitive individuals. Ingestion of large quantities can cause nausea, vomiting, diarrhea, central nervous system depression and myocardial damage. The airborne TLV is 1 mg/m^3. Tests on minnows have yielded 96-hour LD_{50}s of 5 mg/l in soft water and 43 mg/l in hard water.

Treatment Methods:

(1) Adsorption on Magnetite – Laboratory tests have demonstrated up to 86% removal of nickel.
(2) Reverse Osmosis – Tests on nickel plating wastes with a nickel concentration of 93.6 g/l have demonstrated up to 99.6% reduction.

(3) Precipitation/Coagulation – $Al_2(SO_4)_3$ treatment can reduce wastewater nickel concentrations from 16 to 11.8 mg/l. $Ca(OH)_2$ dosages can reduce wastewater nickel concentrations from 16 to 6 mg/l. Lime coagulation and precipitation can reduce nickel concentrations from 160 to 0.08 mg/l. Lime coagulation has been employed to reduce nickel concentrations from 2,935 to less than 1 mg/l. Chitosan polymer can reduce nickel concentrations from 50 to 45 mg/l.
(4) Ion Exchange – Wofatit MC 50 chelating resin has demonstrated virtually complete removal of nickel from solutions containing 11 to 12 mg nickel.
(5) Distillation.
(6) Electrodialysis.

Tungsten

Occurrence: The presence of tungsten in water can be attributed primarily to industrial discharge. Tungsten is commonly used for incandescent lamp filaments and electron tubes in electronic communication apparatus and as a toughener in the manufacture of steel. It exists primarily as the tungstate anion (WO_4^-) in aqueous solution.

Toxicology: Little data is available on the toxicity of tungsten. In experiments on chickens, ingested tungsten prompted the excretion of molybdenum, the loss of which adversely affected their growth metabolism. The LD_{50} for soluble tungsten is 1,190 mg/kg.

Treatment Methods:

(1) Ion Exchange – DEAE cellulose is capable of reducing tungsten concentrations of 0.004 to 2 mg/l by 98% or more.
(2) Precipitation – Chitosan polymer has reduced tungsten concentrations from 500 to 170 mg/l in laboratory tests.
(3) Distillation.
(4) Electrodialysis.

Vanadium

Occurrence: Vanadium in water is usually associated with industrial discharge. It can occur naturally in concentrations up to 15 mg/l, but is usually in significantly smaller quantities. It is used in the production of rust-resistant steel and in glassware, dyes, inks, paints, varnish dryers, alloys, pesticides, and fuel oils. Vanadium pentoxide (V_2O_5) is used as a catalyst in the manufacture of H_2SO_4. In solution, it may be found as V^{+3}, V^{+5}, metavanadate (VO_2^-), VO_4^{-3}, or $V_2O_7^{-4}$.

Toxicology: There is no strong evidence that vanadium is carcinogenic to man or animal. It inhibits the synthesis of cholesterol. When ingested, it can cause gastric intestinal disturbances; inhibit the enzyme cholinesterase; cause a decrease in serum albumin; and cause fatty degeneration, cirrhosis of the liver, kidney damage, anemia, and muscular atrophy. The TLV for V_2O_5 dust is 0.5 $\mu g/m^3$, and for its fumes is 0.05 mg/m^3. The 96-hour TL_m for vanadium as vanadyl sulfate is 4.8 mg/l in soft water, and 30 mg/l in hard water. The 96-hour TL_m for V_2O_5 is 13 mg/l in soft water, and 35 mg/l in hard water.

Treatment Methods:

(1) Coagulation/Precipitation – Coagulation with iron or aluminum salts is capable of reducing concentrations from 249 to 1 mg/l. Chitosan polymer reportedly precipitates vanadium quite effectively. Numerical data were not found, however.
(2) Ion Exchange – De-acidite FF can reduce vanadium in solutions containing up to 5 mg by 96%. Five grams of Dowex1-X8 can produce a 97% vanadium-free effluent from 2 liters of seawater containing 5.58 μg vanadium per liter.
(3) Distillation.
(4) Electrodialysis.

PROCESS DESCRIPTIONS

[Note: Oxidation processes are discussed in Chapter 4, carbon adsorption processes in Chapters 5 and 6.]

Coagulation/Precipitation

Process Description: Coagulation and precipitation are processes which change the physical or chemical properties of dissolved, colloidal, or suspended solids to such an extent that they will settle out of solution by gravity. A coagulant either changes surface charge properties to the extent that particles will tend to agglomerate or it enmeshes particles in a polymeric suspension. In either case, the final products are large, agglomerated particles (or flocs) which more readily settle under the influence of gravity. A precipitant is a chemical which will react with a soluble species to form an insoluble compound which precipitates out of solution. Often, coagulated or precipitated particles are also removed by filtration after sedimentation. See Table 10.10 for a summary of solubilities of inorganic salts.

TABLE 10.10: SOLUBILITIES OF PARTICULAR INORGANIC SALTS, mg/l

	Cations								Anions			
	Sb	Be	Co	Li	Mo	Ni	W	V	SbO_3^-	$MoO_4^=$	WO_4^-	VO_3^-
Carbonate	–	I	I	10^4	–	I	–	–	–	–	–	–
Chloride	10^5	S	10^6	10^6	I	10^6	–	S	–	–	–	–
Ferrocyanide	–	–	I	S	–	I	–	–	–	–	–	–
Hydroxide	–	I	3.2	10^5	10^3	130	–	–	–	–	–	–
Sulfate	I	10^5	10^5	10^5	–	10^6	–	S	–	–	–	–
Sulfide	1.7	S	I	S	I	I	I	I	–	–	–	–
Calcium	–	–	–	–	–	–	–	–	–	I	6.0	–
Iron	–	–	–	–	–	–	–	–	–	I	–	I
Magnesium	–	–	–	–	–	–	–	–	I	10^5	I	–

S = Soluble, solubility ≥100 mg/liter but unknown.
I = Insoluble, solubility <10 mg/liter but unknown.
No entry denotes no compound or no data.
Numbers are solubilities in mg/liter at 20°C and 760 mm of Hg.

Particle coagulation results from the destabilization of stable suspended and dissolved solids and colloids. Stability is due to electrical charge and solvation effects. Destabilization is accomplished by the addition of cationic polyelectrolytes or cationic electrolytes such as Al^{+3}, Fe^{+2}, or Fe^{+3}; the formation of charged hydrous oxides such as $Al_x(OH)_y^{+n}$; the natural coagulation of anionic and cationic polyelectrolytes; or the entrainment in hydrous oxide flocs. Destabilization is most effective when the contaminant solids come in contact with small charged microflocs of hydrous oxide. When followed by flocculation to generate larger particles, the solids can be removed by sedimentation, flotation, or filtration.

To accomplish coagulation or precipitation, chemical coagulants or precipitants are added to the raw water in solid, slurry, or solution form. The chemicals and influent are mixed for 1 to 2 minutes in a rapid mix tank (see Figure 10.1) followed by a slow mix of about 30 minutes to stimulate particle growth. Several hours in a clarifier (see Figures 10.2 and 10.3) are usually sufficient to separate the bulk of the solids from the water although filtration is often necessary (see Figures 10.4, 10.5, and 10.6). Dewatered solids can be recovered and recycled, incinerated, landfilled, used as fertilizer, or simply dumped in the ocean, depending on the nature of the waste, the chemicals added, and local regulations.

A number of chemical compounds are suitable for use as coagulants or precipitants. Common coagulants include alum $[Al_2(SO_4)_3 \cdot 14\ H_2O]$, ferric chloride ($FeCl_3$), lime $[Ca(OH)_2]$, ferrous sulfate ($FeSO_4$), clays, and natural and synthetic polymers. Precipitants include chlorine (Cl_2), potassium permanganate ($KMnO_4$), potassium ferrocyanide $[K_4Fe(CN)_6]$, hydrogen sulfide (H_2S), sodium sulfide (Na_2S), soda ash (Na_2CO_3), caustic soda (NaOH), activated silica (SiO_2), and certain polymeric materials like chitosan. These categories are not rigid and some of these chemicals are classed as both coagulants and precipitants in the literature. This is due to a certain overlap and similarity in the physical processes involved.

Exact water treatment parameters will vary depending on the chemicals used and the nature of the water treated. Water with a total dissolved or suspended solids concentration of less than 50 mg/l is more difficult to treat effectively than more concentrated waters because of the difficulty in forming flocs in such a dispersed system. Similarly, settling is greatly hindered when treating a water with greater than a 2,000 mg/l concentration of solids.

The control of pH can be very important, especially when dealing with ions whose chemical nature is pH dependent, such as carbonates or hydroxides. Alum coagulation treatment is favored at pHs of 5-7; alkaline pHs favor lime treatment. Usually chemical treatment must be followed by some sort of neutralization to stabilize chemically the water and return it to a pH near 7.

There are several technical problems associated with coagulation and precipitation. Both methods produce considerable quantities of sludge solids. For instance, alum produces 0.45 kg per 100 to 200 g of aluminum added. Lime produces 450 to 600 kg of sludge per 1,000 m^3 of soft water. Anhydrous ferric chloride produces about 50 kg of sludge per 1,000 m^3 of soft water. These large sludge quantities present considerable disposal problems. In designing a treatment process involving chemicals, the resulting sludge disposal problem must be taken into account. With large quantities of solids present, scaling can also be a problem. If allowed to continue, scaling requires periodic plant shutdown for cleanup to prevent complete clogging. One way to avoid scaling is to use construction materials which are not readily susceptible to scaling, such as stainless steel or epoxy-reinforced fiberglass.

FIGURE 10.1: TYPES OF FLASH MIXERS

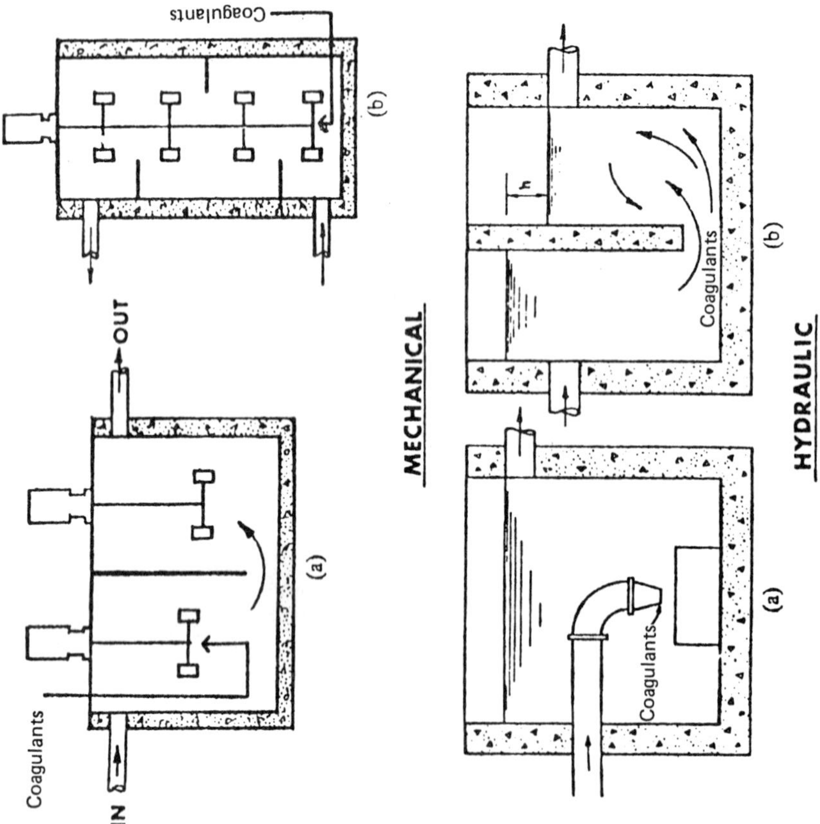

FIGURE 10.2: TYPICAL CIRCULAR CENTER-FEED SEDIMENTATION TANK

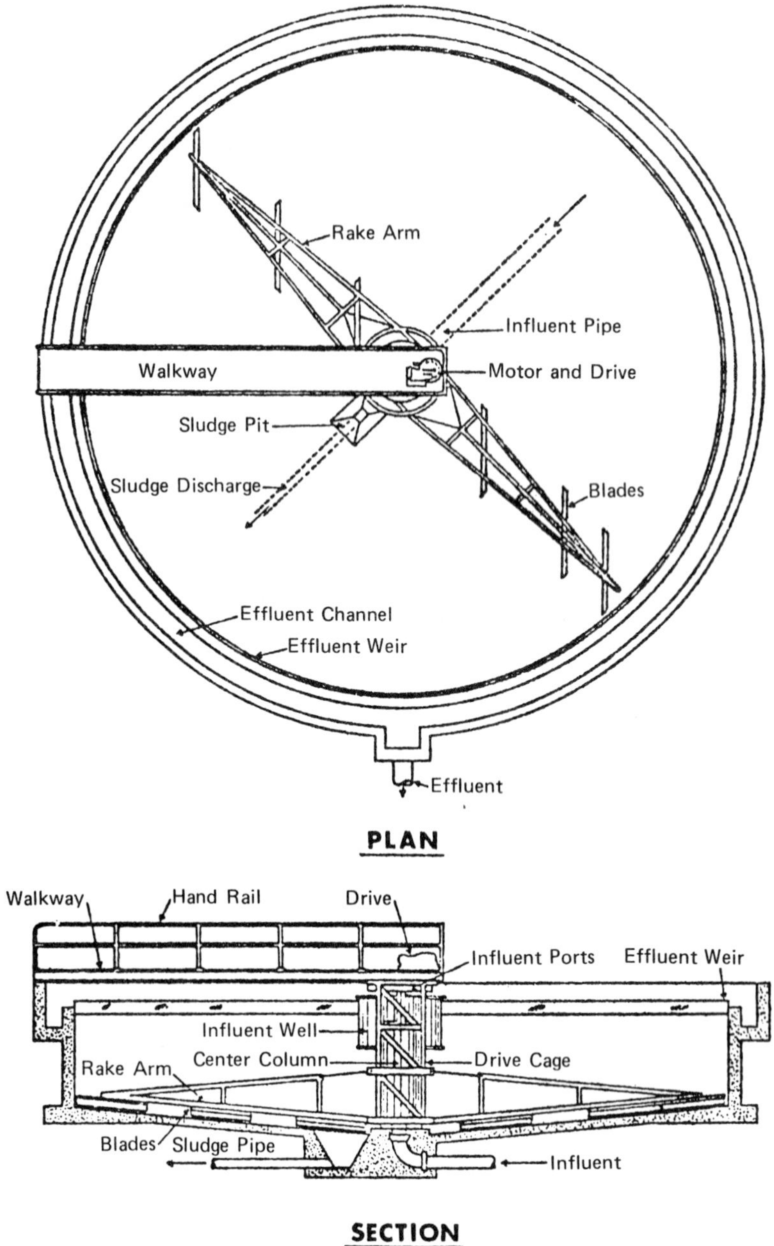

FIGURE 10.3: TYPICAL RECTANGULAR SEDIMENTATION TANK

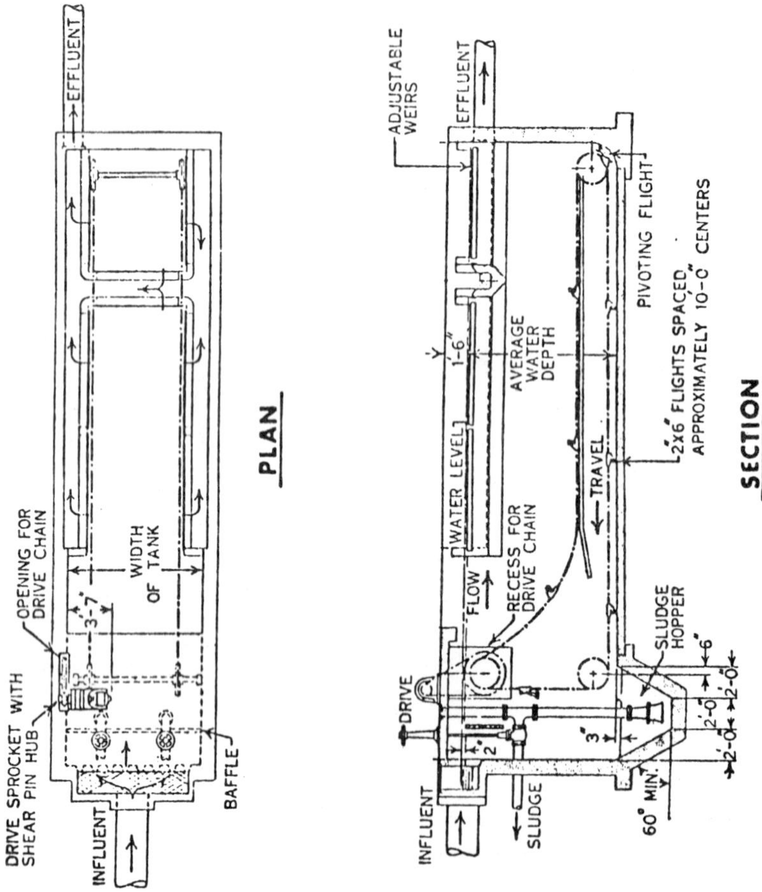

FIGURE 10.4: CUTAWAY VIEW OF TYPICAL RAPID SAND FILTER

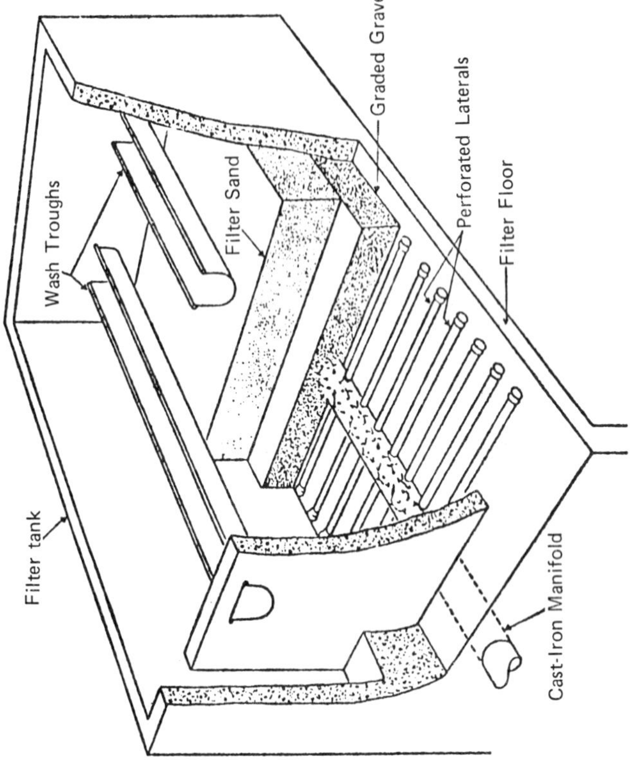

FIGURE 10.5: GENERAL ARRANGEMENT OF A RAPID SAND FILTER PLANT

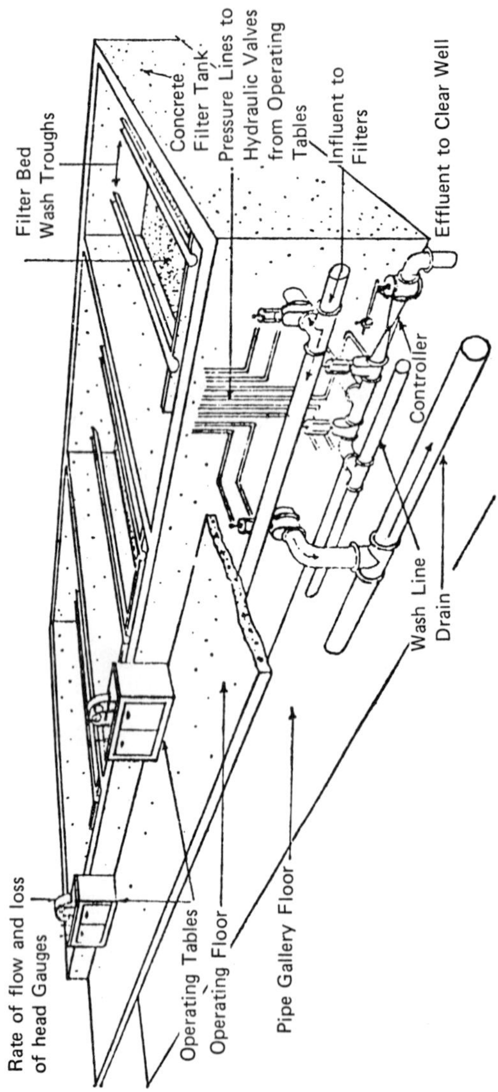

FIGURE 10.6: TYPICAL PRESSURE FILTER

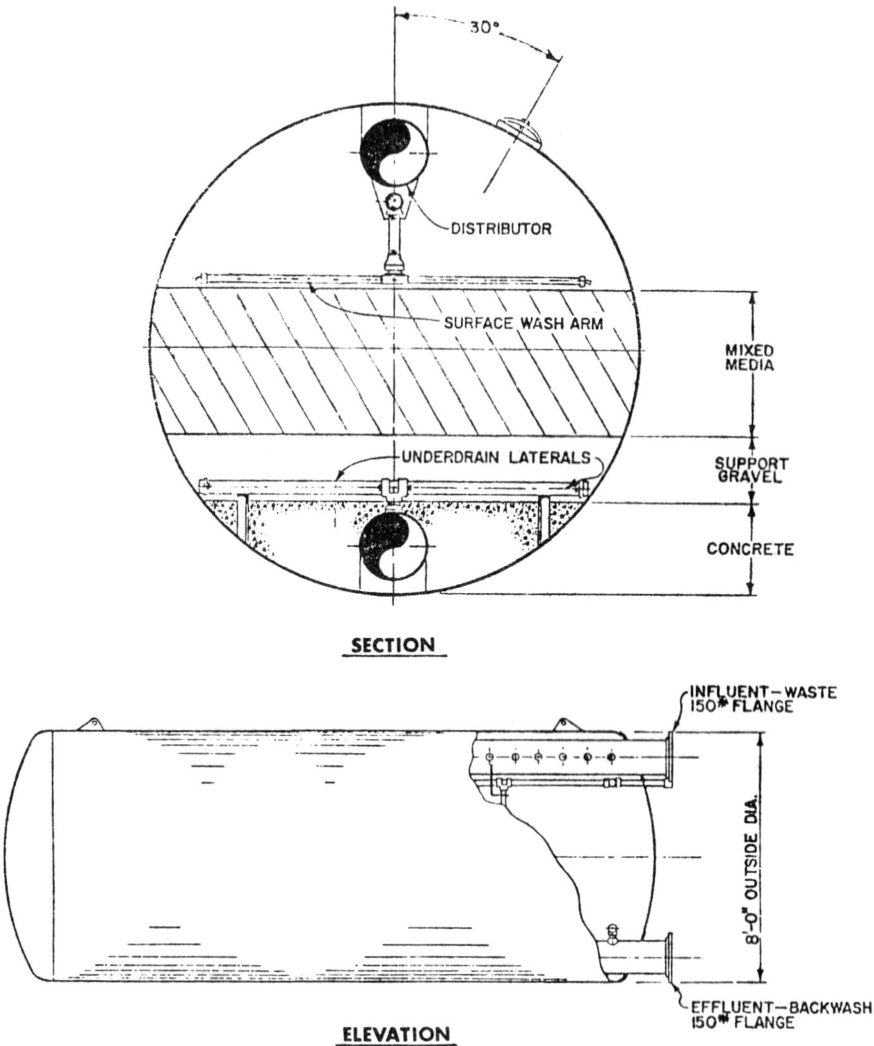

Application: All the coagulation and precipitation processes and required chemicals so far discussed have been used in full-scale treatment plant applications. The same is not true for some of the specific applications to be discussed below. Many of these have been examined on a laboratory scale only. The basic processes, however, are well known and conventional.

Numerous tests have been conducted on several of the subject contaminants, for example, antimony and nickel, using lime as the coagulant. Alum is another commonly used coagulant, and seems to give good results in nickel and vanadium removal.

A new polymer called chitosan is showing promise of being an effective water and wastewater treatment chemical. It is a natural polymer derived from crab shells, but can be synthesized on a commercial scale relatively cheaply and easily. Experiments conducted with a variety of metals have shown excellent results. Chitosan is insoluble in water and is added to the water to be treated in an acetic acid solution. As soon as it is added to the water, the chitosan precipitates in large flocs which chelate and adsorb transition metals. Lab-scale tests have been conducted on cobalt, nickel, molybdenum, and tungsten with good results.

There are several more specialized, less common precipitants in use, such as potassium ferrocyanide, soda ash, sodium or hydrogen sulfide, and caustic soda. Hydrogen sulfide dissolved in hydrochloric acid and added to a water sample will precipitate antimony to the solubility of antimony sulfide. Sulfides will precipitate most metals present in a cationic form. Waters treated with sulfides require some sort of posttreatment, such as the addition of ferrous sulfate, to remove excess sulfides.

Distillation

Process Description: Distillation involves the separation of water from the nonvolatile components of a solution by conversion of the water from a liquid to a vapor followed by condensation of the vapor apart from the influent. Ideally, the process should provide a product water 100% free of nonvolatile contaminants. Realistically, product waters can range from 500 to less than 1 mg/l dissolved solids, depending on the type of distillation system.

Distillation is an expensive method of water purification. Although scale economies do exist, the process is economically unfeasible except when:

(1) Only highly brackish or salt water is available for potable use,
(2) Inexpensive waste heat is available,
(3) A very high degree of treatment is required, or
(4) Contaminants cannot be removed by any other method.

Several full-scale plants are presently in operation. The major problems associated with distillation relate to power requirements, corrosion, scaling, and cost. For example, an influent stream of 3.5% (36 g/l) NaCl at 25°C theoretically requires at least 90 kwhr per 1,000 liters of pure water. Furthermore, because of many thermodynamic irreversibilities in the distillation process, actual processes operate at less than 10% of the optimum thermodynamic efficiency.

As water evaporates during the distillation, dissolved salts become increasingly concentrated in the remaining brine. Eventually, the concentration exceeds the solubility of the dissolved salts and calcium carbonate, calcium sulfate, and magnesium hydroxide scales form. These can plug piping and reduce heating efficiency and must be removed; pH control can minimize the problem. Concentrated solutions of inorganic salts can also be very corrosive. Consequently, only certain corrosion-resistant metals can be used in their construction, such as cupronickel alloys, aluminum, and titanium.

Minor problems include carryover of volatile contaminants, especially ammonia and organic compounds. Ammonia carryover can be avoided by lowering the pH of the feed to less than 3, but this can increase corrosion. Volatile organic compounds can foul heat transfer surfaces and cause taste and odor problems in the effluent. Lowering the distillation temperature helps. Both carryover problems can be fairly well dealt with by various conventional pre- and posttreatment methods.

Three distillation methods are commercially important: (1) boiling type with long tube vertical evaporator, (2) flash evaporation, and (3) forced circulation with vapor compression. In multistage flash evaporation (Figure 10.7), influent water is pumped through heat transfer units in each of the several stages. Steam from the influent water condenses on the outside of the tubes and is collected in trays. The concentrated wastewater cascades from one stage to the next as a result of the pressure differential until it reaches the lowest stage, where it is pumped out.

Although this process is less efficient thermodynamically than ordinary evaporation, construction costs are lower. The influent water must be treated for removal of suspended solids and deaerated before it can be pumped through the system.

In vapor compression evaporators, influent water is heated to less than boiling and the evaporated vapor is compressed and returned to serve as the heating medium. This process requires partial vacuum conditions. The commercial applicability of the configuration is limited by the capital cost of the vapor compressors.

The third type, vertical tube evaporation, holds the most promise for future large-scale desalination plants. Tests have indicated 15 to 20% higher performance ratios than other designs with fewer problems of scale formation.

In a three-stage unit (Figure 10.8), pretreated influent water enters the heat exchanger in the last stage and continues to warm up as it passes through the other heat exchangers. Influent water moving through the heat exchangers condenses the steam from the various heating units. When the hot influent water reaches the first stage, it flows down through vertical steam-heated tubes. The wastewater from the bottom of the first heating unit becomes the feed to the second.

Application: Although no data were located concerning the use of distillation for removing any of the twelve subject contaminants, the very nature of the process makes it obvious that nonvolatile contaminants can be removed up to 100%, especially when dealing with influents of low concentrations such as would be expected with these contaminants. The process would be most applicable to treating water containing cobalt, lithium, molybdenum, nickel, tungsten and vanadium.

FIGURE 10.7: MULTISTAGE FLASH EVAPORATION DISTILLATION

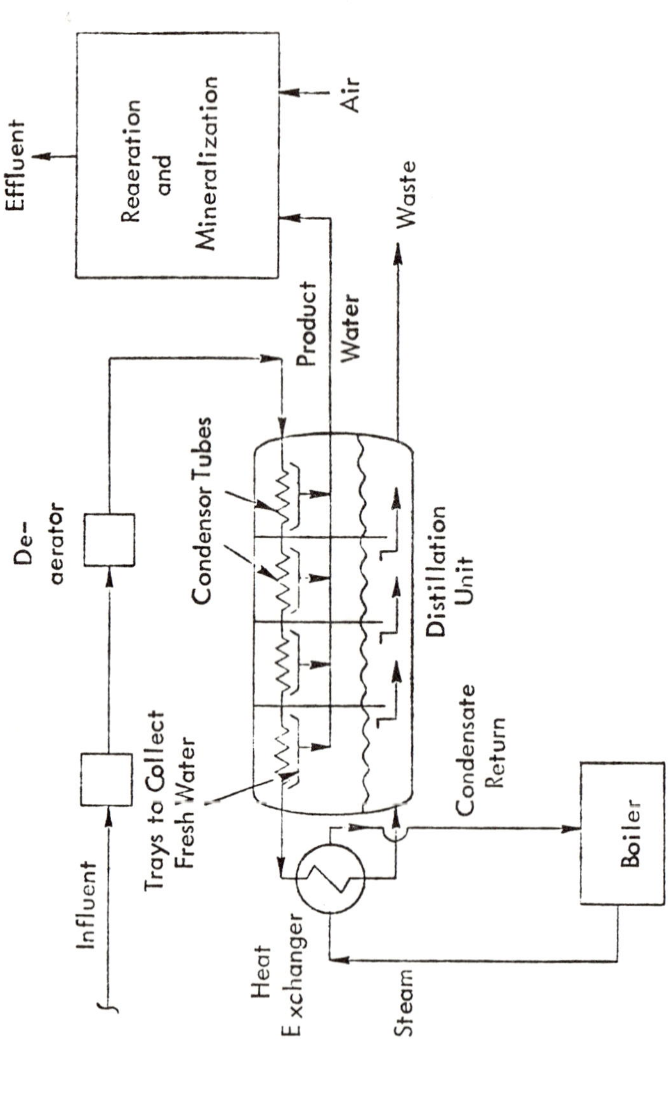

FIGURE 10.8: VERTICAL TUBE EVAPORATOR

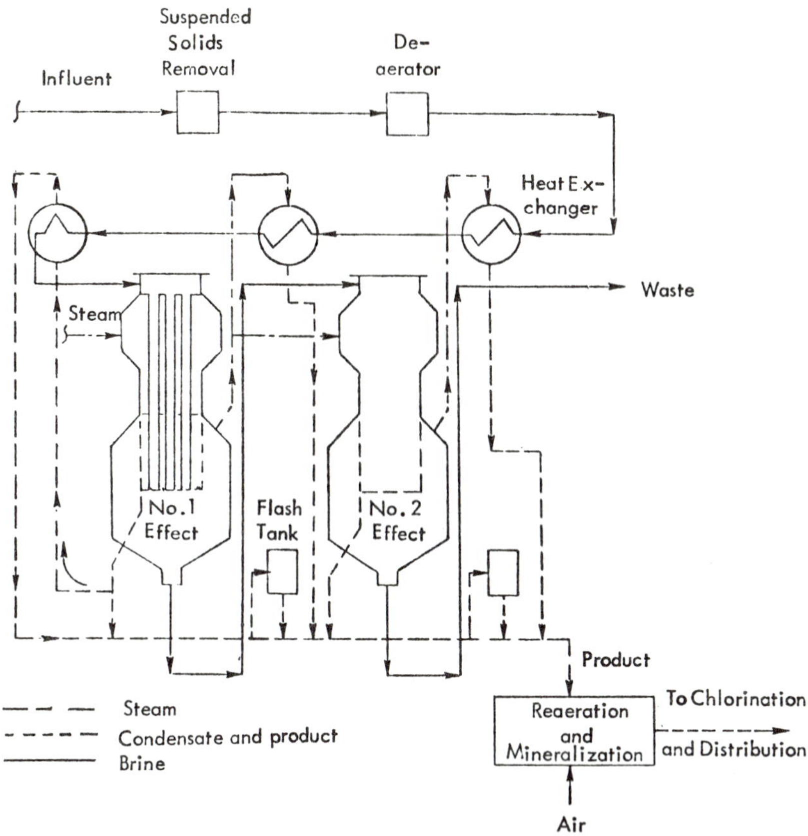

Antimony and beryllium are volatile enough that there might be a carryover problem. All of the organic compounds would present carryover problems.

Electrodialysis

Process Description: Electrodialysis (illustrated in Figure 10.9) utilizes electromotive forces to transport ionized materials across semipermeable membranes separating solution chambers. An operating electrodialysis unit consists of several chambers (formed by alternating anionic and cationic membranes) situated between two electrodes. Electrodialysis membranes are typically made of conventional ion-exchange materials: cationic membranes of sulfonic acid derivatives and anionic membranes of quaternary amine compounds. The contaminated influent passes through the chambers and, under the influence of the electromotive forces, the cations move toward the cathode, and the anions toward the anode. The ions are unable to pass through more than one membrane because the alternated anionic and cationic membranes allow the passage of only one type of ion.

FIGURE 10.9: SIMPLE ELECTRODIALYSIS SYSTEM

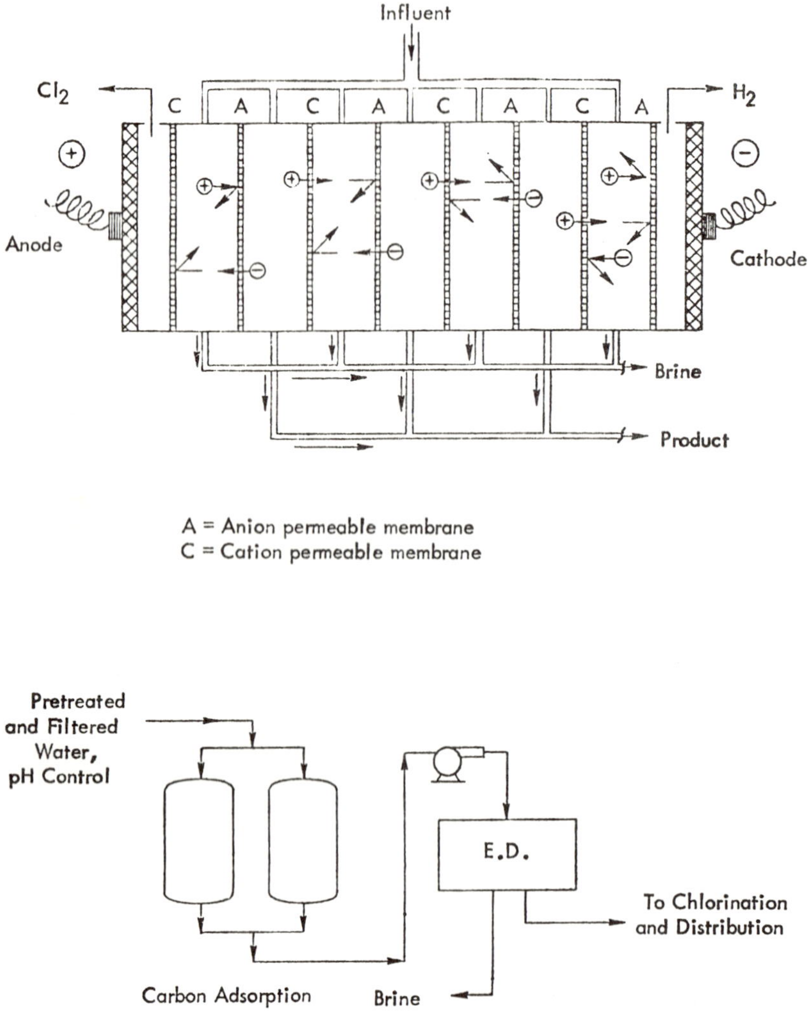

Consequently, every other chamber contains concentrated waste or purified water. There are several efficiency-reducing problems associated with the process. For instance, the passage of an ion from one chamber to another increases the electrical resistance of that chamber. Eventually, the resistance becomes great enough that no further ion migration is possible. This establishes a lower limit of salinity in the product water. An increase in system operating temperature will overcome this to some extent. Ion buildup at the membrane (concentration polarization) will also increase resistance. Ions, which are irreversibly adsorbed by the membranes or which precipitate as a result of changes in solution composition, can foul the membranes. Finally, the concentrated waste brine must be disposed of.

There are several process configurations possible. The continuous process, particularly suited to large-scale applications, consists of a series of electrodialysis units, each of which receives the effluent from the one preceding. This arrangement achieves greater demineralization than modular units, but does have the problem that changing feed water conditions from one unit to another requires continuous system adjustment.

The batch recirculation process achieves the desired effluent quality by pumping a fixed volume of feed from a reservoir, through the electrodialysis unit, and back to the reservoir until the desired effluent quality is achieved. This process requires considerably more hardware in terms of piping, controls, and reservoirs than the continuous process. Higher power is also needed, and since the feed quality is constantly changing, conditions at the membranes vary widely. The process yields a constant effluent quality regardless of changes in influent and membrane properties.

In the feed-and-bleed process, a portion of the effluent water is continuously recycled with the fresh feed, which allows the feed quality to be kept constant. The process does consume considerable power, though, and some fairly sophisticated controls are needed.

In the internally staged process, the effluent makes several passes in series between a single set of electrodes. It is a continuous process that produces a high degree of demineralization with a constant power demand. This process requires more membrane per unit of product water than the other configurations.

Application: No data have been found on the use of electrodialysis for treating any of the subject contaminants. The process is used on a moderately large scale for the removal of general dissolved solids, however. The nature of the process dictates that some degree of removal could be accomplished for the metallic contaminant species. The magnitude of this removal must remain conjectural, though, until further, more comprehensive data are available. The process description was included only because the process is a valuable addition to the list of potential treatment methods and should not be ignored due to lack of concrete data.

Ion Exchange

Process Description: Ion exchange is a common type of chemical reaction. Recently, the ion exchange technique has been associated with the process of causing water to flow through beds or columns of zeolite-type materials. An ion-exchange material is a solid consisting of a molecular network, usually an organic polymer or inorganic silicate, containing electrically charged sites of one particular charge sign and mobile ions of the opposite sign. In a cation exchange resin, the immobile molecular network is anionic; the mobile ions are cations. In anion exchanges, the molecular charges or signs are reversed.

Ion exchange is used to substitute one ion for another in a solution. For example, if water containing Ca^{+2} ions flows through a column of cation-exchange material containing Na^+ ions as the mobile groups, displacement occurs with the Ca^{+2} binding to the molecular network and Na^+ ions appearing in the effluent. This particular process is the one used to convert hard water to soft water.

The nature of the ionic groups on the ion-exchanger material largely determines the behavior of the ion exchanger. The group type affects both the ion-exchange equilibrium and ion selectivity; the total number of groups per unit volume of resin dictates the exchange capacity.

There are two main operating modes for ion exchange: batch and column applications. Batch operation involves adding a quantity of ion-exchange resin to water to be treated and mixing the slurry until a condition of equilibrium is attained for the ion-exchange solution. This mode of operation is used for ionic solutions with high selectivity under the equilibrium conditions. Furthermore, the ion being released by the resin should be relatively harmless or easily removed (e.g., by precipitation from water). Column operation (see Figures 10.10, 10.11, and 10.12) is, strictly speaking, a series of batch operations.

In practice, column operation involves passing water through a packed column or open bed filled with ion-exchange resin. It can be applied for ions with low selectivity toward the resin, and requires that the water be free of settleable and suspended solids before it flows through the column (to prevent bed clogging and fouling).

There are several types of ion exchangers available–strongly and weakly acidic cation exchangers and strongly and weakly basic anion exchangers. Strongly acidic cation exchangers include those resins containing functional acid groups such as sulfonic or sulfuric. Weakly acidic cation exchangers include those resins containing functional weak acid groups such as phenolic, carbonic, or carboxylic. Strongly basic anion exchangers contain such functional groups as primary amine, secondary amine, and tertiary amine.

Weakly basic anion exchangers contain functional groups derived from quaternary ammonium compounds such as $RR'_3N^+OH^-$. R is the organic network of the resin and R' is an organic radical such as an alkyl group.

The efficiency of an ion-exchange unit depends on the characteristics of the exchange material, the operating parameters (temperature, pressure, flow rate, solution pH, etc.), and whether there is one column or several operated in series. The various types of exchange resins and the various functional groups strongly affect the selectivity of a resin and its affinities for different ionic species. For instance, to cite a rather obvious example, an anion-exchange resin could not be used to remove cations from solution. Any attempt to do so would be completely inefficient. The operating parameters generally affect the equilibrium constants and the speeds of reaction. Consequently, a change in any of these parameters would change the efficiency of the unit.

Ion-exchange resins have a finite capacity: as the exchange sites become filled with contaminant ions, the operating efficiency will eventually fall below acceptable levels. The saturated resin must then be removed from operation and regenerated. The details of regeneration vary depending on the resin and the contaminant ions but, in general, regeneration involves the removal of the contaminant ion and its replacement by mobile ions of the same type as were originally on the resin prior to its use. This is accomplished by flooding the resin with an extremely concentrated solution of the original ion, reversing the equilibrium, and exchanging the ions again. The resulting concentrated waste brine of the contaminant ion must be disposed of.

FIGURE 10.10: TYPICAL ION-EXCHANGE BED

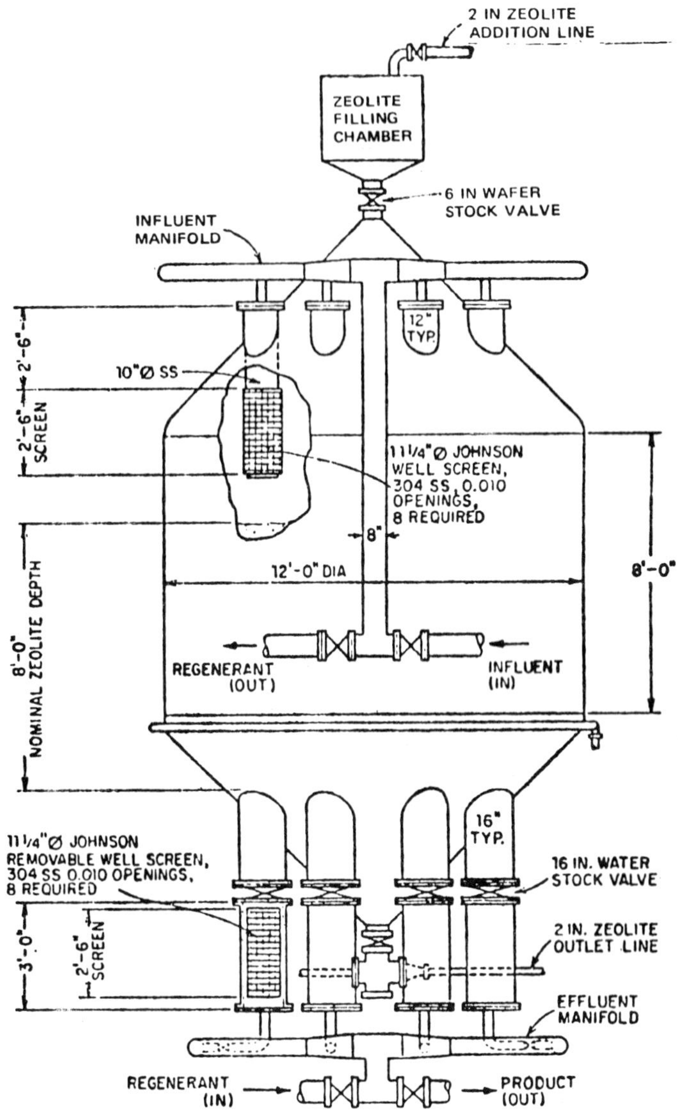

FIGURE 10.11: TWO BED DEMINERALIZER

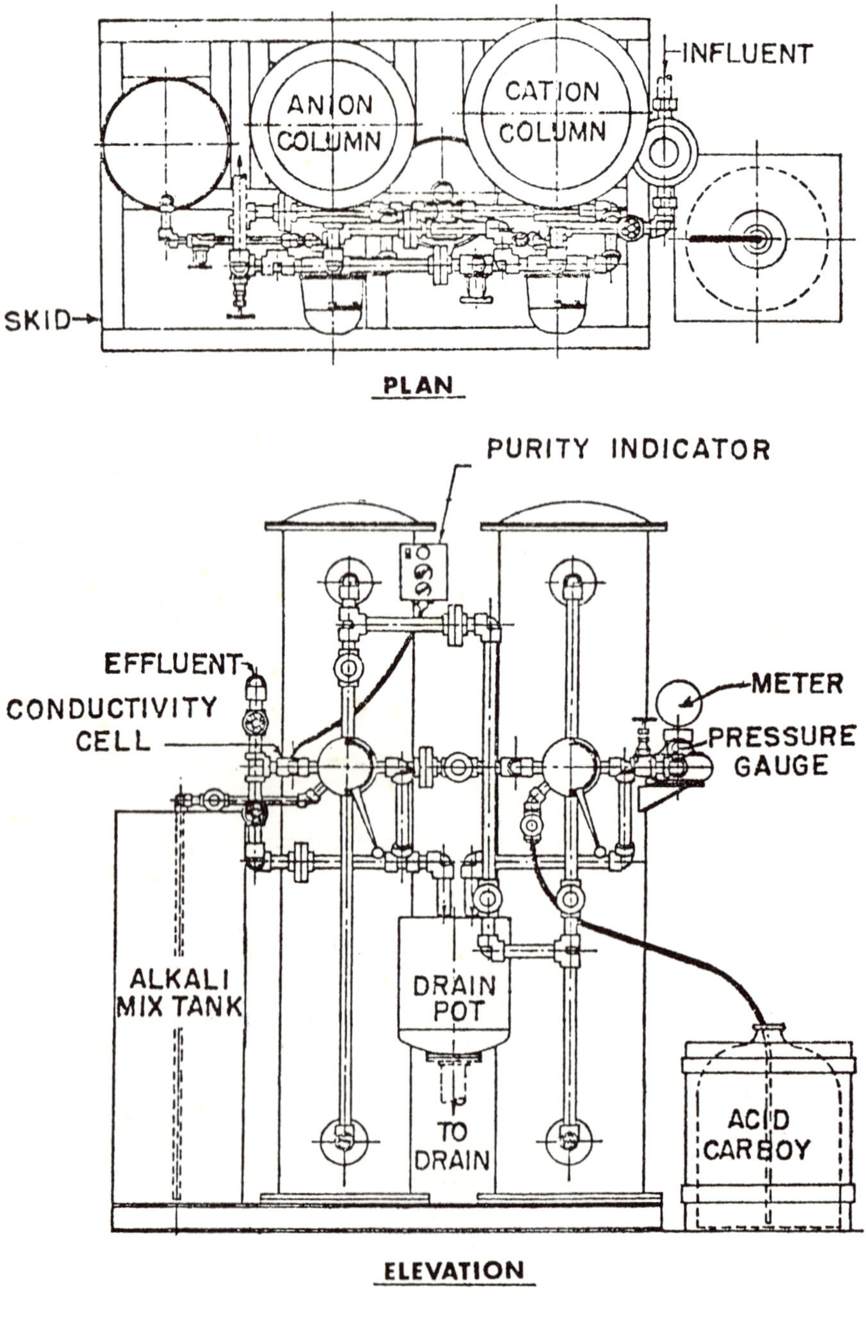

FIGURE 10.12: ION-EXCHANGE COLUMN SERIES

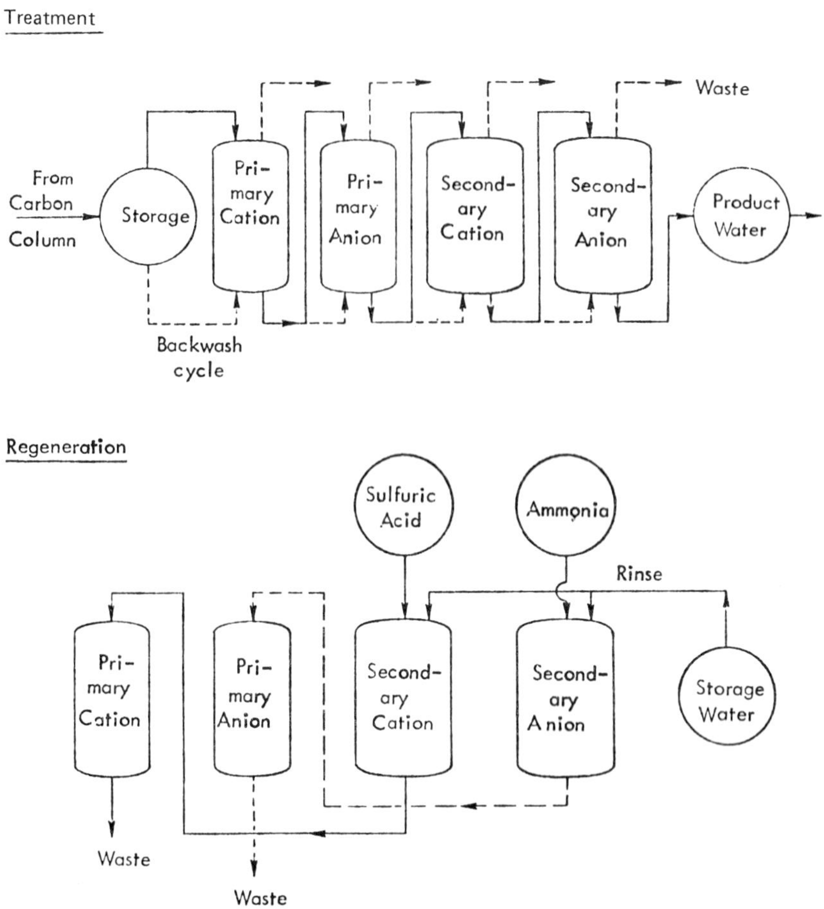

Application: Ion exchange is most commonly used to soften or demineralize water, to concentrate and recover useful by-products from industrial wastes, or in analytical separations. For instance, a simple demineralizer is the two-bed batch operation. This consists of a cation and an anion exchanger operated in series. The contaminant cations and anions are removed and replaced by hydrogen and hydroxide ions. The exchangers are regenerated with concentrated strong acid and concentrated sodium hydroxide, respectively.

All eight of the metal contaminants discussed in this section can be removed to some degree by ion exchange. Specific resins used have been mentioned under the headings of the various metals.

List of Abbreviations

ADI	acceptable daily intake
AWWA	American Water Works Association
BAC	biological granular activated charcoal
CCE	carbon-chloroform extract
DCP	dichlorophenol
DO	dissolved oxygen
DOCl	dissolved organic chlorine
EBCT	empty bed contact time
EMS	emission scan
FA	fulvic acid
GAC	granular activated carbon
GC	gas chromatography
HA	humic acid
MCL	maximum contaminant level
MIB	2-methylisoborneol
MS	mass spectrometry
NAS	National Academy of Sciences
NCI	National Cancer Institute
NIOSH	National Institute for Occupational Safety and Health
NIPDWR	National Interim Primary Drinking Water Regulations
NOMS	National Organic Monitoring Survey
NORS	National Organics Reconnaissance Survey
NPTOC	nonpurgeable total organic carbon
NVTOC	nonvolatile total organic carbon
OSHA	Occupational Safety and Health Administration
PAH	polynuclear aromatic hydrocarbon
POC	purgeable organic carbon
RFM	rapid fluorometric method
RO	reverse osmosis
SAB	Science Advisory Board
SDWA	Safe Drinking Water Act
SPC	standard plate count
TCP	trichlorophenol

TermTHM	terminal trihalomethanes
THM	trihalomethanes
THMFP	trihalomethane formation potential
TOC	total organic carbon
TOCl	total organic chlorine
TOH	total organic halogens
TON	threshold odor number
TTHM	total trihalomethanes
TU	turbidity units
USEPA	United States Environmental Protection Agency
USPHS	United States Public Health Service
WSRD	Water Supply Research Division, USEPA

Bibliography

Federal Register, Vol. 42, No. 132, July 11, 1977.

Federal Register, Vol. 43, No. 28, February 9, 1978.

Federal Register, Vol. 43, No. 130, July 6, 1978.

Statement of Basis and Purpose for an Amendment to the National Interim Primary Drinking Water Regulations on Trihalomethanes, prepared by the Office of Water Supply Criteria and Standards Division of the Environmental Protection Agency, Washington, D.C., January 1978.

Drinking Water and Health, prepared for the U.S. Environmental Protection Agency by the Safe Drinking Water Committee, Advisory Center on Toxicology, Assembly of Life Sciences, National Research Council; published by the National Academy of Sciences, Washington, D.C., June 1977.

Interim Treatment Guide for Controlling Organic Contaminants in Drinking Water Using Granular Carbon, Symons, J.M., Ed., prepared by the Water Supply Research Division, Municipal Environmental Research Laboratory, Office of Research and Development, Cincinnati, Ohio, January 1978.

Interim Treatment Guide for the Control of Chloroform and Other Trihalomethanes, Symons, J.M., Ed., prepared by the Water Supply Research Division, Municipal Environmental Research Laboratory, Office of Research and Development, Cincinnati, Ohio, June 1976.

Economic Impact Analysis of a Trihalomethane Regulation for Drinking Water, prepared by Temple, Barker & Sloane, Inc., Wellesley Hills, Mass., for the U.S. Environmental Protection Agency, Office of Water Supply, Washington, D.C., August 1977.

Economic Analysis of Proposed Regulations on Organic Contaminants in Drinking Water, prepared by Temple, Barker & Sloane, Inc., Wellesley Hills, Mass., for the U.S. Environmental Protection Agency, Office of Water Supply, Washington, D.C., December 1977.

Treatment Effectiveness for the Removal of Selected Contaminants from Drinking Water, prepared by Ralph Stone and Company, Inc., Los Angeles, California, for the Environmental Protection Agency, Division of Water Supply, Washington, D.C., July 1975 (PB-258 271).

Brass, H.J., Feige, M.A., Halloran, T., Mello, J.W., Munch, D. and Thomas, R.T., "The National Organics Monitoring Survey: Samplings and Analyses for Purgeable Organic Compounds," *Drinking Water Quality Enhancement Through Source Protection,* R.B. Pojasek, Ed., Ann Arbor Science (Ann Arbor, 1977).

Symons, J.M., Bellar, T.A., Carswell, J.K., DeMarco, J., Kropp, J.L., Robeck, G.C., Seeger, D.R., Slocum, C.J., Smith, B.L. and Stevens, A.A., "National Organics Reconnaissance Survey for Halogenated Organics," *J. Am. Water Works Assoc.* 67:634 (1975).

Sorg, T.J., "Treatment Technology to Meet the Interim Primary Drinking Water Regulations for Inorganics," *J. Am. Water Works Assoc.,* February 1978, p 105.

Sorg, T.J. and Logsdon, G.S., "Treatment Technology to Meet the Interim Primary Drinking Water Regulations for Inorganics. Part 2," *J. Am. Water Works Assoc.,* July 1978.

Notice

The information contained in this book is based upon publications issued by various government agencies as indicated on page 346. While every effort has been made to correct any obvious errors in the original documents, the nature of the subject matter precludes any further representation of accuracy or certainity. Noyes Data Corporation expressly disclaims any warranty based upon the use of this volume for obtaining water pure enough to drink, and represents only that it has used its best efforts to convey the data or information obtained from the sources cited in an accurate manner.

TOXIC ORGANIC CHEMICALS DESTRUCTION AND WASTE TREATMENT 1978

by E. Ellsworth Hackman III

Pollution Technology Review No. 40
Chemical Technology Review No. 107

This book has been written with several types of readers in mind. Manufacturing personnel are given a basic understanding of a wide variety of potential methods, followed by an exposition of current methods considered best at present.

Pollution control engineers, consultants, R&D investigators, and facilities planners or designers are given approaches for waste treatment and separation of difficult-to-destroy toxic organic chemicals.

Local, regional and national environmental control personnel in government are furnished insights into the scope and magnitude of toxic and refractory organic chemicals wastes problems and their refractoriness.

At the outset it will become obvious that there is no way in which modern life can function without the vast majority of these compounds which play an important part in materials, food, and energy technology. Bans on use and precipitous searches for substitutes were found to be poor alternatives in most cases. This author maintains that effective pollution control and waste treatment is much to be preferred.

It is necessary to walk a very fine line to achieve success in protecting the general public, the workers, the environment and future generations from toxic organics while at the same time trying to expand the job market in chemical products.

This book concentrates on those toxic organic compounds which are considered the most toxic and refractory substances, but having many useful properties.

Part I (chapters 1 to 5) identifies and describes the problems of toxic organic chemical pollution, and then shows all the technical avenues of approach to solving these problems, whether they are already in practice or forecast for use in the future.

Part II (chapters 6 to 12) categorizes the industrial sources of toxic organic chemical pollution, and presents the waste treatment systems found most practical and effective. Part II also places emphasis on cost factors.

Each compound that has been examined in this book has very powerful reasons for its existence and use. When substitutes for certain chemicals are sought, the virtues of the original compound come into clear focus. At times the substitutes are found to be hazardous. Some very effective fire retardants and noncombustible fluids are being removed from use because of their toxic hazards. Benefits and costs of substitution are weighed carefully.

Therefore emissions control should be the keystone of all future industrial efforts. In general, the air, water and solid waste emissions from the plant processes we design will have to be such that none of us would mind having the plant in our own neighborhood.

ISBN 0-8155-0700-3 **317 pages**

RADIOACTIVE WASTE DISPOSAL 1977

Low and High Level

Edited by W. R. Gilmore

Pollution Technology Review No. 38

The increasing use of nuclear energy for electric power generation, as well as the expanding applications of radioisotopes in various fields, are inevitably associated with growing amounts of radioactive wastes.

This book describes the technology being developed to concentrate and confine both high level and nonhigh level radioactive wastes so that they may be stored or otherwise gotten rid of in a comparatively safe and compact mode according to governmental guidelines.

Present radioactive waste management practices at U.S. nuclear facilities and commercial power plants are outlined in detail. Careful attention is also given to proposed schemes for radioactive waste management resulting from advances in this particular technology.

Most important is a long-term solution to the problem of accumulation of wastes from the nuclear fuel cycle. Processes are under development to convert the wastes into solid and virtually insoluble forms (the so-called vitrification processes). There are also various proposals for permanent storage ("disposal") of solidified wastes under the ocean bed or in deep geological formations where they will be virtually inaccessible.

This book is based mainly on federally-financed studies. The Table of Contents (partially reproduced here in condensed form) is organized in such a way as to serve as a subject index. A valuable glossary and list of abbreviations of nuclear and technical terms serves to familiarize the reader with this timely topic.

ISBN 0-8155-0684-8

356 pages

HOW TO REMOVE POLLUTANTS AND TOXIC MATERIALS FROM AIR AND WATER A PRACTICAL GUIDE 1977

by Marshall Sittig

Pollution Technology Review No. 32

This book is designed to provide a one volume ready reference for the handling of noxious materials emerging into the air and water as a result of industrial processes or from running machinery of any kind.

The descriptions are based almost entirely on U.S. patents dealing with practical environmental control methods and systems. The book surveys some 500 patents in the 1973 to 1976 period with exhaustive coverage up to November 1st, 1976. Since environmental patents are given priority handling by the U.S. Patent Office, some applications for these patents were filed only in late 1975 and even early 1976. This book therefore contains substantial technical information on very late developments in their field.

This book is addressed to industrialists who must keep abreast of the latest pollution removal techniques, to legislators and public health officials intent upon sensible rules and regulations, to interested conservationists and to those eager students who can foresee permanent and brilliant careers in the fields of pollution abatement and engineering.

Subject entries are in alphabetical sequence. Because of the encyclopedic nature of the book, only those entries which are at the beginning of the alphabet can be shown here without bias.

Introduction

ACETONE CYANOHYDRIN
Removal from Water

ACID MINE WATERS

ACROLEIN PROCESS EFFLUENTS
Removal from Water

ACRYLIC RESIN EMISSIONS
Removal from Water

ACRYLONITRILE EFFLUENTS
Removal from Air
Removal from Water

ADIPIC ACID EFFLUENTS
Removal from Water

ALDEHYDES
Removal from Air or Water
Removal from Water only

ALKALI
Removal from Air

ALKALI CYANIDES
Removal from Air

ALKYL IODIDES
Removal from Air

ALKYLATION PROCESS EFFLUENTS
Removal from Air

ALUMINUM
Removal from Water

ALUMINUM CELL EFFLUENTS
Removal from Air

ALUMINUM CHLORIDE EFFLUENTS
Removal from Air

ALUMINUM ETCHING LIQUORS
Removal from Water

ALUMINUM REFINING EFFLUENTS
Removal from Air

AMINES
see **Foundry Casting Effluents**

AMMONIA
Removal from Water

AMMONIA-SODA EFFLUENTS
Removal from Water

AMMONIA SYNTHESIS EFFLUENTS
Removal from Water

AMMONIUM PHOSPHATE EFFLUENTS
Removal from Air

AMMONIUM SULFATE
Removal from Air

AMMONIUN SULFIDE
Removal from Water

AMMONIUM SULFITE
Removal from Water

ANTIFOULING PAINTS
Disposal of Residues

ASBESTOS
Removal from Air
Removal from Water

ASPHALT VAPORS
Removal from Air

AUTOMOTIVE EXHAUST
Removal from Air

BATTERY CHARGING EFFLUENTS
Removal from Air

ETC.

The book contains a total of 293 subject entries. The subject name refers to the polluting substance and the text underneath each entry tells how to combat pollution by said substance.

ISBN 0-8155-0654-6

621 pages

WASTE TREATMENT WITH POLYELECTROLYTES AND OTHER FLOCCULANTS 1977

by Sidney J. Gutcho

Pollution Technology Review No. 31

The recycling of wastewater to a purity acceptable and useful in population centers and industry can be accomplished through the judicious use of polyelectrolytes and other flocculants, such as natural latexes.

Polyelectrolytes are high molecular weight polymers and are classified into three types—anionic, cationic and nonionic. They promote the process whereby suspended solids and colloidal materials in the water are agglomerated into masses sufficiently large to settle. Being polyelectrolytes, they neutralize any static charges on dirt particles, allowing them to come together.

Correctly applied, organic flocculants are practically nontoxic, biodegradable, small in volume, easily incinerated, and effective under varied pH and temperature conditions. This makes them ideal for clarification of water from reservoirs, lagoons, etc. They are being used at an increasing rate for the removal of solids from various industrial wastewaters including the effluents of the mining and papermaking industries.

This book presents a broad spectrum of the manufacturing technologies and practical applications, as they have been elaborated by many industries and municipalities. A partial and condensed table of contents follows here. Numbers in parentheses indicate the number of processes per topic.

ISBN 0-8155-0648-1

274 pages

PARTICULATES AND FINE DUST REMOVAL 1977
Processes and Equipment

by Marshall Sittig

Pollution Technology Review No. 34

Particulates and fine dust created by man's activities contribute significantly to all major aspects of air pollution. While the generation of natural fine dusts (fugitive dusts) is also very large in some parts of the earth, industrially generated, particle-loaded air emissions may push the particulate level to a point where acceptable air quality standards are exceeded continuously.

How to reduce such emissions at the source, and what processes and equipment to use, is the subject of this book, which is based mainly on reports of federally-financed air pollution studies.

Particulate removal, conventionally called "deliberate dust collecting," involves the removal of coarse and fine particles from an air stream. Although generally limited to that required for controlling industrially generated emissions into the atmosphere, particulate control equipment also has wide application for air cleaning and air filtration inside the plant, as well as for recovery of usable products (primarily in the processing industries).

The Table of Contents (partially reproduced here in condensed form) is organized in such a way as to serve as a subject index. It provides detailed guidance to the information contained in this book. Bibliographic reference lists follow each of the 28 chapters so as to provide the reader and user of this volume with more easy access to complete information on this timely topic.

ISBN 0-8155-0664-3 **605 pages**

SEWAGE SLUDGE TREATMENT AND DISPOSAL 1976

by Richard W. James

Pollution Technology Review No. 29

Sludge is the major by-product of wastewater treatment processes, regardless of whether the wastewater is from industrial or household sources or both.

In any water pollution control program where semisolid effluents must be dealt with, the most crucial operation is the sludge conditioning process used prior to dewatering.

Regardless of the mechanical method used later on for dewatering, it has become standard practice to condition the sludge first. This conditioning is almost always based on chemical oxidation, whether such a reaction is obvious or not. The operational details of a variety of such processes are described in this book. The activated sludge process, for instance, uses a heterogenous microbial population for oxidizing soluble and colloidal organic matter to carbon dioxide and water in the presence of molecular oxygen.

This chemical conditioning facilitates mechanical dewatering dramatically. Dewatering may be done by gravity filtration, vacuum filtration, centrifugation, or by flotation and sedimentation.

Incineration is all important and constitutes the largest chapter in this book, but another two chapters are devoted to practical end uses for treated sludge, viz. addition to fertilizers, coal and cement, and conversion to synthesis gas.

This book presents over 190 processes described in the U.S. patent literature. These processes include large scale community systems as well as special compact units for boats, etc. A partial and condensed table of contents follows here. Numbers in parentheses indicate the number of processes per topic.

ISBN 0-8155-0630-9 **339 pages**

MUNICIPAL SOLID WASTE MANAGEMENT 1975

by David Rimberg

Pollution Technology Review No. 26

This book has been prepared with the intention to provide sufficient technological and administrative guidance for designing practical and economical solid waste management systems in accordance with the *Federal Solid Waste Disposal* and *Resource Recovery Acts.*

The reader is given ample background and current information pertinent to the subject as obtained from recent studies performed mostly under the auspices of the EPA (U.S. Environmental Protection Agency). Armed with this information the reader can endeavor to plan or modify a local government's solid waste management system.

The book is logically arranged into three sections. The first section is devoted to the technological aspects of solid waste management and thus comprises the first four chapters. Discussed in these chapters are the sources and characteristics of solid waste, methods of storage, collection and disposal, and a survey of the newest equipment and architectural designs. The next two chapters (the second section of the book) deal with the financial, economic, and labor aspects of solid waste management. The last part of the book is a description of several prototype municipal installations in various parts of the United States as built or modified rather recently. A partial and condensed table of contents follows here. Chapter headings are given, and some of the more important subtitles are included also.

ISBN 0-8155-0596-5 **380 pages**